Methods in Cell Biology

VOLUME 39
Motility Assays for Motor Proteins

Series Editors

Leslie Wilson
Department of Biological Sciences
University of California
Santa Barbara, California

Paul Matsudaira
Whitehead Institute for Biomedical Research and
Department of Biology
Massachusetts Institute of Technology
Cambridge, Massachusetts

Methods in Cell Biology

Prepared under the Auspices of the American Society for Cell Biology

VOLUME 39
Motility Assays for Motor Proteins

Edited by

Jonathan M. Scholey

Section of Molecular and Cellular Biology
University of California
Davis, California

ACADEMIC PRESS, INC.
A Division of Harcourt Brace & Company

San Diego   New York   Boston   London   Sydney   Tokyo   Toronto

This book is printed on acid-free paper. ∞

Academic Press, Inc.
1250 Sixth Avenue, San Diego, California 92101-4311

United Kingdom Edition published by
Academic Press Limited
24–28 Oval Road, London NW1 7DX

International Standard Serial Number: 0091-679X

International Standard Book Number: 0-12-564139-7 (Hardcover)
International Standard Book Number: 0-12-638920-4 (Paperback)

PRINTED IN THE UNITED STATES OF AMERICA
93 94 95 96 97 98   EB   9 8 7 6 5 4 3 2 1

CONTENTS

Contributors   xi

Preface   xv

1. *In Vitro* Methods for Measuring Force and Velocity of the Actin–Myosin Interaction Using Purified Proteins

Hans M. Warrick, Robert M. Simmons, Jeffrey T. Finer, Taro Q. P. Uyeda, Steven Chu, and James A. Spudich

    I. Introduction   2
   II. Preparation of *in Vitro* Motility Assay Components   4
  III. *In Vitro* Assay for Myosin Velocity   7
  IV. *In Vitro* Assay for Myosin Force in the Motility Assay   12
   V. Components of the Optical Trap System   13
  VI. Future Directions   18
     References   20

2. Myosin-Specific Adaptations of the Motility Assay

James R. Sellers, Giovanni Cuda, Fei Wang, and Earl Homsher

    I. Introduction   24
   II. Description of Equipment   25
  III. Basic Assay Procedure   27
  IV. Adaptations of the Assay   31
   V. Quantitation of Rate of Actin Filament Sliding   42
     References   48

3. Motility of Myosin I on Planar Lipid Surfaces

Henry G. Zot and Thomas D. Pollard

    I. Introduction   51
   II. Methods   53
  III. Evaluation of the Method   59
     References   61

4. Microtubule and Axoneme Gliding Assays for Force Production by Microtubule Motor Proteins

Bryce M. Paschal and Richard B. Vallee

    I. Introduction   65
   II. Microtubule Gliding Assay   67

III. Polarity of Force Production   70
References   73

5. Analyzing Microtubule Motors in Real Time

S. A. Cohn, W. M. Saxton, R. J. Lye, and J. M. Scholey

I. Introduction   76
II. Configuration of the Computer–Video Microscope Setup   76
III. The Microtubule Motility Assay   81
IV. Overview of Typical Results Obtained with the Real-Time Motility Assay   83
V. Conclusions   87
Addendum   87
References   88

6. Assays of Axonemal Dynein-Driven Motility

Winfield S. Sale, Laura A. Fox, and Elizabeth F. Smith

I. Introduction   89
II. Dark-Field Video Microscopy   90
III. Procedures and Analysis of ATP-Induced Microtubule Sliding in
Flagellar Axonemes   91
IV. The Second Assay: Dynein-Driven Microtubule Translocation Using
Purified Proteins   97
V. Summary: Future   103
References   103

7. Preparation of Marked Microtubules for the Assay of the Polarity of
Microtubule-Based Motors by Fluorescence Microscopy

Jonathon Howard and Anthony A. Hyman

I. Introduction   105
II. Preparation of Labeled Microtubules   106
III. Reagents   106
IV. Preparation of the Bright Microtubule Seeds   106
V. Preparation of the Polarity-Marked Microtubules   109
VI. Visualization of Marked Microtubules in Motor Polarity Assays   110
References   113

8. Expression of Microtubule Motor Proteins in Bacteria for
Characterization in *in Vitro* Motility Assays

Rashmi Chandra and Sharyn A. Endow

I. Introduction   115
II. Construction of Plasmids for Expression in Bacteria   116
III. Vectors Used for Expression of Motor Proteins   118
IV. Expression of Proteins in *Escherichia coli*   119
V. Partial Purification of Proteins from Bacteria for *in Vitro* Motility Assays   120

VI. Brief Description of Motility Assays for Testing the
Functional Properties of the Expressed Proteins   123
VII. Microtubule Binding Assays   124
VIII. Measurement of ATPase Activity as a Determinant of Motor
Protein Function   124
IX. Problems Encountered with Prokaryotic Expression Systems
and Some Solutions   125
References   125

9. Tracking Nanometer Movements of Single Motor Molecules

Michael P. Sheetz and Scot C. Kuo

I. Introduction   129
II. Movements by Single Motor Molecules   130
III. Microtubule Movements on Motor-Coated Glass   130
IV. Conclusion   135
References   135

10. Assay of Microtubule Movement Driven by Single Kinesin Molecules

Jonathon Howard, Alan J. Hunt, and Sung Baek

I. Introduction   138
II. Protein Preparation   139
III. Flow Cells and Microscopy   139
IV. Microscopy and Motion Analysis   143
V. *In Vitro* Motility Assay   144
VI. Evidence That Single Molecules Suffice for Microtubule-Based Motility   145
References   147

11. *In Vitro* Motility Assays Using Microtubules Tethered to
Tetrahymena Pellicles

Vivian A. Lombillo, Martine Coue, and J. Richard McIntosh

I. Introduction   149
II. Isolation of Assay Components   151
III. *In Vitro* Motility Assay I: Inducing Chromosome and Vesicle Motions
with Microtubule Depolymerization   158
IV. *In Vitro* Motility Assay II: Testing Microtubule Sliding Activity of
Motor Enzymes   160
V. Other Procedures   163
VI. Additional Remarks   164
References   165

12. Use of ATP Analogs in Motor Assays

Takashi Shimizu, Yoko Y. Toyoshima, and Ronald D. Vale

I. Introduction   167
II. Purity Check and Purification of ATP Analogs   168

III. Chemistry of ATP Analogs   169
IV. Additional Precautions in Using ATP Analogs   174
 V. Uses of ATP Analogs   174
     References   176

13. Myosin-Mediated Vesicular Transport in the Extruded Cytoplasm
     of Characean Algae Cells

Bechara Kachar, Raul Urrutia, Marcelo N. Rivolta, and Mark A. McNiven

  I. Introduction   179
 II. Extruded Algal Cytoplasm Preparations   180
III. Methods for the Direct Visualization of Myosin-Mediated
      Organelle Movements   181
IV. Observation of Myosin-Mediated Vesicular Movements *in Vitro*   183
 V. Identification of a Soluble Pool of Myosins in *Nitella* Cytoplasm 187
VI. Concluding Remarks   187
     References   190

14. Assay of Vesicle Motility in Squid Axoplasm

Scott T. Brady, Bruce W. Richards, and Philip L. Leopold

  I. Extrusion of Isolated Axoplasm   192
 II. Physiological Buffers   194
III. Video Microscopy   195
IV. Pitfalls   197
 V. Summary   201
     References   201

15. Assay of Membrane Motility in Interphase and Metaphase *Xenopus* Extracts

Viki J. Allan

  I. Introduction   203
 II. Preparation of *Xenopus* Extracts and Membrane Fractions   205
III. Motility Assay   213
IV. Data Acquisition, Handling, and Analysis   220
 V. Conclusions and Prospects   223
     References   224

16. Microtubule Motor-Dependent Formation of Tubulovesicular
     Networks from Endoplasmic Reticulum and Golgi Membranes

James M. McIlvain, Jr., Carilee Lamb, Sandra Dabora, and Michael P. Sheetz

  I. Introduction   228
 II. Materials and Methods   228
III. Discussion   233
     References   235

17. Cytoplasmic Extracts from the Eggs of Sea Urchins and Clams
for the Study of Microtubule-Associated Motility and Bundling

Neal R. Gliksman, Stephen F. Parsons, and E. D. Salmon

    I. Introduction   238
   II. Sea Urchin Egg Cytoplasmic Extracts   238
  III. Clam Egg Cytoplasmic Extracts   243
  IV. Visualizing the Microtubule-Associated Motility and Crosslinking   245
   V. Example Results for Cytoplasmic Extracts   247
      References   250

18. Kinesin-Mediated Vesicular Transport in a Biochemically Defined Assay

Raul Urrutia, Douglas B. Murphy, Bechara Kachar, and Mark A. McNiven

    I. Introduction   253
  II. Assay Reagents and Components   254
 III. Motility Assay Procedure   257
 IV. Kinesin-Mediated Dynamics *in Vitro*   258
  V. Biological Implications of Kinesin-Supported Motility Assays   260
 VI. Summary   264
     References   265

19. An Assay for the Activity of Microtubule-Based Motors on the
Kinetochores of Isolated Chinese Hamster Ovary Chromosomes

A. A. Hyman and T. J. Mitchison

    I. Introduction   267
  II. Methods for Assaying Motor Activity on the Kinetochore   268
 III. Conclusions   276
    References   276

20. The Diatom Central Spindle as a Model System for Studying
Antiparallel Microtubule Interactions during Spindle Elongation *in Vitro*

Christopher J. Hogan, Patrick J. Neale, Manlin Lee, and W. Zacheus Cande

    I. Introduction   278
  II. Diatom Cultures   279
 III. *In Vitro* Models   283
 IV. Concluding Remarks   290
   References   291

Index   293
Volumes in Series   301

CONTRIBUTORS

Numbers in parentheses indicate the pages on which the authors' contributions begin.

Viki J. Allan (203), Structural Studies Division, Medical Research Council Laboratory of Molecular Biology, Cambridge, CB2 2QH, United Kingdom

Sung Baek (137), Department of Physiology and Biophysics, University of Washington, Seattle, Washington 98195

Scott T. Brady (191), Department of Cell Biology and Neuroscience, University of Texas Southwestern Medical Center, Dallas, Texas 75235, and Marine Biological Laboratory, Woods Hole, Massachusetts 02543

W. Zacheus Cande (277), Departments of Molecular and Cell Biology, University of California, Berkeley, Berkeley, California 94720

Rashmi Chandra (115), Department of Microbiology, Duke University Medical Center, Durham, North Carolina 27710

Steven Chu (1), Departments of Physics and Applied Physics, Stanford University, Stanford, California 94305

S. A. Cohn (75), Department of Biological Sciences, DePaul University, Chicago, Illinois 60614

Martine Coue[1] (149), Department of Molecular, Cellular and Developmental Biology, University of Colorado, Boulder, Colorado 80309

Giovanni Cuda (23), Laboratory of Molecular Cardiology, National Heart, Lung, and Blood Institute, National Institutes of Health, Bethesda, Maryland 20892

Sandra Dabora (227), Department of Cell Biology, Washington University, St. Louis, Missouri 63130

Sharyn A. Endow (115), Department of Microbiology, Duke University Medical Center, Durham, North Carolina 27710

Jeffrey T. Finer (1), Departments of Biochemistry and Developmental Biology, Beckman Center, Stanford University School of Medicine, Stanford, California 94305

Laura A. Fox (89), Department of Anatomy and Cell Biology, Emory University School of Medicine, Atlanta, Georgia 30322

Neal R. Gliksman[2] (237), Department of Biology, University of North Carolina, Chapel Hill, North Carolina 27599

Christopher J. Hogan (277), Departments of Molecular and Cell Biology, University of California, Berkeley, Berkeley, California 94720

Earl Homsher (23), Department of Physiology, School of Medicine, Center for Health Sciences, University of California, Los Angeles, Los Angeles, California 90024

---

[1] *Present address*: Institute Jacques Monod, University of Paris, Paris, France.

[2] *Present address*: Department of Cell Biology, Duke University Medical Center, Durham, North Carolina 27710.

Jonathon Howard (105, 137), Department of Physiology and Biophysics, University of Washington, Seattle, Washington 98195

Alan J. Hunt (137), Department of Physiology and Biophysics, University of Washington, Seattle, Washington 98195

A. A. Hyman (105, 267), Department of Pharmacology, University of California, San Francisco, San Francisco, California 94143

Bechara Kachar (179, 253), Laboratory of Cellular Biology, National Institute for Deafness and Other Communication Disorders, National Institutes of Health, Bethesda, Maryland 20892

Scot C. Kuo (129), Department of Cell Biology, Duke University Medical Center, Durham, North Carolina 27710

Carilee Lamb (227), Department of Cell Biology, Duke University Medical Center, Durham, North Carolina 27710

Manlin Lee (277), Institute of Life Science, National Tsing Hua University, Hsingchu, Taiwan 30043, Republic of China

Philip L. Leopold[3] (191), Department of Cell Biology and Neuroscience, University of Texas Southwestern Medical Center, Dallas, Texas 75235, and Marine Biological Laboratory, Woods Hole, Massachusetts 02543

Vivian A. Lombillo (149), Department of Molecular, Cellular and Developmental Biology, University of Colorado, Boulder, Colorado 80309

R. J. Lye (75), Department of Genetics, Washington University School of Medicine, St. Louis, Missouri 63110

James M. McIlvain, Jr. (227), Department of Cell Biology, Duke University Medical Center, Durham, North Carolina 27710

J. Richard McIntosh (149), Department of Molecular, Cellular and Developmental Biology, University of Colorado, Boulder, Colorado 80309

Mark A. McNiven (179, 253), Center for Basic Research in Digestive Diseases, Mayo Clinic, Rochester, Minnesota 55905

T. J. Mitchison (267), Department of Pharmacology, University of California, San Francisco, San Francisco, California 94143

Douglas B. Murphy (253), Department of Cell Biology and Anatomy, The Johns Hopkins Medical School, Baltimore, Maryland 21205

Patrick J. Neale[4] (277), Department of Plant Biology, University of California, Berkeley, Berkeley, California 94720

Stephen F. Parsons (237), Department of Biology, University of North Carolina, Chapel Hill, North Carolina 27599

Bryce M. Paschal (65), Cell Biology Group, Worcester Foundation for Experimental Biology, Shrewsbury, Massachusetts 01545, and Department of Cell Biology, University of Massachusetts Medical School, Worcester, Massachusetts 01605

---

[3] *Present address*: Department of Pathology, Columbia University College of Physicians and Surgeons, New York, New York 10032.

[4] *Present address*: Smithsonian Environmental Research Center, Edgewater, Maryland 21037.

Thomas D. Pollard (51), Department of Cell Biology and Anatomy, The Johns Hopkins Medical School, Baltimore, Maryland 21205

Bruce W. Richards (191), Department of Cell Biology and Neuroscience, University of Texas Southwestern Medical Center, Dallas, Texas 75235, and Marine Biological Laboratory, Woods Hole, Massachusetts 02543

Marcelo N. Rivolta (179), Laboratory of Cellular Biology, National Institute for Deafness and Other Communication Disorders, National Institutes of Health, Bethesda, Maryland 20892

Winfield S. Sale (89), Department of Anatomy and Cell Biology, Emory University School of Medicine, Atlanta, Georgia 30322

E. D. Salmon (237), Department of Biology, University of North Carolina, Chapel Hill, North Carolina 27599

W. M. Saxton (75), Department of Biology, Indiana University, Bloomington, Indiana 47405

J. M. Scholey (75), Section of Molecular and Cellular Biology, University of California, Davis, Davis, California 95616

James R. Sellers (23), Laboratory of Molecular Cardiology, National Heart, Lung, and Blood Institute, National Institutes of Health, Bethesda, Maryland 20892

Michael P. Sheetz (129, 227), Department of Cell Biology, Duke University Medical Center, Durham, North Carolina 27710

Takashi Shimizu (167), Research Institute of Bioscience and Human-Technology, Higashi, Tsukuba, Ibaraki 305, Japan

Robert M. Simmons (1), MRC Muscle and Cell Motility Unit, Randall Institute, Kings College London, London WC2B 5RL, England

Elizabeth F. Smith (89), Department of Anatomy and Cell Biology, Emory University School of Medicine, Atlanta, Georgia 30322

James A. Spudich (1), Departments of Biochemistry and Developmental Biology, Beckman Center, Stanford University of School of Medicine, Stanford, California 94305

Yoko Y. Toyoshima[5] (167), Department of Biology, Ochanomizu University, Ohtsuka, Bunkyo-ku, Tokyo 112, Japan

Raul Urrutia (179, 253), Laboratory of Cellular Biology, National Institute for Deafness and Other Communication Disorders, National Institutes of Health, Bethesda, Maryland 20892, and Center for Basic Research in Digestive Diseases, Mayo Clinic, Rochester, Minnesota 55905

Taro Q. P. Uyeda (1), Departments of Biochemistry and Developmental Biology, Beckman Center, Stanford University School of Medicine, Stanford, California 94305

Ronald D. Vale (167), Department of Pharmacology, University of California, San Francisco, San Francisco, California 94143

Richard B. Vallee (65), Cell Biology Group, Worcester Foundation for Experimental Biology, Shrewsbury, Massachusetts 01545

[5] *Present address*: Department of Pure and Applied Sciences, College of Arts and Sciences, University of Tokyo, Meguro-ku, Tokyo 113, Japan.

Fei Wang (23), Laboratory of Molecular Cardiology, National Heart, Lung, and Blood Institute, National Institutes of Health, Bethesda, Maryland 20892

Hans M. Warrick (1), Departments of Biochemistry and Developmental Biology, Beckman Center, Stanford University School of Medicine, Stanford, California 94305

Henry G. Zot (51), Department of Physiology, University of Texas Southwestern Medical Center, Dallas, Texas 75235

PREFACE

Motor proteins that hydrolyze ATP to generate force for movement along cytoskeletal fibers are now thought to have a variety of important cellular and developmental functions in eukaryotes, such as controlling the steady-state structural organization of cytoplasm, moving and positioning intracellular macromolecules or organelles, as well as driving cell movement, cell division, flagellar movement, and muscular contraction.

It would be difficult to overstate the importance of the development, during the mid-1980s, of light microscopic "motility assays" for monitoring the activity of these motor proteins; the application of such assays has led to the identification, purification, and characterization of many novel motor proteins, and is illuminating the precise molecular mechanisms by which motor proteins generate force and motion.

Thus, the editors of *Methods in Cell Biology* recognize that a volume in this series should contain a comprehensive sample of methods for performing microscopic motility assays on purified myosins, dyneins, and kinesins (Chapters 1–12) and on crude cell extracts capable of supporting organelle transport and mitotic movements (Chapters 13–20). Each chapter represents a practical guide for any researcher who may wish to perform a particular type of motility assay. Consequently it is hoped that the volume will prove useful for a large number of investigators of the cytoskeleton and related areas of cell biology.

Jonathan M. Scholey

CHAPTER 1

In Vitro Methods for Measuring Force and Velocity of the Actin–Myosin Interaction Using Purified Proteins

Hans M. Warrick,★ Robert M. Simmons,† Jeffrey T. Finer,★ Taro Q. P. Uyeda,★ Steven Chu,‡ and James A. Spudich★

★Departments of Biochemistry and Developmental Biology
  Beckman Center, Stanford University School of Medicine
  Stanford, California 94305

†MRC Muscle and Cell Motility Unit
  Randall Institute, Kings College London
  London WC2B 5RL, England

‡Departments of Physics and Applied Physics
  Stanford University
  Stanford, California 94305

  I. Introduction
 II. Preparation of in Vitro Motility Assay Components
     A. Preparation of Myosin
     B. Preparation of N-Ethylmaleimide–Heavy Meromyosin
     C. Quick Freeze of Myosin and Heavy Meromyosin
     D. Preparation of Actin
     E. Labeling of Actin with Polystyrene Beads
     F. Motility Assay Surface Preparation
 III. In Vitro Assay for Myosin Velocity
     A. Basic Features
     B. Digital Image Analysis
     C. Extraction of Motor Events from the Position Data
 IV. In Vitro Assay for Myosin Force in the Motility Assay
  V. Components of the Optical Trap System
     A. Laboratory Requirements
     B. Microscopes
     C. Lasers
     D. Laser Beam Modulation and Deflection

E. Objectives
F. Illumination and Fluorescence
G. Condensers
H. Detection System
I. Microscope Stage
J. Feedback Control
VI. Future Directions
References

I. Introduction

Myosin is a class of molecular motor that causes unidirectional movement of actin filaments, using the chemical energy obtained from the hydrolysis of ATP. In both muscle and nonmuscle cells, myosins play critical roles in various forms of cellular movement and shape changes. Cytokinesis, directed cell migration by chemotaxis, capping of ligand-bound cell surface receptors, developmental changes in cell shape, and muscle contraction are only a few examples of events in which molecular motors of the myosin class are involved. In spite of decades of investigation, the molecular basis of the conversion of the chemical energy into mechanical work remains an enigma. Studies with muscle fibers have been important in defining many aspects of the contractile process; however, such studies are complicated by problems associated with the large number of motor molecules working simultaneously and asynchronously, as well as by the presence of a large number of components whose functions are in many cases unknown. *In vitro* motility assays provide an important approach to investigate myosin function using only a small number of purified components. Several *in vitro* motility assays have been used to quantitate velocity as a parameter of myosin function, and recently assays have been developed that measure force production. In the future, additional *in vitro* assays will need to be developed to probe other aspects of motor function (e.g., cooperativity, efficiency).

A number of *in vitro* movement experiments with extracts containing actin and myosin were reported in the 1970s, which formed the foundation for all subsequent work in this area (for review, see Kamiya, 1986). Early reports that purified actin and myosin can produce directional movement *in vitro* involved measuring the streaming of an actin- and myosin-containing solution in glass capillaries (Oplatka and Tirosh, 1973), movement of bundles of actin as measured by dark-field microscopy (Higashi-Fujime, 1985), and rotation of cylinders or pinwheels coated with actin in a solution of myosin (Yano, 1978; Yano *et al.*, 1982). A quantitative assay was later developed that used the oriented polar cables of actin filaments that are found in the giant internodal cells of the alga *Nitella axillaris*. Polystyrene beads coated with purified myosin were observed to move when placed on the *Nitella* actin array (Sheetz and Spudich, 1983). The velocity of the bead movement was shown to be dependent on the type of myosin used, the ATP concentration, pH, and ionic strength (Sheetz *et al.,* 1984). Two-headed proteolytic fragments of myosin were prepared (long heavy meromyosin and short heavy meromyosin) and found to be sufficient to move beads in this assay (Hynes *et al.,* 1987). A more defined *in vitro* assay was developed (Kron and Spudich, 1986) in which individual fluorescently labeled actin filaments were observed to move over a glass surface coated with myosin. Using this myosin-coated surface assay, it was possible to show that single-headed myosin molecules will support movement (Harada *et al.,* 1987) and, later, that the subfragment 1 (S1) of myosin is all that is needed to generate movement of actin filaments (Toyoshima *et al.,* 1987). S1 was also shown to be able to produce force by a microneedle assay developed by Kishino and Yanagida (1988). These experiments defined the minimum enzymatic unit of myosin capable of motor function.

One intrinsic property of the myosin enzyme is its step size, which is defined as the average distance that a myosin moves an actin filament per ATP hydrolyzed. A tightly coupled model for myosin action depends on a one-to-one relationship between the release of ATP hydrolysis products and a force-producing conformational change in myosin while bound to actin. This model places an obvious constraint on the maximum step size in that the displacement between binding and release of the myosin head from a moving filament cannot exceed a value twice the ~20-nm chord length of the myosin head. There are, however, other models in which no direct tight coupling is required between ATP hydrolysis and movement (e.g., Vale and Oosawa, 1990). The value of the step size could help to differentiate between these different models of myosin function, and therefore several approaches using *in vitro* motility assays have been used to estimate it. Although there is not yet general agreement regarding the value of the step size (reported experimental values range from 5 to 100 nm), the methods that have been used to date are summarized below.

There is a minimum length of actin filament, dependent on the density of myosin on the surface, for continuous movement in the *in vitro* motility assay. Filaments longer than this minimum length move continuously at the maximum speed, whereas shorter filaments dissociate from the surface in the presence of ATP. If one assumes that the shortest filaments undergoing continuous movement are the filaments that are held onto the surface by at least one strongly bound (stroking) myosin head, then simultaneous measurement of the total displacement of a population of minimum-length filaments and the resultant total amount of ATP hydrolyzed in the *in vitro* motility assay can be used to estimate the step size. The step size is calculated by dividing one parameter by the other to obtain the distance moved per ATP hydrolyzed (Toyoshima *et al.,* 1990; Harada *et al.,* 1990).

Another approach to estimating the step size was taken by Uyeda *et al.* (1990). With the assumption of a tightly coupled model, step size can be expressed as the product of t_s, the time of the strongly bound state in one ATP hydrolysis cycle, and v_0, the maximum velocity of movement.

$$d = v_0 t_s$$

Provided that the heads are cycling at their maximum rate, V_{max}, the relationship can be expressed as a function of f, the ratio of the strongly bound state time to the total cycle time (t_c).

$$d = v_0 * f * t_c$$

where $f = t_s/t_c$ and $t_c = 1/V_{max}$. The proportion, f, could be determined by fitting the observed velocity-versus-filament length data to yield another estimate of step size.

A third approach to estimating the step size used methylcellulose in the *in vitro* movement assay buffer, allowing observation of the movement of individual actin filaments on a small number of motor units (Uyeda *et al.*, 1991). Frequency analysis of the observed velocities showed that sliding speeds distribute around integral multiples of a unitary velocity. This discreteness was interpreted to result from differences in the numbers of motors interacting with each actin filament, where the unitary velocity reflects the activity of one motor unit. When this unitary velocity is combined with the cycle time of the myosin ATP hydrolysis another estimate of step size can be obtained.

A fourth approach to measuring the step size involves the use of optical tweezer technology (Ashkin *et al.*, 1986; Block, 1990; Chu, 1991). An optical tweezer, or laser trap, allows the manipulation of a single actin filament (via inert beads that are attached to the filament), so that the interaction of that filament with myosin molecules can be examined with much greater resolution and precision than heretofore possible (Simmons *et al.*, 1992). Besides the ability to more accurately measure displacements produced by the actin–myosin interaction, optical tweezers also allow measurement of the resultant motive force. Measurement of force *in vitro* (Kishino and Yanagida, 1988; Oiwa *et al.*, 1990, 1991) can generate considerable insight into motor function.

II. Preparation of *in Vitro* Motility Assay Components

Most of the components of the motility assay can be prepared using protocols described in two recent reviews (Kron *et al.*, 1991a,b). The descriptions that follow include modifications or extensions of those protocols. Protocols not covered in those reviews are also described.

A. Preparation of Myosin

The key to good movement in the *in vitro* motility assay is the quality of the myosin used. We have used full-length rabbit skeletal myosin, prepared by the standard procedure of Margossian and Lowey (1982), and its proteolytic fragments heavy meromyosin (HMM) and subfragment 1 (S1) (Kron *et al.*, 1991a). Gel electrophoresis is routinely used to monitor the purity of the preparations. Gels stained with Coomassie should show only the bands corresponding to the myosin heavy and light chains. Myosin can also be used from a wide variety of organisms including the cellular slime mold *Dictyostelium discoideum*, which has the particular advantage that it can be used to create recombinant myosins (Ruppel *et al.*, 1990; Uyeda *et al.*, 1992; Ruppel and Spudich, 1992).

An actin affinity step can greatly improve the quality of movement by reducing the number of partially inactivated motors in the myosin preparation. Myosin (30–100 μg/ml) is mixed with filamentous actin (final concentration 150 μg/ml) and with ATP (final concentration 2 mM). If the myosin used is competent to form filaments then a high salt concentration (200 mM) is required. After 5 to 10 minutes on ice, the mixture is centrifuged at 75K rpm for 10 minutes in a Beckman TL100 rotor to sediment the actin and the subset of myosin that is irreversibly bound to actin in the presence of ATP. If highly purified actin (Pardee and Spudich, 1982) is not used, the actin preparation may have an associated contaminating proteolytic activity. It is therefore advisable to use the supernatant containing the actin-affinity-purified myosin within a few hours. Additional purification of the myosin fragments using a sizing column (e.g., Superose 6, Pharmacia) can greatly extend the lifetime of the myosin fragment preparation by removing residual protease activity.

B. Preparation of N-Ethylmaleimide–Heavy Meromyosin

Myosin can be modified with *N*-ethylmaleimide (NEM) in the same step as the HMM cleavage (described in Kron *et al.*, 1991a) with the following modifications. The concentration of dithiothreitol (DTT) in the buffers is reduced to 0.1 mM. Immediately following the 10-minute incubation with α-chymotrypsin, the reaction is stopped by addition of phenylmethylsulfonyl fluoride (PMSF) to 100 μM. NEM is added to a final concentration of 1.5 mM and the mixture is incubated for an additional 20 minutes at 25°C. The modification is then stopped by addition of 9 vol of cold stop buffer [100 μM NaHCO$_3$, 100 μM ethylene glycol bis(β-aminoethyl ether) N,N^1-tetraacetic acid (EGTA), 10 mM DTT, 3 mM MgCl$_2$, and 100 μM PMSF] and clarified by high speed centrifugation.

C. Quick Freeze of Myosin and Heavy Meromyosin

Myosin has traditionally been stored in 50% glycerol at -20°C. The quality of *in vitro* movement can be preserved for longer periods by preparing HMM from myosin that was originally stored at -80°C following quick freezing in liquid nitrogen. Myosin is stored at a concentration of 20–30 mg/ml in the absence of glycerol and quickly frozen dropwise directly into liquid nitrogen. Immediately prior to use, myosin is thawed quickly in a 40°C water bath just to the point at which thawing is complete. It is then transferred to ice. HMM and NEM–HMM can also be stored by this technique.

D. Preparation of Actin

Filamentous actin is prepared from acetone powder of rabbit muscle (Pardee and Spudich, 1982). It can be stored on ice at relatively high concentrations (4 mg/ml) in 0.2 mM ATP for several weeks. If long filaments are desired, we recommend slight modifications in the actin purification protocol as follows: Use buffer A containing a somewhat higher ATP concentration (0.5 mM instead of 0.2 mM), use higher salt in the high-salt wash step (0.8 M instead of 0.6 M), and dialyze for a shorter period to depolymerize the F-actin (5 hours instead of overnight; this results in loss of actin that did not depolymerize, but the G-actin obtained is particularly good at polymerizing into long, stable filaments).

To label actin fluorescently, tetramethylrhodamine–phalloidin (Molecular Probes) is applied using the procedure described by Kron *et al.* (1991a). Considerable variations in filament labeling have been observed between different batches of tetramethylrhodamine–phalloidin. The labeled actin solution is stable several weeks if stored on ice, in the dark.

E. Labeling of Actin with Polystyrene Beads

Beads of diameter 0.2–1.0 μm are attached to actin filaments using NEM–HMM as the linking agent. Covaspheres (Duke Scientific MX Covaspheres) are mixed with an equal volume of NEM–HMM (~0.25 mg/ml) prepared as described and incubated for 1.5 hours at room temperature (22°C). The mixture is then diluted with 2 vol of bead wash buffer (10 mM Tris–Cl, pH 8, 1 mg/ml bovine serum albumin) and the beads are sedimented in a microcentrifuge. The beads are washed two to three more times by removing the supernatant, resuspending in bead wash buffer, and sedimenting by centrifugation. After the final wash the NEM–HMM beads are resuspended in the motility assay buffer (Kron *et al.*, 1991a), and they may be stored on ice for several days. The NEM–HMM beads are attached to rhodamine–phalloidin-labeled actin filaments just prior to use in the motility assay. Mixing ratios and conditions have to be determined empirically.

F. Motility Assay Surface Preparation

The surface to which the myosin units are attached plays an important role in determining the observed velocity in the assay. The exact nature of the contact of the myosin with the surface is largely uncharacterized. This connection must be flexible to allow the myosin to reach the actin filament in the solution above and yet have enough resistance so that the force generated by the motor unit can be transferred to the substratum. Very clean glass, nitrocellulose-covered surfaces, and siliconized surfaces have all been used to observe myosin-induced motility. Nitrocellulose-coated surfaces are easy to prepare and are very reliable, and nitrocellulose solutions from many different sources can be applied in different ways and all seem to work. Yanagida and co-workers (Harada *et al.*,

1987, 1990) have used primarily siliconized surfaces. We have examined a large number of siliconizing reagents and had until recently found them unsatisfactory, producing very variable and poor movement. The siliconizing agent dichlorodimethylsilane (DowCorning, Z1219) appears to be one exception to this rule. The coverslips are prepared by dipping them in a freshly prepared solution of 2% dichlorodimethylsilane in chloroform for a few seconds and then allowing them to dry completely before use. The observed filament velocities on this surface appear to have a tighter distribution than observed on nitrocellulose surfaces.

III. *In Vitro* Assay for Myosin Velocity

A. Basic Features

Two velocities appear characteristic of movement of actin filaments along molecules of a particular myosin. In a simple model of myosin function (Uyeda *et al.*, 1991), the maximum velocity, which is the velocity obtained over large numbers of active motor units, is related to the duration of the tightly bound step of the actin-activated myosin ATPase cycle. The minimum velocity, which is observed under very low motor concentrations, is related to the activity of a single motor and is limited by the duration of the total actin-activated myosin ATPase cycle time. At low myosin concentrations the addition of methylcellulose (0.5–1.75%) in the motility assay buffer is essential because it suppresses lateral movement of actin filaments and keeps the filaments from diffusing away from the myosin-coated surface.

A flow cell is prepared as described in Kron *et al.* (1991a). In brief, a surface-treated glass coverslip is placed face down on two parallel beads of grease (Lubraseal, Thomas) on a microscope slide, supported along two sides by strips of glass coverslip. If the grease lines are about 10 mm apart a flow cell of 50-μl volume is created. A smaller flow cell (10 μl) can be made if the grease lines are placed closer together; this allows the use of minimal amounts of myosin solution, which is important when the myosin is difficult to obtain. To use methylcellulose in the assay buffers, thicker spacers (coverslip No. 2) are required because of viscosity limits.

Images of rhodamine–phalloidin-labeled actin filaments are observed with a fluorescence microscope combined with a low-light-level camera. Very sensitive video cameras and image intensifiers generally suffer from substantial lag and persistence problems which could affect high-resolution velocity determinations. The images are best preserved if the output of the video camera is directly fed to an optical disk recorder (Panasonic, TQ-2028F). More economically, the images can be recorded in real time on $\frac{3}{4}$-in. video tape (U-matic, Sony) and selected images that contained filaments that are well separated and in good contrast are then transferred to the optical disks. The optical disk player has the ability to display stable images of successive frames under computer control.

Each video frame (every 1/30th second) from the record is sent to a frame grabber capable of resolving the image into a 640×512 matrix of intensities (Imaging Technology, PC-Vision Plus), and the digitized information is analyzed by computer. We find that each transfer of data from camera to recording device or recording device to computer benefits from the use of a time base corrector (Sony, MPU-100).

B. Digital Image Analysis

One of the better methods for determining the position of an object, such as the fluorescent image of an actin filament in a video frame, is to determine its centroid (Gelles *et al.*, 1988). The fluorescent image of a filament is more intense in the center than at its edges, and information about the position of the filament can be greatly improved if the digitized image is first filtered using a low-pass filter. In the work of Uyeda *et al.* (1991), the value of the averaging filter was determined empirically using data from a stationary image. At the magnification used, a 7×7 convolution was found to be the best for removing high-frequency noise. The convolution was calculated by determining the value of each pixel in the image and then replacing it with the average of its own value and the 48 nearest neighbors. This filter, when applied to the image, made the background more uniform. The next step was to determine the perimeter of the filament. A threshold intensity value was selected that separates filaments from other filaments as well as from the background. The absolute value of the threshold varied from record to record because of differences in illumination but usually appeared quite constant within a single record (more than 600 frames).

The location of the pixels that exceed the threshold can be used to define the location of the filament. The set of pixel locations that determine the edge of the object can be averaged to determine the centroid of that object. By use of this filter, the position of a nonmoving filament can be determined to a standard deviation of ± 14 nm. An even better centroid calculation could, however, be made if each point that exceeded the threshold value was weighted by its intensity and then averaged. A number of commercially available image processing software packages are available that can do the calculations involved in the above procedures (e.g., Bioscan; Optimas, Jandel Scientific; Java), but one must be careful because some programs round calculations to the nearest pixel, which could lead to the appearance of artificial displacements.

C. Extraction of Motor Events from the Position Data

The centroid of a filament when calculated every thirtieth of a second appears to contain noise from several sources. Filters can be applied to the data to reduce specific components of this noise. The directed movements of the filaments resulting from the action of motor molecules, at very low motor densities, are very close in magnitude to those expected from random Brownian movements. A number of filters can be used to separate the random events from those of directed motor actions.

The movement of a filament by a motor molecule occurs longitudinally, determined by the actin filament polarity. Therefore, much of the lateral movement of the filament probably results from noise in the image from a variety of sources. The position of the filament could be affected by changes in overall illumination, the stochastic nature of the fluorescent image, or Brownian motion of the filament when not strongly bound to motor units. Lateral movements could occur even if filaments are tightly bound to a motor unit because of elastic elements contained within the actin or myosin structure or the elastic nature of their interaction. One way to reduce the lateral component of the movement is to project the position-versus-time information onto a curve that best fits all the points. In the work of Uyeda *et al.* (1991) a fourth-order polynomial fit of the position data was used. Each point was then projected perpendicularly onto the fitted curve and given a new point. It should be noted that the application of this filter to the data set appeared to change the position of the data points only slightly, indicating that lateral displacement noise does not form a large component of the data.

The next step in filtering the data was to evaluate the change in position of the centroid and score it as to whether it occurred in the general direction of travel of the filament or opposed to it. These data, when summed over the time of the observations, were plotted as a displacement-versus-time curve.

The last filter applied to the data relied on the observation that when a filament moved, it appeared to move at a relatively constant velocity for a specific interval and then changed to a new one. The simplest interpretation of that observation would be that as a filament moved across a specific area of the surface that was sparsely coated with myosin molecules, it was in contact with a defined set of active motor units. While interacting with the same number of motor units, the filament should move at constant speed. When the filament advances to a new area containing a different number of active motors the velocity will change. It therefore seemed appropriate to draw lines through the regions of constant movement on the displacement-versus-time curve. In the work of Uyeda *et al.* (1991), when a segment of seven data points could be identified whose standard deviation after linear least-squares regression was less than a set limit (0.0175 μm), it was retained in the data set. The window of seven points was extended at either or both ends until the standard deviation of the fit of the line exceeded the limit. When that occurred the last point was deleted from both ends of the fit, and the regression line slope and length were stored. The window was advanced and the process repeated until the filament movement record was exhausted. The data from many filaments were pooled and displayed in histograms such as those shown in Figs. 1 and 2. It was observed that the set limit for linear fitting could be varied over a wide range ($4\times$) without significantly changing the appearance of the final histogram. The

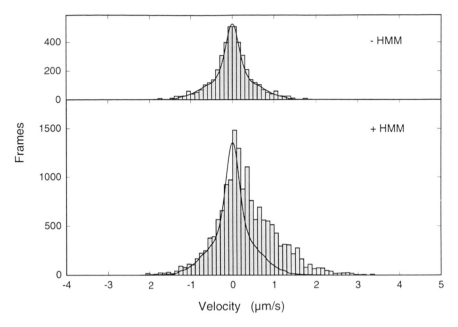

Fig. 1 Actin filament velocity distributions pooled from many experiments. Top: In the absence of myosin motor fragment (HMM) Brownian movements are observed. The direction of the movements could not be determined without motors so the velocities were accumulated on one side and reflected on the other, as Brownian movements would be expected in both directions with equal probability. Bottom: In the presence of low densities of HMM motors, the distribution contains additional positive velocities peaked at regular intervals (for details, see Uyeda *et al.*, 1991). The solid curved lines were drawn to fit the negative-velocity distribution and were then made symmetrical for the positive-velocity distribution. The purpose is to draw attention to the extra positive velocities apparent in the presence of HMM.

principal change with making the limit more restrictive was the reduction in the number of data points in the histogram.

Obviously from such an analysis it is essential to exclude as many artifacts as possible. A critical control involved the analysis of the movements of filaments observed in the absence of motors. Under these conditions, no pattern in the velocity peaks could be observed (Fig. 1, top). In addition, data sets containing active movements analyzed using two different objective lenses (63X and 100X) gave the same distribution of velocity peaks (Fig. 2) (Uyeda *et al.*, 1991). It appears unlikely, therefore, that the instrumentation could have created the observed quantized distribution seen when actin filaments moved on a sparse lawn of HMM molecules (Fig. 1, bottom).

Some distributions appeared to contain overrepresented negative velocities (Figure 2, top and middle), but as the number of observations increased the distribution of negative velocities became smoother. This effect is already seen in Fig. 2 (bottom), but is very apparent in Fig. 1 (botttom). Pooled velocities

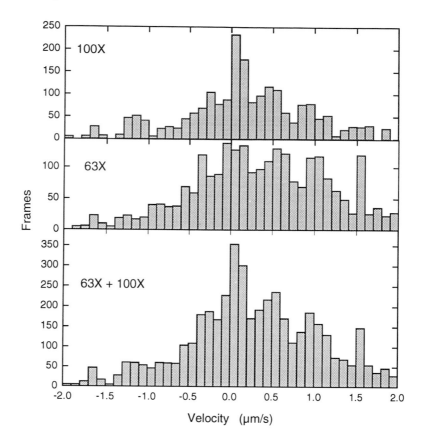

Fig. 2 Actin filament velocity distributions in the presence of low densities of HMM show similar peaks when different microscope objectives are used. Top, 100× objective; middle, 63× objective; bottom, pooled data from top and middle panels. For details, see Uyeda *et al.* (1991).

from many experiments show specific positive velocities which are overrepresented at regular intervals (Figs. 1 and 2, bottom). These quantized velocities could reflect the action of a small number of motor units operating on each filament. The minimum velocity peak would then represent the activity of a single HMM molecule and could be an intrinsic parameter of that motor. One motor preparation generated a slightly longer spacing in the distribution of its velocity peaks (Fig. 2), perhaps as a result of a small difference in the preparative protease digestion. In the majority of experiments a primary peak of velocity is seen at about 0.3 μm/s. If this were due to the activity of one HMM molecule and if each head of HMM was acting independently to drive the actin filament movement, then the step per ATPase cycle can be estimated as $d = t_{c}v$, where v is one-half the measured velocity for the two-headed molecule, or 0.15 μm/s, and t_{c} is about 30 milliseconds. The step size d would therefore be about 5 nm. If more HMM molecules were involved in producing the peak in velocity at 0.3 μm/s, then the step size would be even smaller. Note that a large step size of, for example, 50 nm should result in a peak in the histogram at about 3 μm/s, which was not seen in the experiments of Uyeda *et al.* (1991) (see Fig. 1, bottom).

IV. *In Vitro* Assay for Myosin Force in the Motility Assay

The *in vitro* motility assay can be extended to allow the measurement of force on single actin filaments by combining it with the technique of optical trapping. In the optical trap (or optical tweezers) technique, a laser beam is introduced into the back aperture of a microscope objective and brought to a focus in a diffraction-limited spot in solution. A refractile particle that comes into the vicinity of the focus is drawn into the laser beam and becomes trapped so that it can be manipulated in three dimensions (for a good introduction, see Ashkin *et al.*, 1986; Block, 1990; Chu, 1991). Particles that may be easily trapped range in size from about 0.1 to 10 μm and include latex beads and cells. This technique can be used to measure force on single actin filaments by attaching beads to the filaments as described earlier. The laser can be used to trap beads moving on the surface, and the force necessary to do so can be measured by lowering the laser power until the bead and filament resume movement.

The main requirement for such a system, in addition to a light microscope, is a continuous-wave laser. Using a Nd-YAG laser with a wavelength of 1.064 μm, we have found that a beam with power of 10 mW introduced into the back aperture of a microscope objective produces a force of 1–10 piconewtons (pN) on a 1-μm polystyrene bead. Forces of at least 100 pN can be obtained by using a higher-power laser. These force levels (1–100 pN) are well matched to what would be predicted for force on a single actin filament in a motility assay. The laser beam is reflected into the optical path of the microscope via a dichroic filter and eventually enters a high-numerical-aperture objective, in our case a Zeiss 1.25 NA Plan-Neofluar. The specimen is placed on a microscope stage and its position is finely controlled over a range of about 30 μm using piezoelectric transducers (PZTs). The PZT movements serve two purposes: they provide a means of steering a moving bead into the light trap, and they provide a means of force calibration.

With a PZT-controlled stage it is possible to steer a moving filament so that its attached bead moves into the vicinity of the trap. When the trap strength is sufficiently high the bead can be trapped in the laser beam. The trap strength can then be lowered by decreasing the laser power via an attenuator until the bead resumes movement. The bead continues moving immediately as expected, but only in about 20–30% of the cases. In the majority of the cases, the bead does not resume directed movement at all or undergoes a short delay. Likely causes of this behavior are detachment of the bead–actin complex from the myosin-coated surface at the moment it is pulled into the trap or breakage of the filament near the bead.

The laser trap force is calibrated by applying a viscous drag to the trapped bead. This is accomplished by moving the surrounding fluid using a PZT-controlled microscope stage. By Stokes' law, the force applied to the bead is directly proportional to the bead radius r, the fluid velocity v, and the fluid viscosity η.

$$\text{Stokes' force} = 6\pi\eta rv$$

To measure the escape force, the fluid velocity can be increased to the point at which a bead in solution can no longer be held by the trap.

The technique so far described gives reliable force values, but it cannot be used to monitor force fluctuations in real time. We have overcome this limitation by observing that when a force less than the escape force is applied to a bead, the bead is displaced from the center of the trap by up to 0.1 μm, but remains trapped. In fact, the amount of displacement from the center of the trap is directly proportional to the applied force up to about half of the escape force. Thus, if the bead position is closely monitored, it can be used to measure force. In our setup, the image of the bead is projected onto a quadrant photodiode detector which gives the subnanometer position of the bead in two dimensions with millisecond resolution. There is a severe potential artifact in this type of experiment: the bead position becomes less stable in the trap when the focus of the laser beam is too close to the surface, so this must be carefully avoided. The main limitation in this type of experiment is that a relatively weak trap has to be used to obtain a satisfactory signal-to-noise ratio, where most of the noise arises from the Brownian motion of the bead. With a weak trap, the stiffness of the trap is on the order of a few piconewtons per micrometer, which allows the bead to undergo considerable myosin-directed movement during force fluctuations. The stiffness of the trap can be enhanced considerably by the use of feedback from the bead position detector to deflect the laser beam. We use acousto-optic modulators to produce the deflection (described in more detail below). When the feedback circuit is operational, if the bead is pulled on by the filament so that the image of the bead moves away from the center of the quadrant detector, the laser beam is deflected, moving the position of the trap in the opposite direction to the direction of pull, so that the bead remains nearly stationary.

V. Components of the Optical Trap System

An excellent introduction to the theory and applications of the optical trap technique is the review by Block (1990). We concentrate here on the additional equipment and precautions needed to make measurements of force at the sub-piconewton level and distance at the subnanometer level. The main components of the optical trap system are the laser and the optical system. The function of the optical system is to provide an image of the preparation to a video camera and to project an image of a trapped bead onto a detector, which gives its x and y positions. The following is a description of the major parts of the system shown in Fig. 3.

A. Laboratory Requirements

For precision measurements, the optical and laser components need to be mounted on a good-quality antivibration table. The laboratory should be as isolated as possible from external noise and from noise within the laboratory, including air-conditioning or ventilation system noise.

B. Microscopes

Our first prototype apparatus is based on an upright microscope (Zeiss Axioplan). Two major problems have been encountered that affect the choice of subsequent instruments. The first is that upright microscopes are inherently subject to vibration, and this is increased when devices such as detectors and

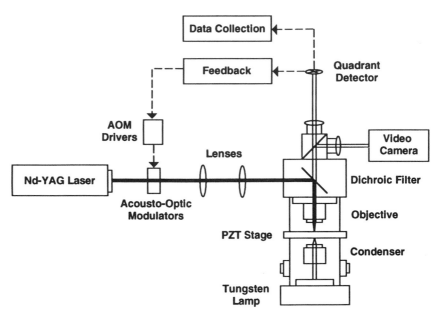

Fig. 3 Schematic drawing of a force-measuring laser trap microscope apparatus. The modified microscope is on the right and the laser on the left. The dark heavy line represents the path of the laser beam. The double line shows the path of the tungsten lamp light that hits the quadrant detector, giving the position of the bead. Dashed lines indicate electrical circuits for system control and data collection. AOM, acousto-optic modulator; PZT, piezoelectric transducer.

video cameras are mounted on them. We have dealt successfully with most of the vibrations by additional supports and by damping the upper part of the microscope and the detector by surrounding them with a column packed with lead shot, but it is likely that an inverted microscope or a purpose-built microscope would be superior. The second problem, which is likely to be encountered in all commercially available microscopes, is the difficulty of placing optical elements, particularly dichroic mirrors, in the optimal positions.

C. Lasers

Most optical traps have so far used continuous-wave Nd-YAG lasers, with a wavelength of 1.064 μm. This is probably not the optimal wavelength. Berns *et al.* (1992) have shown, using a tunable laser (Ti:sapphire, 0.7–1.0 μm), that at longer wavelengths the trap force per watt of laser power is greater and there is less damage to biological tissues. For small forces, single-mode diode lasers are an attractive and inexpensive option (Afzal and Treacy, 1992); however, Nd-YAG lasers, particularly diode-pumped lasers, are extremely stable. In addition, the quality of the beam is very good and they require very little maintenance or alignment. Some of them have a low-intensity noise ($<1\%$), which means that force in the trap can be controlled to this level, though the noise is much less than the equivalent Brownian noise from a bead, so it is not a primary factor. Beam pointing noise and drift are important factors. A change of angle of the beam by 1 microradian has the effect of moving the trap by (very approximately) 1 nm at the level of the specimen. The best lasers are stable over a period of half a minute (long enough for most measurements) to 1 microradian or better, and they have negligible high-frequency noise. Encasing the optical path also helps avoid changes of beam position caused by air currents. Lasers with noisy cooling fans or water supplies are to be avoided. An alternative method to reduce any error caused by pointing instability would be to detect the position of the beam at the microscope and use a feedback circuit via one of the devices described below to correct the error.

D. Laser Beam Modulation and Deflection

Modulation of the beam is needed to control its intensity and to deflect it. Commercially available electro-optic modulators or acousto-optic modulators can be used to modulate the beam intensity, though some of these also deflect the beam slightly, which could be detrimental to an experiment. In principle it would be possible to use intensity modulation rather than deflection to improve the stiffness of the trap, but we have not explored this. Several deflection systems are commercially available. We chose the acousto-optic modulator type (which is essentially a diffraction grating controlled by acoustic power) because it offers a wide bandwidth, high dynamic range, very low noise, and reasonable linearity; however, in the type we use (Isomet 1206C), throughput efficiency is poor, the beam is somewhat degraded in quality, and there is a limit on laser power (though other designs may be better). Other deflection devices are galvanometer-type mirror movements and PZT-operated mirror movements. Galvanometer-type mirror movements have a slower response (bandwidth ~1 kHz) and higher noise than acousto-optic modulators, but have no limit on laser power. PZT mirror movements feature a reasonable response (bandwidth up to 9 kHz), low noise, and high power capability, but they suffer from hysteresis and creep (which could be corrected with a beam detection and feedback system).

E. Objectives

In our experience, it is only the more expensive objectives that trap successfully (e.g., Zeiss: Plan-Neofluar and Plan-Apochromat ranges), possibly because they are better corrected in the infrared than cheaper objectives. The latter often trap to some extent in the xy plane but not in the z direction. Using an objective with an iris diaphragm (required for dark-field illumination), we found the minimum requirement for a numerical aperture to be 1.2. The upper limit to the laser power that an objective will tolerate is not known. The maximum we have used is 200 mW (measured at entry into the back of the objective). Phase-contrast objectives can also be used; however, they may be more susceptible to damage because of possible absorbance of laser light by the phase ring.

F. Illumination and Fluorescence

We have so far used conventional tungsten illumination for the bead detection system. With a low-noise power supply, 60-Hz ripple (or its multiples) is not apparent in the detector output, possibly in part because such variations in illuminating intensity are subtracted out. We have also tried mercury and xenon arc lamps, and these certainly give a more intense illumination, although they can suffer from flicker and oscillations of intensity (which might be counteracted by feedback control or appropriate normalization of the detector outputs). An even more promising approach, however, is use of a multimode diode laser with a stabilized power supply.

Epifluorescence is necessary for visualization of actin filaments, and if simultaneous fluorescence and bead detection are to be achieved, there is a restriction on the wavelength of illuminating light that can be used. With a dichroic mirror for rhodamine, light is not transmitted through the microscope optical path below 565 nm. Furthermore, the source for bead illumination must have a wavelength that is well clear of the rhodamine emission (600–650 nm). Simultaneous fluorescence and detection could be achieved by filtering the tungsten illumination; however, this would result in a considerable loss of intensity. Again, laser diode systems are an attractive option. Fortunately, the cheapest laser diode systems with reasonably high power have wavelengths around

800 nm. At this wavelength range, the illuminating light and the fluorescence imaging light could easily be separated with a dichroic mirror.

G. Condensers

We use a conventional bright-field condenser (Zeiss 1.2–1.4 NA). An alternative, which we have tried, is dark-field illumination, which increases the contrast of the bead image considerably, although with reduced intensity, so that with conventional illumination there is little to choose between bright-field and dark-field. Using bright-field illumination, we increase the contrast of a bead by using the low focus position of the microscope so that the bead appears dark on a light background and by adjusting the aperture of the condenser to maximize the detector output for a given bead movement. This is simply done by trapping a bead and applying a square wave to the laser beam deflection system.

H. Detection System

We chose to measure bead position using a photodiode quadrant detector, because such detectors are simple to use and give a better signal-to-noise ratio than other devices (e.g., CCD arrays). Light levels are high, so there is no advantage in using photomultipliers. The main restriction is on useful range of movement, which is limited to about half the diameter of the bead. The x and y positions of the centroid of the bead are derived by appropriate additions and subtractions of the outputs from the four photodiode amplifiers. Noise on the bead position signals arise from several sources: the illuminating light, dark noise from the photodiodes, Johnson noise in the feedback resistor of the photodiode current-to-voltage amplifier, and noise in the amplifier itself. Modern operational amplifiers (e.g., Burr Brown, OPA 121) are essentially noise free, and Johnson noise can be counteracted (at least up to the point where the amplifier is saturated) by increasing the size of the feedback resistor (R), as gain is proportional to R but noise to $\sqrt{R}$. Dark noise can be reduced by choosing the smallest quadrant detector (e.g., Hamamatsu S1557), as dark noise is proportional to photodiode area. An additional gain in using a small detector is that the magnification of the microscope can be lower, leading to a more compact design and less liability to vibration. Extraneous noise in the illumination system is dealt with above, but an irreducible factor is the shot noise caused by the quantum nature of light. Thus, if the other sources of noise are kept to a minimum, signal-to-noise ratio depends on the intensity of the illumination light and also on the extent to which "empty" illumination can be reduced: illumination of the photodetector that does not change in intensity when the bead changes position contributes only noise. With conventional tungsten illumination, the root-mean-square noise level on detection of a 1-μm bead is a few nanometers for a bandwidth of 1 kHz. It is likely that this can be improved by an order of magnitude or more by the use of a more powerful source of illumination.

Equally, this would allow the use of smaller beads. So far we have used the simplest system available, although it seems that there are considerable advantages in a more sophisticated approach based on differential interference contrast (Denk and Webb, 1990).

I. Microscope Stage

We use two piezoelectric transducer blocks (Physik Instrumente, P-771.00) to provide movement over distances of about 50 μm when tracking a bead or when calibrating force. These have very low noise and little long-term drift and the frequency response is about 0.5 kHz, but they suffer from creep and it would be preferable to replace them for tracking by a motor-controlled movement. So far we have not located a commercially available system with a noise level that is guaranteed to be below 1 nm and that can be conveniently mounted in place of a microscope stage.

J. Feedback Control

We use purpose-built feedback amplifiers with integral and direct feedback to control bead position. Force on a bead is derived from the difference between the bead position and the trap position, with a correction for viscous force. Additional circuitry deals with switching on the feedback system using sample and hold circuitry to avoid switching artifacts when a bead is already under force. The application of feedback can be seen in Fig. 4, where fluctuations in the position of a bead can be controlled by feedback circuits dynamically changing the position of the trap.

VI. Future Directions

Much research remains to be done with the existing techniques for measuring force and velocity. In particular, there is a need to rapidly characterize the mutated forms of myosin molecules that are now becoming available (Ruppel and Spudich, 1992; Uyeda *et al.*, 1992). With improved software for image analysis and fluctuation analysis, it should become possible to characterize several new mutants a day, at least with respect to maximum velocity, calculated unitary force, and probably the major rate constants affecting attachment and detachment. In addition, experiments need to be done to clarify some fundamental issues of *in vitro* assays, for example, the effect on filament velocity of varying the ATP concentration at low myosin density (Pate and Cooke, 1989) and direct determination of the number of molecules that can bind to actin filaments under different surface conditions. Further work is required to understand the theory of the optical trap for a range of experimental conditions (i.e., size of bead, strength of trap), and fluctuation analysis needs to be extended to more complex schemes of the actin–myosin interaction.

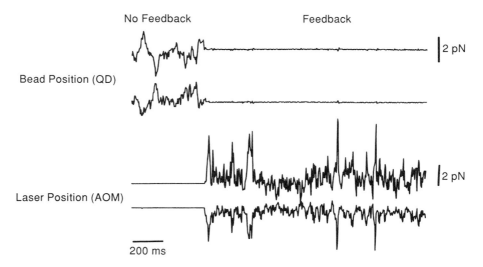

Fig. 4 The laser trap microscope can be used to examine the force applied to a bead carrying an actin filament that is interacting with a myosin-coated surface. The upper pair of traces show the displacement of a bead in the x and y directions, respectively. The lower pair of traces show the position of the trap in the x and y directions, respectively. When feedback circuits are activated, the position of the trap can be changed so that the bead is held in a constant position. QD, quadrant detector; AOM, acousto-optic modulator.

There remain, however, some fundamental obstacles to the quantitative interpretation of results from the present motility assay technique. When an actin filament is free to interact with a number of myosin molecules in a random array on a surface, there are questions concerning the effect of myosin heads in the "wrong" orientation and also concerning the interactions of the actin filament with the surface itself. Some of these difficulties can be avoided by the use of native or synthetic thick filaments rather than isolated molecules; however, just as in membrane channel noise analysis, new developments of the technique are likely to lie in recording and analyzing single events. Partly for technical reasons, it is difficult to achieve a satisfactory arrangement for recording single-molecule interactions with the methods used so far. One reason is that (as already mentioned above) at very low myosin head densities, actin filaments diffuse away from a myosin-coated surface. This could in principle be dealt with by the use of methylcellulose (Uyeda *et al.,* 1990), but in practice, it has been found that filaments with beads attached will not move on a surface when the beads are more than 0.1 μm in diameter, presumably because the larger beads become snagged by the methylcellulose gel (J. Finer, unpublished results). Beads of this diameter are at present outside of the range of detection for measurements at adequate bandwidth. Our approach is to avoid these difficulties, and also minimize the problems over interaction between the trapped actin filament and the surface, by attaching two beads to a filament and trapping each bead in a separate optical trap. In preliminary experiments, we have succeeded in trapping such filaments, stretching them to reduce lateral movement of the filament, and then lowering the filament onto a bump (glass microsphere) on a surface coated with myosin (J. Finer and R. Simmons, unpublished results). In this configuration, the actin filament and associated beads remain free from the surface except by way of interaction with one or a few myosin molecules. There seems to be no technical reason why this approach should not lead to the recording of single-molecule interactions, using surfaces that are sparsely populated with myosin or with synthetic myosin thick filaments.

Recording of unitary mechanical events should lead rapidly to an unambiguous determination of the force and step size of a myosin molecule. Beyond that, it should be possible to undertake a kinetic analysis of the myosin cycle, in much the same way as the kinetics of membrane channel events have been analyzed. Particular points of interest will be: Does the myosin molecule have one, or more than one, force producing step? Is there a rapid equilibrium, as often assumed, between the attached states underlying force generation? How does the coupling of force, displacement, and the kinetic events occur?

References

Afzal, R. S., and Treacy, E. B. (1992). Optical tweezers using a diode laser. *Rev. Sci. Instrum.* **63**(4), 2157–2163.

Ashkin, A., Dziedzic, J. M., Bjorkholm, J. E., and Chu, S. (1986). Observation of a single-beam gradient force optical trap for dielectric particles. *Opt. Lett.* **11**, 288–290.

Berns, M. W., Aist, J. R., Wright, W. H., and Liang, H. (1992). Optical trapping in animal and fungal cells using a tunable, near-infrared titanium-sapphire laser. *Exp. Cell Res.* **198**, 375–378.

Block, S. M. (1990). Optical tweezers: A new tool for biophysics. *In* "Noninvasive Techniques in Cell Biology" (J. K. Foskett and S. Grinstein, eds.), Vol. 9, pp. 375–402. Wiley-Liss, New York.

Chu, S. (1991). Laser manipulation of atoms and particles. *Science* **253**, 861–866.

Denk, W., and Webb, W. W. (1990). Optical measurement of picometer displacements of transparent micropic objects. *Appl. Opt.* **29**, 2382–2391.

Gelles, J., Schnapp, B. J., and Sheetz, M. P. (1988). Tracking kinesin-driven movements with nanometre-scale precision. *Nature (London)* **331**, 450–453.

Harada, Y., Noguchi, A., Kishino, A., and Yanagida, T. (1987). Sliding movement of single actin filaments on one-headed myosin filaments. *Nature (London)* **326**, 805–808.

Harada, Y., Sakurada, K., Aoki, T., Thomas, D. D., and Yanagida, T. (1990). Mechanochemical coupling in actomyosin energy transduction studied by in vitro movement assay. *J. Mol. Biol.* **216**, 49–68.

Higashi-Fujime, S. (1985). Unidirectional sliding of myosin filaments along the bundle of F-actin filaments spontaneously formed during superprecipitation. *J. Cell Biol.* **101**, 2335–2344.

Hynes, T. R., Block, S. M., White, B. T., and Spudich, J. A. (1987). Movement of myosin fragments in vitro: Domains involved in force production. *Cell (Cambridge, Mass.)* **48**, 953–963.

Kamiya, N. (1986). Cytoplasmic streaming in giant algal cells: A historical survey of experimental approaches. *Bot. Mag. (Tokyo)* **99**, 441–467.

Kishino, A., and Yanagida, T. (1988). Force measurements by micromanipulation of a single actin filament by glass needles. *Nature (London)* **334**, 74–76.

Kron, S. J., and Spudich, J. A. (1986). Fluorescent actin filaments move on myosin fixed to a glass surface. *Proc. Natl. Acad. Sci. U.S.A.* **83**, 6272–6276.

Kron, S. J., Toyoshima, Y. Y., Uyeda, T. Q. P., and Spudich, J. A. (1991a). Assays for actin sliding movement over myosin-coated surfaces. *In* "Methods in Enzymology" (R. B. Vallee, ed.), Vol. 196, pp. 399–416. Academic Press, San Diego.

Kron, S. J., Uyeda, T. Q. P., Warrick, H. M., and Spudich, J. A. (1991b). An approach to reconstituting motility of single myosin molecules. *J. Cell Sci.* **14s,** 129–133.

Margossian, S. S., and Lowey, S. (1982). Preparation of myosin and its subfragments. *In* "Methods in Enzymology" (D. Frederiksen and L. Cunningham, eds.), Vol. 85, pp. 55–71. Academic Press, New York.

Oiwa, K., Chaen, S., Kamitsubo, E., Shimmen, T., and Sugi, H. (1990). Steady-state force-velocity relation in the ATP-dependent sliding movement of myosin coated beads on actin cables in vitro studied with a centrifuge microscope. *Proc. Natl. Acad. Sci. U.S.A.* **87,** 7893–7897.

Oiwa, K., Chaen, S., and Sugi, H. (1991). Measurement of work done by ATP-induced sliding between rabbit muscle myosin and algal cell actin cables in vitro. *J. Physiol. (London)* **437,** 751–763.

Oplatka, A., and Tirosh, R. (1973). Active streaming in actomyosin solutions. *Biochim. Biophys. Acta* **305,** 684–688.

Pardee, J. D., and Spudich, J. A. (1982). Purification of muscle actin. *Methods Cell Biol.* **24,** 271–289.

Pate, E., and Cooke, R. (1989). A model of crossbridge action: The effects of ATP, ADP, and P_i. *J. Muscle Res. Cell Motil.* **10,** 181–196.

Ruppel, K. M., and Spudich, J. A. (1992). Mutagenesis of the ATP- and actin-binding sites and other highly conserved regions of *Dictyostelium* myosin. *Mol. Biol. Cell* **3,** 191a.

Ruppel, K. M., Egelhoff, T. T., and Spudich, J. A. (1990). Purification of a functional recombinant myosin fragment from *Dictyostelium discoideum. Ann. N. Y. Acad. Sci.* **582,** 147–155.

Sheetz, M. P., and Spudich, J. A. (1983). Movement of myosin-coated fluorescent beads on actin cables in vitro. *Nature (London)* **303,** 31–35.

Sheetz, M. P., Chasan, R., and Spudich, J. A. (1984). ATP-dependent movement of myosin in vitro: Characterization of a quantitative assay. *J. Cell Biol.* **99,** 1867–1871.

Simmons, R. M., Finer, J. T., Warrick, H. M., Kralik, B., Chu, S., and Spudich, J. A. (1992). Force on single actin filaments in a motility assay measured with an optical trap. *In* "Mechanism of Myofilament Sliding in Muscle Contraction" (H. Sugi and G. H. Pollack, eds.) Plenum, New York.

Toyoshima, Y. Y., Kron, S. J., McNally, E. M., Niebling, K. R., Toyoshima, C., and Spudich, J. A. (1987). Myosin subfragment-1 is sufficient to move actin filaments in vitro. *Nature (London)* **328,** 536–539.

Toyoshima, Y. Y., Kron, S. J., and Spudich, J. A. (1990). The myosin step size: Measurement of the unit displacement per ATP hydrolyzed in an in vitro assay. *Proc. Natl. Acad. Sci. U.S.A.* **87,** 7130–7134.

Uyeda, T. Q. P., Kron, S. J., and Spudich, J. A. (1990). Myosin step size. Estimation from slow sliding movement of actin over low densities of heavy meromyosin. *J. Mol. Biol.* **214,** 699–710.

Uyeda, T. Q. P., Warrick, H. M., Kron, S. J., and Spudich, J. A. (1991). Quantized velocities at low myosin densities in an in vitro motility assay. *Nature (London)* **352,** 307–311.

Uyeda, T. Q. P., Ruppel, K. M., and Spudich, J. A. (1992). Site directed mutagenesis of the 50K/20K junction domain of Dictyostelium myosin. *Mol. Biol. Cell* **3,** 44a.

Vale, R. D., and Oosawa, F. (1990). Protein motors and Maxwell's demons: Does mechanochemical transduction involve a thermal ratchet? *Adv. Biophys.* **26,** 97–134.

Yano, M. (1978). Observation of steady streamings in a solution of Mg-ATP and acto-heavy meromyosin from rabbit skeletal muscle. *J. Biochem. (Tokyo)* **83,** 1203–1204.

Yano, M., Yamamoto, Y., and Shimizu, H. (1982). An actomyosin motor. *Nature (London)* **299,** 557–559.

CHAPTER 2

Myosin–Specific Adaptations of the Motility Assay

James R. Sellers, ⋆ **Giovanni Cuda,** ⋆ **Fei Wang,** ⋆
and Earl Homsher†

⋆ Laboratory of Molecular Cardiology
National Heart, Lung, and Blood Institute
National Institutes of Health
Bethesda, Maryland 20892

† Department of Physiology
School of Medicine, Center for Health Sciences
University of California, Los Angeles
Los Angeles, California 90024

I. Introduction
II. Description of Equipment
    A. Microscope
    B. Video Imaging System
    C. Image Processor
    D. Recording System
    E. Quantitation System
III. Basic Assay Procedure
    A. Buffers
    B. Preparation of Coverslip Surfaces
    C. Construction of the Flow Cell
    D. Preparation of Rhodamine–Phalloidin-Labeled Actin
    E. Setup of Motility Assay
    F. Improvement of the Quality of Movement
    G. Use of Methylcellulose
IV. Adaptations of the Assay
    A. Assays with Synthetic Thick Filaments
    B. Assays with Soluble Myosin or Its Subfragments
    C. Assays with Native Thick Filaments
    D. Assays of Smooth Muscle and Nonmuscle Myosin II

E. Assays with Very Slow or Very Fast Myosins
         F. Assays for Thin-Filament Regulatory Systems
         G. Antibody-Assisted Assays
         H. Methods to Introduce a Load onto the Moving Actin Filaments
   V. Quantitation of Rate of Actin Filament Sliding
         A. Description of the Expert Vision System
         B. Methods of Reporting Data
      References

I. Introduction

Muscle shortening is due to the sliding of actin filaments toward the center of bipolar myosin filaments. The ability of purified myosin to actively propel actin filaments can be directly visualized and studied using an *in vitro* motility assay system in which myosin bound to a glass surface translocates fluorescently labeled actin filaments (Harada *et al.,* 1987; Kron and Spudich, 1986). The initial discovery that aided the development of this assay was the ability to image single actin filaments that were complexed with a fluorescent rhodamine derivative of phalloidin, a mushroom toxin that binds stoichiometrically to actin molecules within a filament (Yanagida *et al.,* 1984). Rhodamine–phalloidin serves two important functions in this respect. In addition to providing the fluorescence required to visualize the actin filaments, it also greatly decreases the critical concentration for actin polymerization. This allows for the extreme dilution of actin necessary to observe individual actin filaments in the microscope.

The ability to directly observe the behavior of single actin filaments as they are translocated by myosin has spawned a number of variations on the assay which have contributed to a more detailed understanding of the molecular basis for the interaction of actin and myosin (Kishino and Yanagida, 1988; Uyeda *et al.,* 1991; Ishijima *et al.,* 1991; Sellers and Kachar, 1990).

The motility assay is a quick and sensitive method to quantitate the interaction of myosin with actin. Small amounts of myosin are required for the assay (0.3–10 μg myosin per assay). It complements the traditional measurement of the actin-activated MgATPase activity of myosin. There are several distinctive characteristics of the sliding actin *in vitro* motility assay. First, the rate of movement of actin filaments is largely independent of the density of myosin heads bound to the surface (Collins *et al.,* 1990). This means that, over a wide range of myosin concentrations introduced into the flow cell, the velocity is constant. This is not strictly true for the case of very low myosin densities where the rate of movement of actin filaments does decrease dramatically as the density is lowered. This observation was exploited by Uyeda *et al.* (1991) in an attempt to determine the distance that myosin translocates an actin filament per

ATP hydrolyzed. The rate of movement is also not dependent on the length of actin filaments (Harada *et al.*, 1990; Collins *et al.*, 1990; Takiguchi *et al.*, 1990). Second, the direction of movement of an actin filament is determined by the inherent polarity of the actin filaments (Takiguchi *et al.*, 1990; Harada *et al.*, 1990). Third, each myosin isoform has its own characteristic rate of translocation of actin filaments. Reported actin translocation rates measured under the same ionic conditions for various myosins, vary widely ranging from 0.04 μm/s for brush border myosin I or phosphorylated human platelet myosin II in the absence of tropomyosin to a high value of about 6 μm/s for native clam thick filaments (Table I). These dramatic differences in speeds require different approaches in the motility assay. Fourth, the rate at which any given myosin translocates actin filaments is a function of the ionic conditions and temperature of the assay.

II. Description of Equipment

A good discussion of the various components required for the *in vitro* motility assay system is found in Kron *et al.* (1991), and an excellent description of video microscopy in general is available (Inoue, 1986). A schematic of a typical system is shown in Fig. 1.

A. Microscope

The sliding actin *in vitro* motility assay can be conducted with either an upright or an inverted microscope equipped with an epifluorescence illuminator and a C-mount adaptor for coupling the imaging system. For cases where myosin filaments are also to be imaged by differential interference contrast

Table I
Rate of Actin Filament Sliding by Different Myosins[a]

Myosin	Speed (μm/s)	Myosin	Speed (μm/s)
Chicken brush border myosin I	0.04	Rabbit fast skeletal muscle myosin	4.54
Human platelet myosin	0.04	Rabbit slow skeletal muscle myosin	0.81
with tropomyosin	0.08	*Mercenaria mercenaria* pink muscle	6.0[b]
Bovine aorta smooth muscle myosin	0.13	myosin	
Turkey gizzard smooth muscle myosin	0.22	*Mercenaria mercenaria* white muscle	2.5[b]
with tropomyosin	0.50	myosin	
Rabbit cardiac myosin	0.49	*Mytilus* anterion byssus retractor	1.1[b]
		muscle myosin	

[a] Conditions: Buffer M, 25°C.
[b] These samples also included 0.2 mM CaCl$_2$.

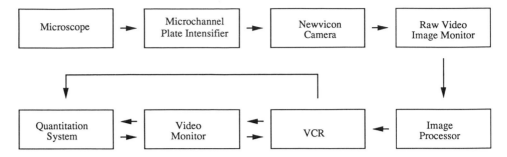

Fig. 1 Schematic diagram of a typical system for recording and quantitating the *in vitro* motility of fluorescently labeled actin filaments by myosin.

(DIC) microscopy, the microscope must be equipped with a high-numerical-aperture condenser (NA = 1.4), Wollaston prisms, a polarizer, and an analyzer.

In addition, it is important for quantitative work to accurately control the temperature of the sample. This can be accomplished in several ways using both commercially available and homemade systems. It should be noted that, when oil immersion is used, the objective becomes a large heat sink and should be thermally jacketed. A water jacket system for the objective and the microscope stage has been described by Kron *et al.* (1991).

B. Video Imaging System

A low-light video imaging system must be selected that is best suitable for the application. The choices available generally fall into two classes. The first type of system uses silicon intensifier target (SIT) or intensifier silicon intensifier target (ISIT) cameras. SIT cameras are minimally sensitive enough for imaging the rhodamine–phalloidin-labeled actin, whereas ISIT cameras offer more than enough sensitivity. These cameras, however, are characterized by a rather poor temporal resolution at low levels of illumination. The other choice of imaging systems is a microchannel plate intensifier coupled with a Newvicon or CCD camera. The microchannel plate-intensified CCD systems offer sufficient sensitivity and high temporal resolution.

The wide range in motility rates shown in Table I places differing demands on the imaging system. The choice, therefore, must involve a compromise between temporal resolution and sensitivity. In addition, spatial resolution and linearity of the image field are important. For some applications, linearity of the light signal must also be considered.

C. Image Processor

Though not strictly required for the assay, the use of an image processor can greatly improve image quality and can provide limited quantitation functions. The basic requirements of an image processor are background subtraction, frame averaging, and contrast enhancement. Most commercial systems can also provide distance scales, measure distances between two points, display a stopwatch, and create legends on the video screen. The more sophisticated programs can overlay images, create collages, display pseudocolor images, quantitate image intensities, and detect motion.

D. Recording System

The most common recording system is a VHS or super VHS video recorder. Another videotape format available is the U-matic or $\frac{3}{4}$-in. video recorder, which offers higher resolution, but is considerably more expensive than the VHS formats and is therefore not as common. The lack of general availability of the U-matic format may create problems in presenting data at remote locations. A higher-resolution but considerably more expensive format is an optical memory disk recorder (OMDR).

E. Quantitation System

The ability to quantitate the movement of actin filaments is the primary goal of most motility assays. Only one commercially available system is in use (Motion Analysis, Santa Rosa, CA). Its use will be described in a later section. Various investigators have written tracking programs in conjunction with commercially available frame grabbers (Uyeda *et al.*, 1991; Work and Warshaw, 1992).

III. Basic Assay Procedure

A. Buffers

Buffer A	4 mM imidazole (pH 7.0), 2 mM MgCl$_2$, 0.1 mM ethylene glycol bis(B-aminomethyl ether) N,N'-tetraacetic acid (EGTA), 3 mM NaN$_3$, 1mM dithiothreitol (DTT)
Buffer B	0.5 M NaCl, 10 mM 4-morpholinepropanesulfonic acid (Mops) (pH 7.0), 0.1 mM EGTA, 1 mM DTT
Buffer C	20 mM KCl, 10 mM Mops (pH 7.2), 5 mM MgCl$_2$, 0.1 mM EGTA, 10 mM DTT
Buffer M	20 mM KCl, 10 mM Mops (pH 7.2), 5 mM MgCl$_2$, 1 mM ATP, 0.1 mM EGTA, 10 mM DTT, 0.7% methylcellulose (4000 cps), 2.5 mg/ml glucose, 0.1 mg/ml glucose oxidase, 0.02 mg/ml catalase (These last four components were purchased from Sigma Chemical Company, St. Louis, MO)

B. Preparation of Coverslip Surfaces

Myosin can be successfully bound to a number of surfaces in preparation for the motility assay. If myosin filaments are to be studied, clean glass coverslip surfaces (typically 18 mm², No. 1) can be used. Uncoated glass surfaces do not work well for monomeric myosin or its subfragments. In these cases, it is better to use surfaces coated with nitrocellulose or silicon.

An easy way to prepare nitrocellulose surfaces is to place one drop of 1% nitrocellulose (superclean grade, EF Fullam, Schenectady, NY) in amyl acetate using a Pasteur pipet onto the surface of water in a 10-cm-diameter round dish. The amyl acetate evaporates in about a minute, leaving a nitrocellulose film floating on the surface of the water. Coverslips (typically 18 mm², No. 1) are carefully placed on the film using forceps and allowed to sit for 1 to 2 minutes. The film between adjacent coverslips is torn away using forceps and a free nitrocellulose-coated coverslip is retrieved by holding it with forceps and pushing it down into the water where it is inverted and lifted out. This motion prevents surface action from damaging the film. The coated coverslips are allowed to air-dry. The surfaces are generally used within a few hours of preparation.

The preparation of silicon-coated coverslips has been described elsewhere (Takiguchi *et al.,* 1990).

C. Construction of the Flow Cell

A simple disposable flow cell can be constructed using slivers (about 2–3 mm in width) of a No. 1 coverslip cut with a diamond scribe as spacers. Two tracks of Apiezon M grease (London, England) are applied parallel to the long axis of a microslide using a syringe with a 21-gauge needle. Two coverslip slivers are laid about 8–12 mm apart on the outside of each track of grease. An 18-mm² coverslip is placed on top of the tracks. If the coverslip is coated with either nitrocellulose or silicon, it is placed film side down. The coverslip is pressed onto the grease track using the forceps. The flow cell typically has a volume of about 40–50 μl. To conserve sample, much smaller cells can be constructed by cutting the coverslip in half, reducing the width of the chamber, and using thinner coverslip slivers (No. 0) as spacers. Flow cells with a sample volume as small as 10 μl can be easily made and used. Fast-drying glue or fingernail polish can be used instead of grease to construct the flow cell.

D. Preparation of Rhodamine–Phalloidin-Labeled Actin

1. Place 60 μl of 3.3 μM rhodamine–phalloidin (Molecular Probes, Eugene, OR) in methanol into an 1.8-ml Eppendorf microcentrifuge tube and dry using a Speed Vac concentrator (Savant, Hicksville, NY). A conventional lyophilizer or nitrogen stream can be used, but these techniques leave the dried rhodamine–phalloidin over a large surface area.

2. Redissolve the rhodamine–phalloidin in 2–3 μl of methanol and add 90 μl of buffer A. Vortex well or place into a sonicating water bath for 5 minutes to dissolve the powder.

3. Add 10 μl of a freshly diluted 20 μM F-actin solution in buffer A. Gently mix the actin using a pipet tip that has been cut with a razor to produce a wider bore, to prevent shearing of the actin filaments. Incubate for at least 2 hours prior to use and store on ice in the dark until needed. The rhodamine–phalloidin-labeled actin can be used for several weeks. Dilute to 20 nM with buffer C prior to use in the motility assay.

E. Setup of Motility Assay

The following example is for myosin that is applied to the flow cell as monomers. This procedure would also be applicable when using rabbit skeletal muscle myosin subfragments or soluble nonmuscle myosin I. Modifications needed to examine the movement of F-actin by myosin in filamentous form are described as needed.

1. Introduce myosin at concentrations between 30 and 200 μg/ml dissolved in buffer B into the flow cell which is inclined at an angle of about 30°.

2. After about 60 seconds, wash the flow cell with 2 to 3 vol of buffer B containing 0.5 mg/ml bovine serum albumin. If myosin filaments are to be applied, dilute the myosin in a lower-ionic-strength buffer (<150 mM KCl) where myosin forms stable filaments. Wash the unbound myosin filaments out and coat the surface with 0.5 mg/ml bovine serum albumin in the buffer used to dilute the myosin.

3. Incubate for 60 seconds and then wash with 2 vol of buffer C. (See Section III,F for notes on how to improve the quality of movement.)

4. Add 2 vol of 20 nM rhodamine–phalloidin-labeled actin in buffer C.

5. After 30–60 seconds start the reaction with the addition of two flow cell volumes of buffer M. Because of the viscosity of the methylcellulose in the solution, it is usually necessary to use a piece of filter paper or a tissue to wick the solution through the flow cell. The presence of methylcellulose in buffer M is optional and is not always necessary. See the discussion below for more information. The combination of glucose, glucose oxidase, and catalase in buffer M serves to scavenge oxygen and prevent photobleaching of the rhodamine–phalloidin-labeled actin filaments (Kishino and Yanagida, 1988). The oxygen-scavenging components are added to buffer M from stock solutions when commencing an experiment and are used within 1 hour of dilution. Degassing of the buffers prior to use can also be used to retard photobleaching.

6. Blot off any solution that has spilled onto the coverslip and apply one drop of immersion oil to this surface. Place the slide onto the stage and equilibrate to temperature.

7. Focus on the fluorescently labeled actin filaments on the underside of the coverslip. Actin filaments should be readily apparent and moving. If too few or too many actin filaments are present on the surface, adjust the concentration of fluorescently labeled actin added in step 4 and/or the time of incubation with fluorescently labeled actin in subsequent slides. In some cases it may be necessary to increase the concentration of myosin added to the coverslip if too few actin filaments are bound.

F. Improvement of the Quality of Movement

Check the "quality" of movement. Ideally, greater than 80% of the actin filaments on the surface should be moving in a constant manner at any given instance. Poor quality of movement is characterized by substantial numbers of actin filaments that are immobile, filaments that are moving in a "stop-and-go" manner, and shearing of actin filaments into small pieces that are difficult to image. The probable reason behind less than optimal movement is the presence of damaged "rigor"-like myosin heads which bind tightly to actin filaments and do not undergo the normal ATP-dependent cycling motion. Actin-binding proteins that may contaminate the myosin preparation, such as tropomyosin, can also contribute to poor-quality movement. This type of behavior makes quantitation of the movement very difficult.

To obtain smooth, consistent movement of actin filaments, high-purity, freshly prepared myosin or its subfragments should be used. The myosin should be stored in 1–5 mM DTT to prevent oxidation and should be periodically clarified to remove aggregates. Freshly prepared myosin typically exhibits good-quality movement for 1–10 days depending on the source. Storage of myosin in 50% glycerol at $-20°C$ or by quick freezing and storage in liquid nitrogen inveritably detract from the quality of movement. We find that chymotryptic heavy meromyosin (HMM) (Margossian and Lowey, 1982) prepared from either freshly prepared or glycerol-stored rabbit skeletal muscle myosin gives very good movement for 3–7 days even if the parent myosin itself does not exhibit good-quality movement. The reason for this is not known, but may be related to proteolytic removal of the rigor-like myosin heads in the preparation.

If myosin exhibits poor-quality movement there are two procedures that should help by effectively removing these rigor-like heads from the myosin preparation. The first of these involves a modification of the procedures described earlier (Section III,E): After washing out the unbound myosin and coating the coverslip with bovine serum albumin (step 2), wash the flow cell with 2 volumes of a solution of 5 μM phalloidin-labeled actin (note *not* rhodamine–phalloidin) in buffer C that was sheared by passage through a syringe with a 26-gauge needle several times. Let this incubate for 1–2 minutes and then wash with two flow cell volumes of buffer C containing 1 mM ATP followed by a wash with buffer C, before continuing on to step 4 as described in Section III,E. This treatment apparently ties up the rigor-like myosin heads with unlabeled actin and allows for much smoother motion of the subsequently added rhodamine–phalloidin actin. This procedure does not adversely affect the velocity of movement if performed on a preparation of myosin that is already moving well. It can be incorporated as a routine part of the assay.

The other modification that can enhance the quality of movement is to treat the soluble myosin in buffer B before commencing the assay with a molar stoichiometric quantity of F-actin in the presence of 2 mM MgATP. The sample is centrifuged for 10 minutes at 100,000 rpm in a TL-100 centrifuge (Beckman, Palo Alto, CA) to sediment the actin and the associated myosin rigor heads. Very little active myosin is lost during this procedure; however, determination of the protein concentration after centrifugation will be complicated by the presence of ATP and a small amount of unsedimented F-actin. This may be a problem in some cases where the concentration of myosin must be known accurately.

G. Use of Methylcellulose

If the actin filaments are moving erratically and tend to dissociate from the surface either entirely or partially via the leading or trailing end, the most likely reason is insufficient interaction between the actin filaments and the myosin bound to the surface. This is most commonly seen at either low myosin surface densities or higher ionic strengths when no methylcellulose is used.

The presence of methylcellulose in the motility buffer decreases the lateral diffusion of free actin filaments and allows the movement of actin filaments to be observed even when the concentration of myosin on the surface is not sufficient to sustain motility in its absence (Uyeda et al., 1990). Methylcellulose is virtually essential to observing the movement of actin filaments by vertebrate smooth muscle and nonmuscle myosin monomers (Umemoto and Sellers, 1990). It has no effect on the rate of movement of actin filaments by myosin when compared with samples that are undergoing smooth and continuous motion in the absence of methylcellulose. In general, methylcellulose is *not* required when observing the motion of actin filaments by myosin filaments, rabbit skeletal muscle myosin monomers, or subfragments of rabbit skeletal muscle myosin. It is useful when assaying the movement of actin filaments over very low surface concentrations of myosin, at higher ionic strengths where the interaction between actin and myosin is weakened, or with myosins that interact more weakly with actin even under low-ionic-strength conditions, such as smooth muscle myosin.

IV. Adaptations of the Assay

The motility assay is a versatile assay subject to many modifications depending on the system being studied. In this section some of the possible variations are discussed.

A. Assays with Synthetic Thick Filaments

Myosin exists primarily as filamentous structures in skeletal muscle and smooth muscle. The ability of myosin to self-assemble *in vitro* is well known and synthetic thick filaments can be easily prepared by dialysis or dilution of soluble myosin at high ionic strength (0.5 M) to a buffer of lower ionic strength (150 mM or lower). Study of the movement of actin filaments by synthetic thick filaments is advantageous in that it might allow for the detection of possible cooperative regulatory effects. In addition, it represents the movement of actin filaments by locally high densities of myosins heads.

The preparation of synthetic thick filaments must be carefully controlled as larger, more aggregated structures than are observed in muscles are usually obtained *in vitro* (Koretz, 1982). In general, it is advisable that short synthetic thick filaments be prepared for use in the assay. When smooth muscle or nonmuscle myosin is used, suitable synthetic thick filaments can be prepared by dialysis of the myosin at 1–2 mg/ml in an initial buffer containing 0.5 M KCl against a buffer containing 150 mM KCl, 10 mM MgCl$_2$, 0.1 mM EGTA, and 1 mM DTT at pH 7.0. Thick filaments can be bound to virtually any surface, including uncoated, clean glass coverslips.

F-actin tends to shear more easily when it is moved by filamentous myosin rather than monomeric myosin. This may be due to the higher local concentration of myosin heads that may exist on myosin filaments or possibly to the more severe changes in angle that may occur when an actin filament switches from one thick filament to another. Another potential problem in working with synthetic thick filaments is knowing whether the myosin has remained bound to the coverslip as filaments during the course of preparing the flow cell and conducting the assay. For rabbit skeletal myosin, this does not seem to be a problem. Smooth muscle and nonmuscle myosin thick filaments are generally more labile *in vitro;* however, this lability does not seem to be a problem in the motility assay where the myosin filaments are bound to glass or coated-glass surfaces (Okagaki *et al.,* 1991). To avoid any potential problem arising from this lability, one study added a monoclonal antibody that was shown to bind to the rod portion of smooth muscle myosin and stabilize the thick filaments to ensure that there was no significant depolymerization of myosin (Warshaw *et al.,* 1990).

B. Assays with Soluble Myosin or Its Subfragments

There are several reasons for conducting the motility assay using monomeric myosin. First, the observed movement of actin filaments is smoother and there is less shearing than when myosin is in the form of synthetic thick filaments. Second, most kinetic studies of actin–myosin interaction have been performed with the subfragments of myosin rather than with the intact myosin molecule. The ability to measure the movement of actin filaments by subfragments of myosin allows for a much more meaningful comparison of the motility with the kinetics of actin–myosin subfragment interaction. In addition, members of the nonmuscle myosin I family do not form filaments and, therefore, must be assayed as monomers.

There is no difference in the rate of translocation of actin filaments by either myosin filaments or monomers when assayed under the same ionic conditions (Table II). Likewise, some myosin subfragments, such as chymotryptic HMM from rabbit skeletal muscle, generate the same rate of movement as the parent myosin molecule (Table III). The movement of actin filaments by some rabbit skeletal muscle myosin subfragments such as tryptic HMM and various subfragment 1 (S1) preparations, however, can be either faster or slower than that of the parent myosin, which means that some caution should be exercised in the selection of subfragments. Little information is available about the behavior of subfragments from other myosins; however, preliminary data suggest that platelet chymotryptic HMM appears to move actin filaments at the same rate as does platelet myosin (G. Cuda and J. R. Sellers, unpublished). Even though monomers and filaments of smooth muscle myosin move actin filaments at the same rate, there is a major difference in the two systems, in that the presence of methylcellulose is virtually essential to obtain consistent movement when using monomers. We have found this to be true of a number of different myosins.

C. Assays with Native Thick Filaments

1. Advantages of Using Native Thick Filaments in the Motility Assay

Some molluscan muscles have exceedingly large thick filaments that can measure up to 50 μm in length and can be visualized by video microscopy (Fig. 2, Table IV). These thick filaments can be easily isolated from molluscan muscles without dissolution. They will be referred to as "native" thick filaments to distinguish them from synthetic thick filaments described above.

The native molluscan thick filaments support movement of actin filaments provided calcium is present in the assay buffer. They present several advantages over monomeric myosin or synthetic thick filaments. The first advantage is that,

Table II
Rate of Actin Filament Sliding by Myosin Filaments or Monomers

Myosin	Conformation	Speed (μm/s)	Reference
Rabbit skeletal muscle myosin	Monomers	3.9	Toyoshima *et al.* (1987)
	Filaments	3.5	Toyoshima *et al.* (1990)
Turkey gizzard smooth muscle myosin	Monomers	0.25	Umemoto and Sellers (1990)
	Filaments	0.28	Umemoto and Sellers (1990)
Human platelet myosin	Monomers	0.17	Umemoto and Sellers (1990)
	Filaments	0.14	Umemoto and Sellers (1990)

Table III
Rate of Actin Filament Sliding by Different Myosins and Their Subfragments

Myosin	Subfragment	Speed (μm/s)	Reference
Rabbit skeletal	Whole myosin	3.9	Toyoshima *et al.* (1987)
	Chymotryptic HMM	4.6	Toyoshima *et al.* (1990)
	Tryptic HMM	7.5	Toyoshima *et al.* (1987)
	Papain–Mg S1	1.8	Toyoshima *et al.* (1987)
	Papain–EDTA S1	1.7	Toyoshima *et al.* (1987)
	Chymotryptic S1	0.9	Toyoshima *et al.* (1987)
Rabbit cardiac	Whole myosin	0.51	Margossian *et al.* (1991)
	Chymotryptic S1	0.37	Margossian *et al.* (1991)

using video-enhanced differential interference contrast (DIC) microscopy, the native thick filaments can be imaged in the light microscope with the same objective and camera as used to follow the movement of the fluorescently labeled actin filaments. Therefore, it is possible to first capture the DIC image of the field of native thick filaments and then to switch to a fluorescence mode and image the actin filaments as they slide over the thick filament whose position is known. This allows for a correlation of the movement of an actin filament with its position on the native thick filament.

The second advantage is that this is perhaps the most structurally appropriate way to study the movement of actin filaments by myosin. The third advantage is a practical one. These thick filaments present a long linear track of myosin heads which should be useful for a variety of techniques such as measurement of the

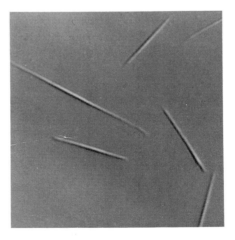

Fig. 2   Video-enhanced DIC light micrograph of native thick filaments from clam pink muscle.

Table IV
Lengths of Various Native Thick Filaments[a]

Muscle	Length (μm)	Reference
Limulus Striated	4	Levine and Kensler (1985)
Scallop striated	2	Vibert and Craig (1983)
Mytilus edulis pedal retractor	17	Castellani *et al.* (1983)
Caenorhabditis elegans	10	Epstein *et al.* (1985)
Mytilus edulis (anterior byssus retractor)	50	Yamada *et al.* (1989)
Mercenaria mercenaria pink	10	Sellers and Kachar (1990)
Mercenaria mercenaria white	40	J. R. Sellers, unpublished

[a] In most cases these values represent the maximal length measured. Many filaments may be shorter as a result of breakage during preparation.

step size of myosin, optical trapping of actin filaments, and measurement of force using microneedle displacement assays. Fourth, the diameter of the native molluscan thick filaments is quite large, up to 0.5 μm depending on the species and muscle, which means that most of the movement will be taking place above the glass surface of the coverslip. This minimizes possible artifactual surface interactions.

2. Preparation of Molluscan Thick Filaments

Buffer D   10 mM ATP, 10 mM MgCl₂, 1 mM EGTA, 20 mM Mops (pH 7.0), 3 mM NaN₃, 1 mM DTT, 0.1 mM PMSF

This preparation was used by Kachar and Sellers (Sellers and Kachar, 1990) and is a modification of the method of Yamada *et al.* (1989). It works with all molluscan muscles tested.

1. Remove muscle (about 1 g) from the animal, dice into small pieces, and place in buffer D. Rinse the muscle twice in buffer D and then homogenize twice in 5 ml of buffer D in an Omnimixer (Sorvall, Newtown, CT) using the small cup at a setting of 7.5, for 7 seconds on ice.

2. Mix the homogenized muscle with an equal volume of buffer D containing 0.1% Triton X-100.

3. After letting it sit 5 minutes on ice, sediment the homogenate at 500*g* for 5 minutes.

4. Carefully remove the supernatant and sediment for 20–30 minutes at 5000*g*. Gently resuspend the pellet in buffer D. The time of the centrifugation is dependent on the length of the thick filaments. Long thick filaments pellet more

rapidly than short ones. The time of centrifugation is optimized to produce a pellet that contains the thick filaments, but that is loose enough to easily resuspend in buffer D.

5. Repeat the two centrifugation steps (3 and 4) twice.

6. Gently resuspend the pellet in 1–2 ml of buffer D, incubate on ice, and sediment at 500*g* for 5 minutes prior to use. The activity is good for about 1 day. Sodium dodecyl sulfate–polyacrylamide gels show that the thick filament preparation consists primarily of myosin, paramyosin, and a few minor bands. There appears to be little actin in the preparation as judged from lack of staining with rhodamine–phalloidin.

This preparation has been tried unsuccessfully on muscles from lobster and *Limulus polyphemus,* the horseshoe crab. In both of these cases a considerable amount of actin was still present and there was no translocation of the actin filaments. It may be necessary to depolymerize actin with DNase I, gelsolin, or severin in these cases to detect movement of fluorescently labeled actin filaments.

3. Modification of the Motility Assay for Use with Native Thick Filaments

Several modifications of the basic assay described in Section III,E are useful for working with native thick filaments.

1. If the movement of actin filaments on single thick filaments is to be observed, it is necessary to dilute the thick filament concentration to about 30 μg/ml.

2. As the thick filaments are so long, it is useful to apply a drop containing the thick filaments to the microscope slide and to build the flow chamber described in Section III,C around the droplet, with the coverslip being placed on top at the end. This gives more randomly oriented thick filaments which help prevent lateral associations which may develop and be difficult to detect when the thick filaments orient under flow.

3. Rhodamine–phalloidin-labeled actin should not be added in the absence of ATP (i.e., at step 4 Section II,D above), but should be present at a concentration of 2 nM when added with buffer M to observe motility. The presence of methylcellulose in buffer M is not required.

4. To obtain video-enhanced DIC images of the native thick filaments, a microscope equipped with a 1.3 or 1.4 NA objective and condenser should be used in the critical illumination mode (Kachar *et al.,* 1987). The DIC video image of the native thick filaments should be background substracted, averaged, and contrast enhanced using a video image processor.

D. Assays of Smooth Muscle and Nonmuscle Myosin II

Vertebrate smooth muscle and nonmuscle myosin II share a number of properties with respect to their ability to move actin filaments in the *in vitro* motility assay. Both require phosphorylation of the 20-kDa light chain for activity, both translocate actin filaments at relatively slow rates (see Table I), and the rate of actin translocation by either of these myosins is increased two- to fourfold by the presence of tropomyosin (Umemoto and Sellers, 1990) (see Table I). To observe the movement of actin filaments by monomeric myosin bound to the surface with either of these myosins, it is necessary to use methlycellulose in the motility buffer.

The 20-kDa light chain of smooth muscle and nonmuscle myosin II can be phosphorylated by myosin light-chain kinase at Ser-19 prior to iniating the assay (Umemoto and Sellers, 1990). Alternatively, the myosin bound to the surface of the coverslip can be phosphorylated by myosin light-chain kinase during the course of the assay. Inclusion of 1 μg/ml turkey gizzard myosin light-chain kinase, 0.1 μM calmodulin, and 0.2 mM CaCl$_2$ into buffer M will result in very rapid phosphorylation of the myosin. The reaction is essentially complete in less than a minute, and the resulting speed of actin filament sliding is identical to that obtained if myosin previously phosphorylated on Ser-19 is applied to the flow cell. It is known that myosin light-chain kinase can also phosphorylate Thr-18 although at a much lower rate; however, myosin phosphorylated at Ser-19 alone moves actin filaments at the same rate as myosin diphosphorylated at Ser-19 and Thr-18 (J. R. Sellers, unpublished data). Thus it is not essential to control for the additional phosphorylation at Thr-18 when conducting the phosphorylation reaction on the surface.

E. Assays with Very Slow or Very Fast Myosins

The wide range of possible velocities when using various myosins presents different imaging problems. In this section the adaptations required to work with a diverse range of speeds are discussed.

1. Slow Myosins

Some myosins move actin filaments very slowly. In particular, phosphorylated human platelet myosin II and chicken brush border myosin I translocate actin filaments at a rate of 0.04 μm/s at 25°C under the ionic conditions stated in the footnote to Table I. Imaging the actin filaments with minimal photobleaching for the 3- to 5-minute observation periods required to accurately track the movement of these slowly moving actin presents problems. Several procedures may help.

1. The use of a sensitive imaging system such as an ISIT camera or a microchannel plate-intensified Newvicon camera is recommended so that the illumination levels can be greatly reduced.

2. Frame averaging using an image processor helps to image the actin filaments that are being illuminated at low light levels. The filaments may be moving so slowly that even a 16- to 32-frame average does not result in significant distortion of the movement.

3. Increasing the magnification helps when computer tracking of the actin filaments is used. In general, the diameter of the field on the video screen should be between 20 and 50 μm.

4. Use of the oxygen scavenging system described in Section III,E is virtually essential to tracking the movement of slow moving actin filaments.

5. Use of an excitation shutter, possibly computer controlled via the image processor and coupled to an optical memory disk recording system to capture a single averaged frame of video data at precise times in coordination with the opening of the shutter, would represent a means to image for a long period with minimal photodamage.

6. The temperature of the assay can be increased. In general, myosin appears to have a Q_{10} for movement of about 2 for temperatures between 20° and 40°C (Anson, 1992; Homsher *et al.*, 1992).

7. Other conditions for the assay that result in faster movement may be employed. For instance, tropomyosin bound to the actin filaments increases the rate of movement of human platelet myosin II by about twofold (see Table I). Tropomyosin inhibits the movement of brush border myosin I, however, so this does not represent a universal cure for slow movement (Collins *et al.*, 1990). Increasing the KCl concentration in the assay to about 100 mM also increases the rate of movement of most myosins by a factor of 2 to 3 compared with that obtained at 20 mM (Umemoto and Sellers, 1990; Takiguchi *et al.*, 1990; Homsher *et al.*, 1992). Again, increasing the ionic strength has no effect on the rate of movement of actin filaments by brush border myosin I (Collins *et al.*, 1990).

2. Fast Myosins

When fast-moving actin filaments are imaged by a SIT camera or a microchannel plate-intensified Newvicon camera the image often exhibits considerable lag, giving the appearance of "comet tails" trailing the bright actin filament. This is due to the poor temporal resolution of these systems. Imaging rapidly moving actin filaments is best accomplished using a microchannel plate intensifier coupled with a CCD camera. CCD cameras have very high temporal resolution. The temporal resolution of the microchannel plate intensifier is a function of the illumination levels. At the high illumination levels that are possible when the actin filaments are only in the field for 15–30 seconds there is little lag in the imaging. Frame averaging of the video data is, in general, not required and may actually distort the position of the actin filament when the actin filaments are moving fast. Lowering the magnification of the system, which, in turn, boosts the apparent brightness of the actin filaments, may help to prevent lag in the imaging system.

F. Assays for Thin-Filament Regulatory Systems

The *in vitro* motility assay can also be used to study thin-filament regulatory systems such as troponin, tropomyosin, caldesmon, and calponin (Shirinsky *et al.*, 1992; Honda and Asakura, 1989; Harada *et al.*, 1990; Okagaki *et al.*, 1991). It is usually necessary to have the thin-filament protein present in the motility buffer at a concentration greater than that of the actin to ensure full saturation of binding, as the actin concentration remaining in the flow cell at the initiation of the reaction is very low.

G. Antibody-Assisted Assays

Some myosin preparations contain a mixture of two or more isoforms which may translocate actin filaments at different rates and which might be difficult to separate by conventional biochemical techniques. One possible way to study the separate functions of the individual isoforms is to make isoform-specific antibodies to the carboxyl-terminal regions of the individual isoforms and use these antibodies to select out the myosin of choice (Fig. 3). This can be accomplished if the sequence of the myosin is known and if unique regions can be identified in the carboxyl terminus that will allow the synthesis of specific peptides to which specific antibodies can be elicited. The isoform-specific antibodies can be affinity purified using the peptide coupled to Affi-Gel columns (Bio-Rad Laboratories, Richmond, CA).

The isoform-specific, affinity-purified IgG is introduced into the motility assay flow cell at a concentration of 0.1 to 0.5 mg/ml in buffer B and allowed to incubate for 20 minutes. The unbound antibody is washed out and the surface is coated with 0.5 mg/ml bovine serum albumin in buffer B. After 1 minute, myosin containing a mixture of isoforms is added at a concentration of 0.1 mg/ml in buffer B. After 5 minutes, the flow cell is washed with buffer B to remove unbound myosin. The motility assay setup then proceeds from step 4 as described in Section III,E. This procedure has been used successfully to study the movement of actin filaments by the slow human skeletal muscle myosin isoform which is purified from human skeletal muscle biopsies as a mixture of fast and slow myosin isoforms (G. Cuda *et al.*, 1993).

H. Methods to Introduce a Load onto the Moving Actin Filaments

The motility assay as described earlier usually operates under unloaded conditions in which actin filaments are being propelled by myosin at V_{max}. From mechanical studies of muscle fibers it is known that velocity of contraction

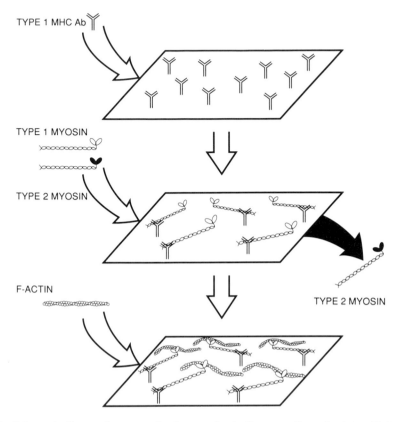

Fig. 3 Schematic diagram demonstrating the use of an isoform-specific antibody to affinity select a particular isoform of myosin from a mixture of isoforms.

decreases as the load imposed on the muscle increases. It is therefore interesting to attempt to make the *in vitro* system do work by introducing a "load" onto the actin filament that must be overcome by myosin to produce movement. There are several ways to accomplish this. Please note that simply increasing the viscosity of the solution does not impose a load on the actin filament.

One of these methods is discussed in Chapter 1 in this volume. One system involves the measurement of force exerted on a single actin filament attached to a glass microneedle as it moves over a surface coated with myosin. The other system uses an optical trap created by an infrared laser beam directed through the microscope. Both of these systems exert force on only a single actin filament at a time. Below, two systems are described that create a load throughout all the actin filaments in the system.

1. Mixtures of Fast and Slow Myosins

When myosins that translocate actin filaments at different rates are mixed in different ratios on the surface of the coverslip, attenuated speeds are obtained. Figure 4 shows such a curve. The hybrid velocities are dominated by the slower moving myosin species. This system can be modeled (Warshaw *et al.*, 1990; Tawada and Sekimoto, 1991) to give information about the relative force–velocity relationships of the individual myosins.

2. Tethering of Actin Filaments through the Use of an Actin–Binding Protein

A number of actin-binding proteins are known, for example, filamin and α-actinin. These two proteins can be used to introduce a load onto the actin filaments by virtue of their biochemical interaction with actin. Premixing of filamin or α-actinin with myosin prior to addition to the flow cell results in a dose-dependent slowing of the sliding velocity of actin filaments, ultimately resulting in complete inhibition of movement (Janson *et al.*, 1992). The presumed mechanism for this is that these proteins bind to the surface in some manner that allows their actin binding site to interact with the actin that is being moved over the myosin molecules which are also bound to the surface. If the

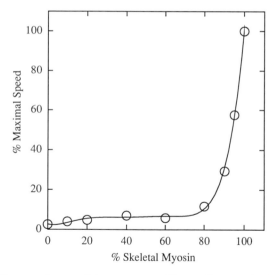

Fig. 4 Slowly cycling myosins retard the movement of fast cycling myosins by creating an internal load. In this experiment rabbit skeletal muscle myosin was mixed at varying ratios with phosphorylated human platelet myosin at a total myosin concentration added to the flow cell of 0.2 mg/ml. The slowly cycling human platelet myosin dominates the speed of actin filament sliding, presumably by creating an internal load on the actin filament.

interaction between the actin-binding protein and the actin filament is sufficiently strong, it presents a load to the myosin interacting with the actin. In these
systems, the on–off rates for binding to actin of the actin-binding proteins would
also be a determinant of the degree of load imposed.

V. Quantitation of Rate of Actin Filament Sliding

The goal of most experiments using the motility assay is to quantitate the
speed of actin filament sliding. This can be accomplished in several ways. The
simplest and least accurate method is to place a piece of plastic wrap onto the
video screen, mark the leading edge of an actin filament at time zero, and, after a
given time increment, and divide the distance by the time to obtain a speed. This
method is obviously prone to significant error and human bias.

A computer-assisted method for tracking moving objects has been described
in detail elsewhere (Sheetz *et al.*, 1986). This system, although obviously better
than the one described above, is still labor intensive and subject to human bias.

A virtually completely computer-driven method, the Expert Vision Motion
Analysis System (Motion Analysis, Santa Rosa, CA), is available. In addition,
several laboratories have developed roughly equivalent systems using commercially available frame grabbers and custom software (Uyeda *et al.*, 1991; Work
and Warshaw, 1992).

A. Description of the Expert Vision System

1. System Components

The Expert Vision system consists of a VP-110 videoprocessor, a 33-MHz
486-AT computer (this was upgraded from the original system) with a 40-MB
hard disk and 1 MB-RAM, a 640W × 400H VGA color monitor, and CellTrak
(Motion Analysis, Santa Rosa, CA) software for two-dimensional image analysis. The input to the video processor is via an EIA RS-170 standard 525-line
video at 60 Hz and a 2 : 1 interlace. The processor's resolution is 256H × 240V
pixels, and it is hardwired for image detection and acquisition, thresholding,
edge mapping, and windowing (the ability to select specific sections of the field
of view for processing).

2. Analysis Procedure

The following procedure to quantitate the movement of actin filaments has
been described previously (Homsher *et al.*, 1992). The first step in the analysis is
definition of the filament edges using standard image processing techniques
(Inoue, 1986). The VP-110 is hardwired to capture fields at user-specified rates
and allows manual adjustment of the gray level threshold (the minimum gray

value defined as being part of a filament) so that the processor-defined filament edges correspond to the image on the screen. If the threshold is set too low, the image outline is much larger than the filament image, and if it is set too high, the filament outline falls within the filament image or disappears altogether. The VP-110 hardware allows the experimenter to constantly monitor the filament outline as the filament moves about the field and thus ensures that the image outline corresponds to the image. Photobleaching should be minimized as it reduces pixel intensity relative to the threshold. Light intensity varies across the visual field (by improper defocusing, improper arc centering, mechanical vibration, arc wander, etc) and can produce spurious changes of the filament edges. The first two of these effects can be controlled, but the latter two can alter the filament edge image.

After the threshold is set, a segment of tape (typically 10–240 seconds) is played to the videoprocessor. In this software-controlled process, the experimenter defines the rate at which fields are acquired and the duration of data acquisition. Typically, 50–150 frames of video data are analyzed from a given segment of tape. Only the coordinates of the pixels forming the filament outline are stored in a file, greatly reducing the amount of data stored. Next a software routine is run in which the characteristics of a filament are defined (i.e., only those images whose length and width exceeded 0.45 μm and whose image area exceeded 0.24 μm^2 were considered filaments), and the centroids of the stored image outlines are calculated by averaging the x–y coordinates of the pixels forming the image edge. The centroid coordinates are then stored in memory. A third software routine is run that defines the path followed by the centroids to represent the filament movement. Here, adjacent stored fields of centroid locations are scanned. A ''search mask'' of a user-defined radius is centered over centroid 1 in field 1 and compared with the locations of centroids in field 2. The centroid in field 2 lying within the search mask area closest to centroid 1 in field 1 is taken as the position of centroid 1 in field 2. Each captured field in a sequence is used to generate a series of centroid paths representing the filament movement in the recorded segment. These paths can be printed out and compared with the observed filament movement in the videotaped sequence. Careful definition of both the frame grabbing rate and the search mask size is important to ensure the computer-determined filament trajectories correspond to the observed movement. Our tests have shown that the correspondence between the computer-generated filament paths and the observed paths in fields containing 20–30 filaments is greater than 95%. For the path determination, the experimenter also defines the minimum consecutive frames a filament centroid must be detected to be scored as a filament. This parameter should be set to correspond to about 20% of the total frames analyzed. This criterion protects against the paths spuriously generated by filaments entering and exiting from the edges of the scanned area and pixel noise. If no centroid is found within the search mask area (e.g., the filament dissociates from the surface, moves out of the field of view, breaks into filaments too small to fulfill the size requirement for a filament), the filament path is defined as ended.

After the path for each moving filament is identified, the speed of each filament between successive frames was calculated, and from these values the mean speed and standard deviation (SD) for each filament were calculated. The total time for the analysis of the mean and SD for each path in a segment of video tape containing 40–50 actin filament paths when 75 frames of video data are analyzed is about 30 seconds. Batch files can be written to process many data sets automatically.

3. Data Filtering

Not all actin filaments move constantly in the course of the assay. Merely presenting the mean and SD as determined in Section V,A,2 of all filaments would be meaningless under some experimental conditions, such as a case where the movement of filaments is being deliberately inhibited by regulatory proteins. Therefore, methods were devised to filter the computer-derived data. A separate routine is available using the Expert Vision Software of the Cell Trak system that allows the path of each individual filament to be displayed on the video screen. The entire filament path or any part of it can be selected by the experimenter for calculation of the filament's speed using a software routine written for that purpose. This routine allows one to select for analysis of only those sections of records during which, in the experimenter's judgment, movement is uniform. There is, however, the danger of bias in this time-consuming selection process, and methods to objectify and expedite this process have been devised.

Taking advantage of the Expert Vision's ability to calculate the mean and standard deviation of a large number of actin filaments, it was noted that filaments that appeared to be moving smoothly by eye typically had a ratio of the SD to the mean of less than 0.3, whereas the ratio for filaments moving erratically or those that were immobile usually greatly exceeded this value. A software program is used to filter the data to eliminate filaments whose SD:mean ratio is greater than 0.3 as the standard for definition of a smoothly moving actin filament. This process does not select on the basis of speed, but rather on the basis of uniformity of movement.

4. Factors Affecting Estimate of Speed

As described in Section V,A,3, the presence of immobile or irregularly moving actin filaments can decrease the estimate of the velocity. A second factor that could affect the estimated velocity in this computer-driven system is the sampling rate (frame grabbing rate). In the absence of noise, the sampling rate should be maximal (60 fields/s or 60 f/s) so as to most closely approximate the filament path. Three features militate against this practice. First, noise is present in the moving filament image, and as the sampling rate increases, the standard deviation of a filament's mean speed increases whereas the mean filament velocity does not. Second, the time required for analysis rises in direct proportion to the number of acquired fields. Third, the noise introduced by stationary filaments becomes a more serious problem at high sampling rates because the calculated rates can overlap those of the moving filaments. There are, however, two potential drawbacks to using sampling rates that are too low for the study of rapidly moving filaments. The first is truncation of curved filament paths and the consequent underestimation of speed. The second is that, at lower sampling rates, the search mask size must be increased and filaments moving into the search mask area interfere with the accurate tracking of filaments. If the sampling rate (in f/s) is roughly equal to the filament velocity (in μm/s) for magnifications of 0.22 μm/pixel, truncation and accurate filament identification are not problems.

B. Methods of Reporting Data

There are several methods of reporting data from motility assays. One method is to display a collage of photographs of video fields at different times during the assay which will demonstrate movement of the actin filaments. This type of analysis is not quantitative and is sometimes time consuming to construct in the absence of a sophisticated image processing system that will assemble the images onto one video screen (Fig. 5). Alternatively, the position of the centroid or that of the leading edge of each actin filament in a video field can be displayed at uniform increments of time. This analysis is visually less cluttered than images of the whole actin filaments and, in general, makes the same point.

As described in Section V,A,3, another method is to determine the mean speed and standard deviation for the movement of the population of actin filaments in the video field that are moving in a uniform manner. This analysis ignores filaments that are not moving or filaments that are moving erratically. It does not, however, allow for an estimation of the quality of movement, that is, how many filaments are stationary or erratically moving. It is sometimes important to know the behavior of the entire population of actin filaments, as some experiments are designed to study inhibition of filament movement by regulatory systems. An excellent way to display this is by constructing histograms of the speed of *all* of the actin filaments paths within the video fields. Such an example is shown in Fig. 6 where the movement of actin filaments by rabbit skeletal muscle heavy meromyosin is inhibited by caldesmon, a thin-filament-binding protein. This type of analysis shows that in the presence of subsaturating caldesmon, the movement of actin filaments is being inhibited in two ways. First, the average velocity of moving actin filaments is decreased, and second, the number of filaments that are moving at an experimentally detectable rate is decreased. The arrowhead marks the position of the mean of the filaments that are moving in a constant manner as determined by the filter routine de-

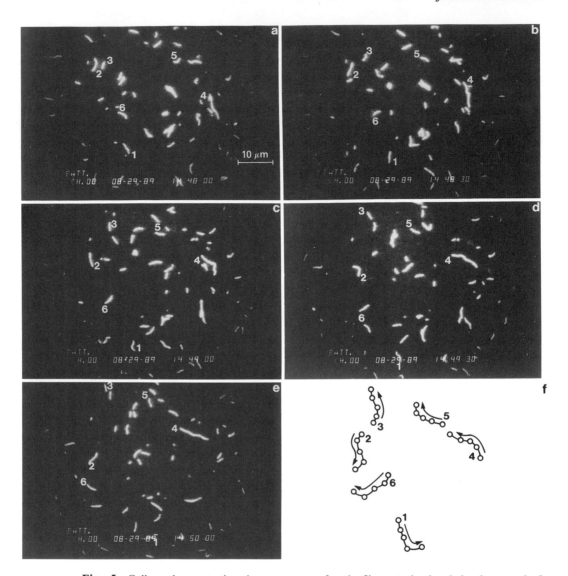

Fig. 5 Collage demonstrating the movement of actin filaments by brush border myosin I. (a–e) Direct photographs of the image from the monitor at 30-second intervals after initiation of the assay. (f) Position of the leading edge of each of the numbered filaments at 30-second intervals. This figure is taken from Collins *et al.* (1990).

scribed in Section V,A,3. The combination of these two graphic procedures adequately describes both the mean velocity of filaments that are moving constantly and the number of filaments that are moving in the assay. It can be seen using the combination of these two graphical techniques that at intermediate

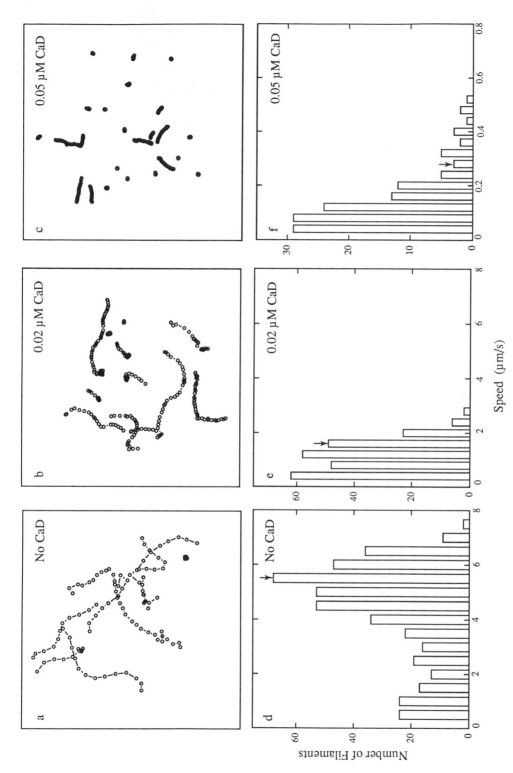

Fig. 6 Two methods of presenting motility data. This experiment is from Shirinsky *et al.* (1992) and involves the inhibition of actin filament sliding over skeletal muscle HMM by caldesmon (CaD) binding to the actin filament. (a–c) Data in the form of centroid plots corresponding to the actin filament paths from a single video record. (d–f) Histograms depicting the speed of all of the actin filaments from 5 to 10 video records. Arrowheads mark the position of the mean speed of the actin filaments that have a ratio of the standard deviation to mean speed of less than 0.3. Data points were taken at 2 f/s in the presence of (a,d) no caldesmon; (b,e) 0.02 μM caldesmon; and (c,f) 0.05 μM caldesmon where actin filament sliding is dramatically slowed, but not completely inhibited.

caldesmon concentrations, the movement of actin filaments is inhibited in a graded manner as opposed to an "all or none" situation.

Acknowledgment

We thank Dr. Robert Adelstein for his comments on the manuscript.

References

Anson, M. (1992). Temperature dependence and Arrhenius activation energy of F-actin velocity generated *in vitro* by skeletal myosin. *J. Mol. Biol.* **224**, 1029–1038.

Castellani, L., Vibert, P., and Cohen, C. (1983). Structure of myosin/paramyosin filaments from a molluscan smooth muscle. *J. Mol. Biol.* **167**, 853–872.

Collins, K., Sellers, J. R., and Matsudaira, P. (1990). Calmodulin dissociation regulates brush border myosin I (110-kD-calmodulin) mechanochemical activity in vitro. *J. Cell Biol.* **110**, 1137–1147.

Cuda, G., Fananapazir, L., Zhu, W., Sellers, J. R., and Epstein, N. D. (1993). *J. Clin. Invest.* (in press).

Epstein, H. F., Miller, D. M., III, Ortiz, I., and Berliner, G. C. (1985). Myosin and paramyosin are organized about a newly identified core structure. *J. Cell Biol.* **100**, 904–915.

Harada, Y., Noguchi, A., Kishino, A., and Yanagida, T. (1987). Sliding movement of single actin filaments on one-headed myosin filaments. *Nature (London)* **326**, 805–808.

Harada, Y., Sakurada, K., Aoki, T., Thomas, D. D., and Yanagida, T. (1990). Mechanochemical coupling in actomyosin energy transduction studied by *in vitro* movement assay. *J. Mol. Biol.* **216**, 49–68.

Homsher, E., Wang, F., and Sellers, J. R. (1992). Factors affecting movement of F-actin filaments propelled by skeletal muscle heavy meromyosin. *Am. J. Physiol.* **262**, C714–C723.

Honda, H., and Asakura, S. (1989). Calcium-triggered movement of regulated actin *in vitro*. A fluorescence microscopy study. *J. Mol. Biol.* **205**, 677–683.

Inoue, S. (1986). "Video Microscopy." Plenum, New York.

Ishijima, A., Doi, T., Sakurada, K., and Yanagida, T. (1991). Sub-piconewton force fluctuations of actomyosin *in vitro*. *Nature (London)* **352**, 301–306.

Janson, L. W., Sellers, J. R., and Taylor, D. L. (1992). Actin-binding proteins regulate the work performed by myosin II motors on single actin filaments. *Cell Motil. Cytoskel.* **22**, 274–280.

Kachar, B., Bridgman, P. C., and Reese, T. S. (1987). Dynamic shape changes of cytoplasmic organelles translocating along microtubules. *J. Cell Biol.* **105**, 1267–1271.

Kishino, A., and Yanagida, T. (1988). Force measurements by micromanipulation of a single actin filament by glass needles. *Nature (London)* **334**, 74–76.

Koretz, J. (1982). Hybridization and reconstitution of thick filament structure. *In* "Methods in Enzymology" (D. Frederiksen and L. Cunningham, eds.), Vol. 85, pp. 20–55. Academic Press, New York.

Kron, S. J., and Spudich, J. A. (1986). Fluorescent actin filaments move on myosin fixed to a glass surface. *Proc. Natl. Acad. Sci. U.S.A.* **83**, 6272–6276.

Kron, S. J., Toyoshima, Y. Y., Uyeda, T. Q. P., and Spudich, J. A. (1991). Assays for actin sliding movement over myosin-coated surfaces. *In* "Methods in Enzymology" (R. B. Vallee, ed.), Vol. 196, pp. 399–416. Academic Press, San Diego.

Levine, R. J. C., and Kensler, R. W. (1985). Structure of short thick filaments from *Limulus* myosin. *J. Mol. Biol.* **182**, 347–352.

Margossian, S. S., and Lowey, S. (1982). Preparation of myosin and its subfragments from rabbit skeletal muscle. *In* "Methods in Enzymology" (D. Frederiksen and L. Cunningham, eds.), Vol. 85, pp. 55–71. Academic Press, New York.

Margossian, S. S., Krueger, J. W., Sellers, J. R., Cuda, G., Caulfield, J. B., Norton, P., and Slayter, H. S. (1991). Influence of the cardiac myosin hinge region on contractile activity. *Proc. Natl. Acad. Sci. U.S.A.* **88,** 4941–4945.

Okagaki, T., Higashi-Fujime, S., Ishikawa, R., Takano-Ohmuro, H., and Kohama, K. (1991). *In vitro* movement of actin filaments on gizzard smooth muscle myosin: Requirement of phosphorylation of myosin light chain and effects of tropomyosin and caldesmon. *J. Biochem. (Tokyo)* **109,** 858–866.

Sellers, J. R., and Kachar, B. (1990). Polarity and velocity of sliding filaments: Control of direction by actin and of speed by myosin. *Science* **249,** 406–408.

Sheetz, M. P., Block, S. M., and Spudich, J. A. (1986). Myosin movement in vitro: A quantitative assay using oriented actin cables from Nitella. *In* "Methods in Enzymology" (R. B. Vallee, ed.), Vol. 134, pp. 531–544. Academic Press, Orlando, FL.

Shirinsky, V. P., Biryukov, K. G., Hettasch, J. M., and Sellers, J. R. (1992). Inhibition of the relative movement of actin and myosin by caldesmon and calponin. *J. Biol. Chem.* **267,** 15886–15892.

Takiguchi, K., Hayashi, H., Kurimoto, E., and Higashi-Fujime, S. (1990). *In vitro* motility of skeletal muscle myosin and its proteolytic fragments. *J. Biochem. (Tokyo)* **107,** 671–679.

Tawada, K., and Sekimoto, K. (1991). A physical model of ATP-induced actin-myosin movement in vitro. *Biophys. J.* **59,** 343–356.

Toyoshima, Y. Y., Kron, S. J., McNally, E. M., Niebling, K. R., Toyoshima, C., and Spudich, J. A. (1987). Myosin subfragment-1 is sufficient to move actin filaments in vitro. *Nature (London)* **328,** 536–539.

Toyoshima, Y. Y., Kron, S. J., and Spudich, J. A. (1990). The myosin step size: Measurement of the unit displacement per ATP hydrolyzed in an in vitro assay. *Proc. Natl. Acad. Sci. U.S.A.* **87,** 7130–7134.

Umemoto, S., and Sellers, J. R. (1990). Characterization of in vitro motility assays using smooth muscle and cytoplasmic myosins. *J. Biol. Chem.* **265,** 14864–14869.

Uyeda, T. Q. P., Kron, S. J., and Spudich, J. A. (1990). Myosin step size. Estimation from slow sliding movement of actin over low densities of heavy meromyosin. *J. Mol. Biol.* **214,** 699–710.

Uyeda, T. Q. P., Warrick, H. M., Kron, S. J., and Spudich, J. A. (1991). Quantized velocities at low myosin densities in an *in vitro* motility assay. *Nature (London)* **352,** 307–311.

Vibert, P., and Craig, R. (1983). Electron microscopy and image analysis of myosin filaments from scallop striated muscle. *J. Mol. Biol.* **165,** 303–320.

Warshaw, D. M., Desrosiers, J. M., Work, S. S., and Trybus, K. M. (1990). Smooth muscle myosin cross-bridge interactions modulate actin filament sliding velocity in vitro. *J. Cell Biol.* **111,** 453–463.

Work, S. S., and Warshaw, D. M. (1992). Computer-assisted tracking of actin filament motility. *Anal. Biochem.* **202,** 275–285.

Yamada, A., Ishii, N., Shimmen, T., and Takahashi, K. (1989). MgATPase activity and motility of native thick filaments isolated from the anterior byssus retractor muscle of mytilus edulis. *J. Muscle Res. Cell Motil.* **10,** 124–134.

Yanagida, T., Nakase, M., Nishiyama, K., and Oosawa, F. (1984). Direct observation of motion of single F-actin filaments in the presence of myosin. *Nature (London)* **307,** 58–60.

CHAPTER 3

Motility of Myosin I on Planar Lipid Surfaces

Henry G. Zot* and Thomas D. Pollard†

* Department of Physiology
University of Texas Southwestern Medical Center
Dallas, Texas 75235

† Department of Cell Biology and Anatomy
The Johns Hopkins Medical School
Baltimore, Maryland 21205

I. Introduction
II. Methods
    A. Preparation of Lipid Vesicles
    B. Formation of Planar Membranes
    C. Motility Recording
III. Evaluation of the Method
    References

I. Introduction

We have combined two evolving technologies, namely, the reconstitution of *in vitro* motility and the formation of substrate-supported planar membranes, to reconstitute the movement of actin filaments relative to lipid membranes with purified myosin I (Zot *et al.*, 1992). This provided a functional test for two of the properties of myosin I: binding to phospholipids attributed to domains in the tail (Adams and Pollard, 1989; Doberstein and Pollard, 1992) and movement of actin filaments attributed to the head of the myosin I molecule (Albanesi *et al.*, 1985; Conzelman and Mooseker, 1987). These separate domains of myosin I potentially enable membrane-mediated movement.

The mechanical properties of myosin are conveniently demonstrated *in vitro* using the gliding filament assay (Kron and Spudich, 1986) in which actin filaments are moved by immobilized myosin. The attractive features of this assay include complete control of the conditions, minimal requirements for materials, and flexible methods of analysis. The assay takes place in a simple flow cell in which myosin is bound to one surface of the flow cell and bathed in a buffer solution containing ATP. Fluorescently labeled actin filaments are added to the flow cell and are moved by the myosin along the plane of the glass surface. Measurements of these movements, normally recorded with a fluorescence microscope and video equipment (Kron *et al.*, 1991), yield the translational velocity of the actin filament. The ability to actually see movement has been a particularly powerful method of demonstrating the flexibility of the myosin head (Toyoshima *et al.*, 1989), the mechanical coupling of heads in a myosin filament (Warshaw *et al.*, 1990; Sellers and Kachar, 1990), and the identity of myosin I as a true myosin (Albanesi *et al.*, 1985; Mooseker and Coleman, 1989; Collins *et al.*, 1990; Zot *et al.*, 1992). Thus, having a visual record may be a strong motivation for the current popularity of *in vitro* assays of motility (cf. other chapters of this volume).

The potential utility of substrate-supported planar membranes is less well appreciated in the cell motility community. The ability to reliably deposit a uniform surface of lipid on a solid support, once limited to those equipped to make Langmuir–Blodgett films, can not be achieved with common laboratory equipment. Brian and McConnell (1984) showed that lipid vesicles fuse spontaneously to form a layer of phospholipid on a glass support. Substrate-supported planar membranes prepared in this way have been used in various fields of investigation including cell surface receptor interactions (Brian and McConnell, 1984; Watts *et al.*, 1984, 1986; Dustin *et al.*, 1988; Poglitsch and Thompson, 1990; Lee and Watts, 1990; Chan *et al.*, 1991; Lawrence and Springer, 1991; Poglitsch *et al.*, 1991; Törzen *et al.*, 1992), lipid–protein interactions (Sui *et al.*, 1988; Frey and Tamm, 1991; Kalb and Engel, 1991; Tendian *et al.*, 1991), biosensor development (Tien and Solaman, 1990), *in vitro* motility (Zot *et al.*, 1992), and membrane structure and dynamics (Merkel *et al.*, 1989; Weisenhorn *et al.*, 1990; Lee *et al.*, 1991). The planar membrane and associated proteins form a surface that can be contained in flow cell and focused with a high-power objective. This enables the reconstitution and visualization of interactions among proteins at the interface between the membrane and the bathing solution.

We found that supported planar membranes of appropriate lipids are excellent substrates for reconstituted movements of actin filaments by myosin I (Zot *et al.*, 1992). Only planar membranes that contain acidic phospholipids support myosin I-based motility. Short movements highlighted by frequent filament detachments are observed with substrates that bind myosin I poorly (Zot *et al.*, 1992) and illustrate the contribution of the substrate to the mechanical activity (Fig. 1). Additional studies are needed to determine the combined properties of the planar membrane and bound protein that target myosin I to specific membrane surfaces.

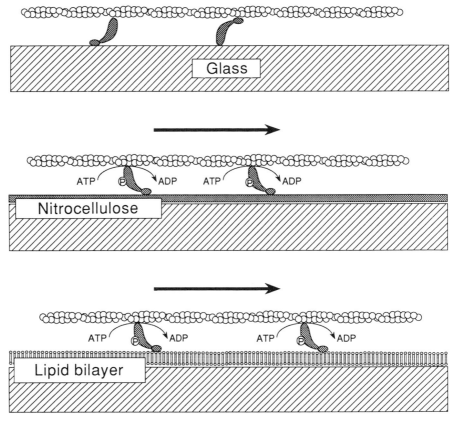

Fig. 1 Schematic summary of substrates that were tested for motility. The three substrates represented here are an unmodified surface of glass, a thin layer of nitrocellulose deposited on glass, and a glass surface covered with a planar membrane composed of phosphatidylcholine and 5–40% phosphatidylserine. A polymer of beads represents filamentous actin; monomeric cross-bridge figures depict myosin I in the phosphorylated state (P), which was found to be necessary for motility in this assay (Zot *et al.*, 1992); and the arrows indicate the movement of the actin filaments. Linear movements of actin filaments were recorded over substrates of nitrocellulose and planar membranes. Detachment but no directed movement of actin filaments was observed with plain glass. The results demonstrate a physiologically relevant set of minimum requirements for motility, which are phosphorylated myosin I, filamentous actin, a membrane bilayer containing acidic phospholipids, and ATP.

II. Methods

Here we describe our most reliable protocol for producing motility on planar lipid membranes. The methods are described in specific terms, calling for myosin I, actin, and certain buffers. Variations on the protocol may become evident

for other applications including other myosins and microtubule motor proteins.

A. Preparation of Lipid Vesicles

1. Materials

nitrogen gas (N_2)

$CHCl_3$: MeOH (10 : 1) (Aldrich Chemicals, Milwaukee, WI)

lipids (Avanti Polar Lipids Inc., Pelham, Alabama)

positive displacement pipets (e.g., Gilson Microman, Rainin Instrument
Company Inc., Woburn, MA)

vacuum desiccator jar

bath sonicator

microcentrifuge

Vortex mixer

glass vials (e.g., Catalog No. 225170, Wheaton, Millville, NJ)

Handling the lipids can be a problem. Lipids are normally dissolved in
$CHCl_3$: MeOH and sealed under N_2. When exposed to air the solvent can
evaporate and the lipids can oxidize. Oxidation may be prevented by purging the
solution with N_2 but vials are difficult to seal, and leakage of the volatile solvent
changes the concentration of the lipid stock. A practical solution to the handling
problem is to order small quantities of lipid in sealed ampules and to open a new
ampule for critical experiments. For consistently pure phospholipids, Avanti is
often mentioned as the best source.

The lipid structures that fuse best with the glass surface are small unilamellar
vesicles (SUVETs). SUVETs are single-bilayer structures of diameters less
than 100 nm formed by simple sonication of lipid sheets in aqueous medium. The
ability of phospholipids to form vesicles is determined by the shape of the head
group and the number and length of the acyl side chains (Israelachvili *et al.*,
1977). It turns out that most phospholipids with two acyl groups have geometries
predicted to favor bilayer formation (Israelachvili *et al.*, 1980). In theory and
practice, phospholipids such as PC, PS, PI, PG, and PA are readily induced to
assemble into stable SUVETs by sonic energy.

2. Protocol for Formation of Lipid Vesicles

1. With the positive displacement pipet, transfer a total of 2.5 μmol of fresh
phospholipids into a glass vial. This is the time to adjust the mixture of lipids to
the desired molar ratios.

2. Dry lipids in a stream of N_2. This can be done with ordinary laboratory
equipment by attaching a 9-in. transfer pipet to a N_2 source and setting the flow
to achieve a cool sensation on a wet lip. Clamp the tip of the pipet inside the vial.

As the solvent evaporates the vial will feel cool. Dried lipids form a film on the glass surface. Cap the vial tightly after drying.

3. To use dried lipids, open the vial, place it in a desiccator containing $CaSO_4$ desiccant, and evacuate with the house vacuum for 30 minutes.

4. While vortexing the vial, add 0.5 ml H_2O and continue vortexing for 60 seconds. The lipid film should disappear from the glass surface and the solution should become opaque. This suspension contains lamellar lipid sheets.

5. Sonicate the suspension until the solution becomes clear. This is conveniently done in a bath sonicator. Adjust the water level in the sonicator so that one or more dimples appear on the surface of the liquid in the bath. Immerse the vial just below the water's surface at one of these sites. Effective sonic action can be felt as vibrations transmitted in the glass vial. Cool the vial in an ice bath for 30 seconds between 30-second bursts of sonication. Typically, 2–3 minutes of sonication is sufficient for clarification; more time may be required for lipid mixtures having a high content of neutral phospholipids. Excessive sonication is not normally a concern as lipid vesicles are stable structures.

6. Microcentrifuge the vesicles at full speed for 10 minuts to sediment remnants of lamellar sheets (a pellet is typically not visible).

7. Store vesicles on ice and use the same day.

B. Formation of Planar Membranes

SUVETs with a wide range of lipid compositions fuse into planar membranes. In our experience, fusion occurs with vesicles composed of pure phosphatidylcholine (PC) as well as PC vesicles containing up to 40% phosphatidylserine (PS). At this high fraction of PS, the vesicles alone do not readily fuse into a planar membrane but require a counterion such as Mg^{2+} in the bathing solution. Whether vesicles fuse is almost certainly dictated by the same factors that govern SUVET formation. One of these factors, the size of the head group, is strongly influenced by electrostatic repulsion. By shielding charged head groups, SUVETs containing PS may be induced to form planar membranes as predicted (Israelachvili *et al.*, 1980) and observed, or SUVETs containing phosphatidic acid (PA) may be induced to form other structures as predicted (Israelachvili *et al.*, 1980).

1. Materials

lipid vesicle stock (5 mM)
coverglasses (e.g., 22 × 40 mm, No. 1)
coverglass rack (e.g., Thomas Scientific Co., Catalog No. 8542)
Linbro detergent (ICN Biomedicals, Costa Mesa, CA), diluted 1 : 4
ion plasma generator (Harrick Scientific, Ossining, NY)

argon gas (Ar)

120°C oven

peristaltic pump

vacuum pump

buffer AB: 25 mM imidazole, pH 7.5, 25 mM KCl, 4 mM MgCl$_2$, 1 mM
ethylene glycol bis(β-aminoethyl ether) N,N'-tetraacetic acid (EGTA)

flow cell (see below)

The flow cell is constructed with the coverglass serving both as the substrate
for planar lipids and the surface for microscopic viewing. A simple and effective
design requires only a standard glass slide, the coverglass, and grease (Kron *et
al.*, 1991). For upright microscopes the flow cell can be constructed with a
square coverglass mounted neatly on the glass slide so that the direction of flow
is parallel to the long axis of the glass slide (Kron *et al.*, 1991). Our modification
for inverted microscopes is to mount a 22 × 40-mm coverglass 90° to the long
axis of a glass slide, allowing the ends of the coverglass to overhang both sides of
the glass slide. This makes the direction of flow 90° to the long axis of the glass
slide. In the inverted position on the microscope, the coverglass is supported
from above by the glass slide and the overextending coverglass forms a trough at
each end of the flow cell. To change solutions at the microscope, fresh buffer is
pooled in one trough and is drawn into the flow cell with filter paper mounted in
the other trough. To change solutions at the bench, the glass slide is held in a
slanted position with large binder clips and buffer added to the trough at the top
runs through the flow cell by gravity. The buffer solutions and their nomencla-
ture are detailed elsewhere (Kron *et al.*, 1991).

2. Protocol for Formation of Supported Planar Membranes

1. Clean coverglasses with detergent. This can be done with the coverglasses
aligned in their holder and the holder placed inside a small beaker. Immerse the
coverglasses in the Linbro detergent. Heat to boiling on hot plate or in microwave. Promptly transfer coverglasses to a beaker containing water. Rinse thoroughly with 10 liters of water for 12 hours by pumping constantly from a
reservoir into the beaker and allowing the overflow to drain from the top. Dry
the coverglasses for 1 hour at 120°C.

2. Just prior to assembling the flow cell, clean a few coverglasses with an Ar
plasma. The preferred method is to purchase an ion plasma generator and
connect it to a vacuum pump and an Ar source. Alternatively, we have found
that a microwave oven can be modified to produce an Ar plasma. To modify the
microwave, drill two holes in the side of the inner chamber only; the outer
covering is left intact. Insert two heat-resistant, nonmetallic tubes (e.g., $\frac{1}{2}$-in. red
rubber vacuum tubing) into the chamber through the holes. Pass the tubing
between the inner chamber and the outer cover to the exterior of the microwave.

Connect one line to a stopcock leading to a vacuum pump and the other to the valve of an Ar source. Inside the microwave, connect the lines to a glass T-connector serving as the port of a suitable flask. We have used a lyophilization flask for this purpose (300-ml Labconco flask, Thomas Scientific, Philadelphia, PA), but virtually any glass flask that holds a vacuum and can withstand heating will work. With the oven off, place the holder containing the coverglasses in the flask and turn on the vacuum pump. Listen for the pump to stabilize; then close vacuum line and open the source of Ar. Close the Ar source and open the line to the vacuum pump. When the pump stabilizes again, turn on the microwave at full power. A glow indicates the presence of the plasma. Expose the coverglasses to the plasma for 1–5 minutes in either the plasma generator or the modified microwave oven.

3. Assemble the flow cell using an Ar^+-cleaned cover glass.

4. Dilute the vesicle stock 1 : 10 in buffer AB and fill the flow cell. Incubate 20 minutes at room temperature in humidified container such as a bell jar containing water-saturated tissue paper.

5. Wash excess vesicles from the flow cell with fresh buffer AB. A planar membrane will be formed on the coverglass.

6. Fluorescence can be used to visualize the membrane. For this purpose, prepare vesicles with 5 mol% rhodamine-labeled phosphatidylethanolamine (PE). The supported planar membrane appears intensely fluorescent in a narrow focal plane. The fluorescence bleaches rapidly but will recover in the dark; recovery of fluorescence is a good test for membrane fluidity. Glass that has not been cleaned with Ar^+ often shows signs of lipid aggregates, which appear blotchy in an otherwise homogenous fluorescent background.

C. Motility Recording

1. Materials

buffer AB/BSA: buffer AB containing 0.5 mg/ml bovine serum albumin (BSA)

glucose (60 mg/ml)

glucose oxidase (200 U/ml)

catalase (0.36 mg/ml)

1 M dithiothreitol (DTT)

0.1 M ATP, pH 7.5

3.3 μM rhodamine–phalloidin (Molecular Probes, Eugene, OR)

filamentous actin (1 mg/ml in buffer AB containing 0.2 mM ATP)

Acanthamoeba myosin I (0.6 mg/ml in 50% glycerol)

Acanthamoeba myosin I heavy-chain kinase (0.3 mg/ml in 50% glycerol)

microscope with video port (e.g., Nikon Diaphot)

epifluorescence illuminator for microscope

lens (NA > 1.2; e.g., Nikon Plan Apochromat 100X)

intensified video camera (e.g., Model SIT 66, Dage-MTI Inc.)

recording device (e.g., retail VCR)

monitor (e.g., 9-in. high-resolution monochrome)

The video microscope is outfitted with the video camera, lens, and fluorescence attachment. A mercury lamp is sufficient for excitation. Both inverted and upright microscopes are suitable, but inverted scope users may find a need to modify the stage to accommodate the flow cell, which itself may be modified for inverted use. A high high-numerical-aperture lens and an oiled coverglass are necessary for the low fluorescence intensity of individual actin filaments. The low light levels require an intensified video camera for recording. Although the quality of the recording system limits the resolution of the image and the precision of the measurement, the most inexpensive format, a retail VCR, offers sufficient resolution for routine measurements.

The proteins constitute the most exotic raw materials. Purification and storage procedures have been described for myosin I and kinase (Lynch *et al.*, 1991) and for actin (Spudich and Watt, 1971; Kron *et al.*, 1991). Actin is labeled with rhodamine–phalloidin as described (Kron *et al.*, 1991) yielding Rh–Ph–actin (0.03 mg/ml) which is diluted 1 : 70 in buffer AB/BSA to make AB/BSA/Rh–Ph–actin. To activate the myosin and prepare AB/BSA/myosin I, dilute myosin I 1 : 20 and myosin I heavy-chain kinase 1 : 1000 in buffer AB/BSA, add 1 mM ATP, and incubate 20 minutes at 25°C.

Glucose, glucose oxidase, catalase, and DTT used together effectively inhibit photobleaching. The antibleaching reagents are included in two solutions for the flow cell: AB/BSA/GOC and AB/BSA/GOC/ATP. To prepare AB/BSA/GOC, dilute the stocks of glucose, glucose oxidase, catalase, and DTT each 1 : 20 in an aliquot of AB/BSA. Prepare AB/BSA/GOC/ATP by diluting ATP 1 : 50 in AB/BSA/GOC.

Dissolved air will promote photobleaching and bubble formation in the flow cell. To avoid these problems, degas AB and AB/BSA prior to use.

2. Protocol for Gliding Filaments

1. Prepare a flow cell containing a planar membrane composed of 20 mol% PS and 80 mol% PC as described above and tilt the slide so that one opening of the flow cell faces downward. Exchange the internal volume of the flow cell by infusion, which consists of pipetting 100 μl buffer AB/BSA into the mouth of the flow cell and allowing the solution to run out of the lower end of the flow cell.

2. Infuse AB/BSA/myosin I and incubate 60 seconds.

3. Infuse AB/BSA twice to remove unbound myosin.

4. Infuse AB/BSA/Rh–Ph–actin and incubate 10 seconds.

5. Infuse AB/BSA twice to remove unbound actin.

6. Infuse AB/BSA/GOC, mount the flow cell on the microscope, and focus on actin filaments. The focal plane can be found through the eyepieces by focusing on the fluorescence emitted by the grease that seals the edges of the flow cell. Focus the fluorescence of the actin filaments using the camera and start the recording device.

7. Draw AB/BSA/GOC/ATP into the flow cell using a filter paper wick at one of the open ends. Take care to keep the filaments in focus, as the solution change sometimes distorts the flow cell.

8. Record the movements of the actin filaments. They start moving immediately and gradually shear into a larger population of smaller fragments. The movements of the fragmented filaments spread to all parts of all fields and persist at least 30 minutes.

III. Evaluation of the Method

The results thus far indicate that supported planar membranes are good models for biological membranes. The phospholipids in supported planar membranes share many important physical properties with the phospholipids in biological membranes, including a lateral diffusion contant of $\sim 10^{-8}$ cm^2 s^{-1} (Brian and McConnell, 1984; Watts $et\ al.$, 1984; Merkel $et\ al.$, 1989; Poglitsch and Thompson, 1990), a packing density of ~ 70 Å^2 per molecule (Zot $et\ al.$, 1992), and the coexistence of liquid and crystal domains (Merkel $et\ al.$, 1989). Membrane fluidity and bilayer structure may be crucial for some applications and the potential importance of these properties dictates that care be taken in the preparation of the supported planar membranes. Thus it may be valuable to note a report claiming that the inner leaflet of the planar membrane next to the glass is not fluid without prior treatment of the glass with Ar$^+$ (Merkel $et\ al.$, 1989).

The lateral movements of myosin I in the plane of the membrane are an issue because a force must be exerted on the membrane surface during the movement of an actin filament. Myosin I, whether bound to the membrane or bound directly to the solid support, moves an actin filament at the same average velocity of 0.2 μm s^{-1} (Zot $et\ al.$, 1992). One may ask, then, how the fluid membrane resists the force acting on its surface during the movement of an actin filament? Posing the question differently, we can ask how much force would be necessary to drag an actin filament, attached to the membrane by myosin, along the surface of the membrane with velocity equal to that of a gliding filament? As shown below, this force is expected to be high relative to the expected drag imposed on an actin filament by the solvent.

The measured surface viscosity of a single layer of lipids (10^{-2} to 10^{-4} g s^{-1}) is higher than expected (Adamson, 1982) and is likely to contribute substantially to different viscosities for the aqueous and membrane phases. The exact solution to the viscosity problem requires some information that is not available, including the geometry of the moving mass, the interactions between the leaflets of the membrane, and compression forces developed in the membrane. But, providing that myosin I associates with the phospholipid head groups on the membrane surface, a conservative estimate of viscosity can be made by assuming that the moving mass has an area equal to the longitudinal section of an actin filament and by ignoring the other interactions and forces. Considering only the surface viscosity of the membrane, the head pressure, P, required to move a rectangularly shaped object along a film of lipids at constant velocity is given in the relationship of Meyers and Harkins (1937) as taken from Adamson (1982):

$$P = \eta^s \frac{12l(dA/dt)}{a^3}$$

where l is the length of the actin filament, a is the diameter of the actin filament, η^s is the surface viscosity, dA/dt is the area of flow given by $2\int_0^{a/2} v \, dx$, and v is the velocity. Using $1.0 \, \mu m$ for l, $7 \times 10^{-3} \, \mu m$ for a, and $0.2 \, \mu m \, s^{-1}$ for v, the area of flow is $1.4 \times 10^{-3} \, \mu m^2 \, s^{-1}$. In this two-dimensional problem $P \times a$ gives a force, F. For $\eta^s = 10^{-2}$ to $10^{-4} \, g \, s^{-1}$, F ranges between 3.5×10^4 to 3.5×10^6 fN. This range is significantly larger than 3 to 5 fN, which is the estimated viscous drag on a microtubule of $1 \, \mu m$ moving at $0.6 \, \mu m \, s^{-1}$ in an aqueous solvent (Howard *et al.*, 1989). Taking the aqueous drag to be less on the smaller and slower actin filament, the opposing forces are expected to differ by at least four orders of magnitude, which should preclude a detectable reduction in gliding velocity as a result of slippage in the membrane.

Being a first and a conservative approximation of forces acting in this system, the preceding exercise illustrates the need for additional information. We are still left with the problem of reconciling a high membrane viscosity with a high membrane fluidity. To address this, a model is needed that can relate the measured diffusion rate of individual molecules in the membrane with the measured bulk property of the surface. Better understanding of the bulk property of the membrane would enable an analysis of the forces ignored in the simplified calculation and, ultimately, a determination of the size of the load that can be moved.

To test theoretical estimates one could take advantage of both established and new trends in reconstitution on the stage of a microscope. Supported planar membranes are paricularly suited for analysis using various laser-related techniques, including epifluorescence, photobleaching, total internal reflectance, and fluorescence correlation spectroscopy (Thompson *et al.*, 1988). Focused infrared laser light, shown to trap objects in a field of light energy (Ashkin and Dziedzič, 1987), is rapidly emerging as a reliable technique to impose a measurable strain on a moving object (Ashkin *et al.*, 1990). Thus far, motility assays using a laser light trap require that the moving object be spherical rather than filamentous. For this purpose, one can reconstitute motility so that free-floating plastic beads with myosin attached move along immobilized filaments of actin (Sheetz and Spudich, 1983). With this assay, myosin I was shown to move not only plastic beads (Albanesi *et al.*, 1985) but also cellular organelles (Adams and Pollard, 1986). The remaining challenges are to demonstrate that myosin I moves vesicles of pure phospholipids over tracks of actin and to determine if additional proteins are required for myosin I to move a measurable load.

Persistent speculation of intrinsic membrane proteins that bind myosin I (Mooseker *et al.*, 1989; Adams and Pollard, 1989; Pollard *et al.*, 1991; Zot *et al.*, 1992) has developed into a possible mechanism operating in the membrane to provide specificity and anchorage for binding and movement. These putative interactions could help produce movement of membranes that are immotile with myosin I alone. Precedence for this view can be found in the study of the microtubule-based movement of vesicles which have been shown to require not only a motor protein, dynein or kinesin, but also supplementary factors (Schnapp and Reese, 1989; Schroer *et al.*, 1988, 1989; Gill *et al.*, 1991). These previous studies purporting exogenous mediators of microtubule-based motility have relied on crude membrane organelles for the reconstitution of motility. Future studies will likely define the conditions which require postulated intrinsic membrane proteins for motility, and supported planar membranes may be appropriate for this purpose.

In summary, motor proteins and lipid molecules can be reconstituted for study using substrate-supported planar lipid membranes. A number of physical techniques reveal both the lateral mobilities of proteins and lipids and the dynamic associations among proteins and lipids in the membrane. Ideally, one would like to demonstrate that a putative receptor or accessory protein is required for movements produced by a motor molecule in the membrane. Interactions resulting in movement are potentially important in determining the modes of regulation and action for proteins such as myosin I, dynein, and kinesin, which are known to bind membranes as well as proteins such as *dilute* (Mercer *et al.*, 1991), p190 (Larson *et al.*, 1990), and MYO2 (Johnston *et al.*, 1991), which will likely be found to bind membranes.

References

Adams, R. A., and Pollard, T. D. (1986). *Nature* (*London*) **322**, 754–756.
Adams, R. A., and Pollard, T. D. (1989). *Nature* (*London*) **340**, 565–588.
Adamson, A. W. (1982). "Physical Chemistry of Surfaces," 4th ed., pp. 117–121. Wiley, New York.
Albanesi, J. P., Fujisaki, H., Hammer, J. A., III, Korn, E. D., Jones, R., and Sheetz, M. P. (1985). *J. Biol. Chem.* **260**, 8649–8652.
Ashkin, A., and Dziedzič, J. M. (1987). *Science* **235**, 1517–1520.

Ashkin, A., Schütze, K., Dziedzič, J. M., Euteneur, U., and Schliwa, M. (1990). *Nature (London)* **348,** 346–348.

Brian, A. A., and McConnell, H. M. (1984). *Proc. Natl. Acad. Sci. U.S.A.* **81,** 6159–6163.

Chan, P. Y., Lawrence, M. B., Dustin, M. L., Ferguson, L. M., Golan, D. E., and Springer, T. A. (1991). *J Cell Biol.* **115,** 245–255.

Collins, K., Sellers, J. R., and Matsudaira, P. T. (1990). *J Cell Biol.* **110,** 1137–1147.

Conzelman, K. A., and Mooseker, M. S. (1987). *J. Cell Biol.* **105,** 313–324.

Doberstein, S. K., and Pollard, T. D. (1992). *J. Cell Biol.* **117,** 1241–1249.

Dustin, M. L., Singer, K. H., Tuck, D. T., and Springer, T. A. (1988). *J. Exp. Med.* **167,** 1323–1340.

Frey, S., and Tamm, L. K. (1991). *Biophys. J.* **60,** 922–930.

Gill, S. R., Schroer, T. A., Szilak, I., Steur, E. R., Sheetz, M. P., and Cleveland, D. W. (1991). *J. Cell Biol.* **115,** 1639–1650.

Howard, J., Hudspeth, A. J., and Vale, R. D. (1989). *Nature (London)* **342,** 154–158.

Israelachvili, J. N., Mitchell, D., and Ninham, B. W. (1977). *Biochim. Biophys. Acta* **470,** 185–201.

Israelachvili, J. N., Marčelja, S., and Horn, R. G. (1980). *Q. Rev. Biophys.* **13,** 121–200.

Johnston G. C., Prendergast, J. A., and Singer, R. A. (1991). *J. Cell Biol.* **113,** 539–551.

Kalb, E., and Engel, J. (1991). *J. Biol. Chem.* **266,** 19047–19052.

Kron, S. J., and Spudich, J. A. (1986). *Proc. Natl. Acad. Sci. U.S.A.* **83,** 6272–6276.

Kron, S. J., Toyoshima, Y. Y., Uyeda, T. Q. P., and Spudich, J. A. (1991). *In* "Methods in Enzymology" (R. B. Vallee, ed.), Vol. 196, pp. 399–416. Academic Press, San Diego.

Larson, R. E., Espindola, F. S., and Espreafico, E. M. (1990). *J. Neurochem.* **54,** 1288–1294.

Lawrence, M. B., and Springer, T. A. (1991). *Cell (Cambridge, Mass.)* **65,** 859–873.

Lee, G. M., Ishihara, A., and Jacobson, K. A. (1991). *Proc. Natl. Acad. Sci. U.S.A.* **88,** 76274–6278.

Lee, J. M., and Watts, T. H. (1990). *J. Immunol.* **145,** 3360–3366.

Lynch, T. J., Brzeska, H., Baines, I. C., and Korn, E. D. (1991). *In* "Methods in Enzymology" (R. B. Vallee, ed.), Vol. 196, pp. 12–23. Academic Press, San Diego.

Mercer, J. A., Seperack, P. K., Strobel, M. C., Copeland, N. G., and Jenkins, N. A. (1991). *Nature (London)* **349,** 709–712.

Merkel, R., Sackmann, E., and Evans, E. (1989). *J. Phys. (Orsay, Fr.)* **50,** 1535–1555.

Meyers, R. J., and Harkins, W. D. (1937). *J. Chem. Phys.* **5,** 601–606.

Mooseker, M. S., and Coleman, T. R. (1989). *J. Cell Biol.* **108,** 2395–2400.

Mooseker, M. S., Conzelman, K. A., Coleman, T. R., Heuser, J. E., and Sheetz, M. P. (1989). *J. Cell Biol.* **109,** 1153–1161.

Poglitsch, C. L., and Thompson, N. L. (1990). *Biochemistry* **29,** 248–254.

Poglitsch, C. L., Sumner, M. T., and Thompson, N. L. (1991). *Biochemistry* **30,** 6662–6671.

Pollard, T. D., Doberstein, S. K., and Zot, H. G. (1991). *Annu. Rev. Physiol.* **53,** 653–681.

Schnapp, B. J., and Reese, T. S. (1989). *Proc. Natl. Acad. Sci. U.S.A.* **86,** 1548–1552.

Schroer, T. A., Schnapp, B. A., Reese, T. S., and Sheetz, M. P. (1988). *J. Cell Biol.* **107,** 1785–1792.

Schroer, T. A., Steur, E. R., and Sheetz, M. P. (1989). *Cell (Cambridge, Mass.)* **56,** 937–946.

Sellers, J. R., and Kachar, B. (1990). *Science* **249,** 406–408.

Sheetz, M. P., and Spudich, J. A. (1983). *Nature (London)* **303,** 31–35.

Spudich, J. A., and Watt, S. (1971). *J. Biol. Chem.* **246,** 4866–4871.

Sui, S. F., Urumow, T., and Sackmann, E. (1988). *Biochemistry* **19,** 7463–7469.

Tendian, S. W., Lentz, B. R., and Thompson, N. L. (1991). *Biochemistry* **45,** 10991–10999.

Thompson, N. L., Palmer, A. G., III, Wright, L. L., and Scarborough, P. E. (1988). *Comments Mol. Cell. Biophys.* **5,** 109–131.

Tien, H. T., and Solaman, Z. (1990). *Biotechnol. Appl. Biochem.* **5,** 478–484.

Törzen, A., Paul Sung, K.-L., Sung, L. A., Dustin, M. L., Chan, P. Y., Springer, T. A., and Chien, S. (1992). *J. Cell Biol.* **116,** 997–1006.

Toyoshima, Y. Y., Toyoshima, C., and Spudich, J. A. (1989). *Nature (London)* **314,** 154–156.

Warshaw, D. M., Desrosiers, J. M., Work, S. S., and Trybus, K. M. (1990). *J. Cell Biol.* **111,** 453–463.

Watts, T. H., Brian, A. A., Kappler, J. W., Marrack, P., and McConnell, H. M. (1984). *Proc. Natl. Acad. Sci. U.S.A.* **81,** 7564–7568.

Watts, T. H., Gaub, H. E., and McConnell, H. M. (1986). *Nature (London)* **320,** 179–181.

Weisenhorn, A. L., Drake, B., Prater, C. B., Gould, S. A., Hansma, P. K., Ohnesorge, F., Egger, M., and Gaub, H. E. (1990). *Biophys. J.* **58,** 1251–1258.

Zot, H. G., Doberstein, S. K., and Pollard, T. D. (1992). *J. Cell Biol.* **116,** 367–376.

CHAPTER 4

Microtubule and Axoneme Gliding Assays for Force Production by Microtubule Motor Proteins

Bryce M. Paschal[*,†] **and Richard B. Vallee**[*]

[*] Cell Biology Group
Worcester Foundation for Experimental Biology
Shrewsbury, Massachusetts 01545

[†] Department of Cell Biology
University of Massachusetts Medical School
Worcester, Massachusetts 01605

I. Introduction
II. Microtubule Gliding Assay
III. Polarity of Force Production
References

I. Introduction

Advances in light microscopic image analysis in the early 1980s (Allen *et al.*, 1981; Inoué, 1981) facilitated the development of *in vitro* motility assays and the discovery of two classes of force-producing enzymes, kinesin (Vale *et al.*, 1985b) and cytoplasmic dynein (Paschal *et al.*, 1987b). Kinesin generates force along microtubules toward their plus ends (Vale *et al.*, 1985c) and is therefore thought to drive the movement of membranous organelles from the nucleus toward the cell periphery. Cytoplasmic dynein is a minus end-directed motor (Paschal and Vallee, 1987) and is thought to be responsible for the retrograde transport of organelles such as lysosomes (Lin and Collins, 1992) and chromosomes (Pfarr *et al.*, 1990; Steuer *et al.*, 1990; Hyman and Mitchison, 1991).

Two general types of assay evolved for the detection of force production by the purified proteins based on the new light microscopic technology. In one case, latex beads were coated with motor protein, and the beads were observed to migrate on the surface of microtubules (see, for example, Vale *et al.*, 1987a) (Fig. 1). This mimicked the topological relationship between organelles and microtubules and could be readily understood in terms of organelle transport in the cell. To determine the direction of force production, microtubules were grown from structures such as centrosomes (Vale *et al.*, 1985c), which established their orientation.

A second type of assay involved adsorption of motor protein to a coverslip. Microtubules were seen to glide over the coverslip surface. This behavior was puzzling when initially detected in squid axoplasm preparations (Allen *et al.*, 1985), though it is now understood in the same terms as bead motility. Although the motor protein is presumed to bind in random orientation to the negatively charged surface of the coverslip, only properly oriented molecules can interact productively with the microtubule (see Fig. 1). The coverslip plays the same role as the coated bead, but is too large to move. As a consequence, it is the microtubule that becomes displaced. In this type of assay, knowledge of the orientation of the microtubule is again necessary to determine the direction of force production. Both flagellar axonemes (Paschal and Vallee, 1987; Lye *et al.*, 1987; Porter *et al.*, 1987) and whole fixed sperm (Yang *et al.*, 1990) have been used for this purpose.

The bead assay is complicated by the presence of large numbers of out-of-focus particles engaged in rapid Brownian movement. In contrast, the microtubule gliding assay is more readily analyzed. It involves smooth movements that

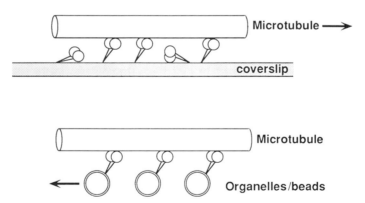

Fig. 1 Diagrammatic representation of microtubule gliding and organelle/bead motility assays. Top: In the microtubule gliding assay, a subset of properly oriented motor molecules is presumed to interact with the microtubule surface. The forces generated by cyclic binding and release result in translocation (gliding) of the microtubule across the coverslip surface in a direction defined by the microtubule surface lattice. Bottom: Beads are predicted to travel in the same direction as organelles, but opposite the direction in which microtubules glide.

persist over long periods, the parameters of which can, therefore, be readily quantified. For these reasons the microtubule gliding assay has proven to be the more straightforward and more commonly used method for assaying force-producing activity.

We describe here the assay in use in our laboratory (Paschal *et al.*, 1987a,b, 1991), which involves the application of microtubules composed of purified brain tubulin to coverslips coated with either cytoplasmic dynein or kinesin. We have used flagellar axonemes prepared from the green alga *Chlamydomonas reinhardtii* as well as from sea urchin sperm. *Chlamydomonas* flagellar axonemes were used to compare the directions of force production by cytoplasmic dynein and kinesin and to demonstrate that cytoplasmic dynein is a minus end-directed (retrograde) motor (Paschal and Vallee, 1987).

II. Microtubule Gliding Assay

The method involves adsorbing purified motor protein to a coverslip, adding taxol-stabilized microtubules and ATP, and monitoring the preparation by video microscopy. The diameter of the microtubule (24 nm) is below the limit of resolution of the light microscope, and the mass is too low to detect reproducibly by conventional light microscopy; however, it is possible to use either differential interference contrast (DIC) microscopy coupled to computer-aided background subtraction or dark-field microscopy to detect these structures. Our laboratory uses a Zeiss IM35 inverted microscope equipped with a 100-W mercury arc lamp, 63× Planachromat objective (oil immersion, NA = 1.4), Zeiss DIC prisms, and a DAGE MTI 67m video camera (transfer factor to camera = 6.25x). Digitization, background subtraction, and image enhancement are performed using a Hughes Aircraft image processor (Model 794 Image Sigma, Hughes Aircraft Co., El Segundo, CA). The final magnification to the video monitor is 8500x. This provides a field of view of approximately 400 μm^2.

The Model 794 Image Sigma is no longer manufactured; however, easy-to-operate image processing computer systems are also available from Dage MTI (Michigan City, MI), Inovision (Durham, NC), Photonics Microscopy (Oakbrook IL), and Universal Imaging (Philadelphia, PA).

1. Glass coverslips should be cleaned prior to use for motility assays. Coverslips (Corning No. 286518, Corning, NY) are mounted in ceramic racks (Thomas Scientific No. 8542-E40, Swedesboro, NJ), sonicated for 20 minutes (50/60 Hz) in a 5% alkaline detergent solution such as Contrad 70 (No. 117-622, Curtin Matheson Scientific, Houston, TX), washed exhaustively (1 hour) with tap water, and rinsed with at least 10 changes of ultrapure water (Milli-Q, Millipore, Bedford, MA). The coverslips are dried at room temperature or briefly in a 60°C oven, and stored in a plastic petri dish between sheets of lint-free lens paper.

2. Microtubules for the gliding assay are prepared as follows. MAP-free

tubulin subunits are purified from third-cycle microtubule protein by ion-exchange chromatography, quick-frozen in liquid nitrogen, and stored in aliquots at −80°C (Vallee, 1986). Microtubule "seeds" (<1 μm in length) are first prepared by incubating 20 μl of tubulin (5 mg/ml stock) and 1 μl taxol [100 μM in dimethylsulfoxide (DMSO)] in 200 μl buffer containing 50 mM KCl, 5 mM MgSO$_4$, 0.5 mM ethylenediaminetetraacetate (EDTA), and 20 mM Tris–Cl, pH 7.6 (Tris–KCl), at 37°C for 5 minutes. Microtubules are then assembled by supplementing the seeds with an additional 20 μl of 5 mg/ml tubulin and incubating for 10 minutes at 37°C. The resulting ∼1 mg/ml stock of microtubules (2–8 μm in length) is sufficient for numerous motility assays and is stable for at least 8 hours at room temperature.

3. Approximately 8 μl of purified cytoplasmic dynein or kinesin (40–100 μg/ml) is applied to the surface of a coverslip, and the preparation is incubated for 5 minutes in a humid chamber. The nonadsorbed protein can be removed by a buffer rinse, though this is optional. Next, 1 μl of 10 mM MgATP and 1 μl of 1 mg/ml taxol-stabilized microtubules are added to a final volume of 10 μl. The coverslip is inverted onto a glass slide (cleaned as described above), and the edges are sealed with VALAP, a 1:1:1 mixture of Vaseline, lanolin, and paraffin. The slide is mounted on the microscope and viewed by DIC or dark-field microscopy. A single video field (400 mm²) should reveal several gliding microtubules (Fig. 2).

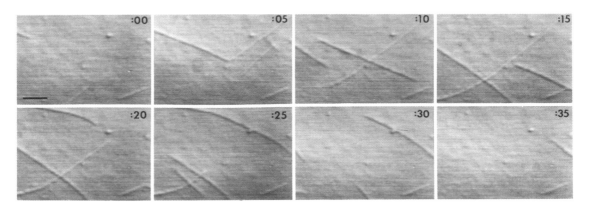

Fig. 2 Microtubule gliding induced by cytoplasmic dynein. Cytoplasmic dynein was purified from calf brain cytosol as described previously (Paschal *et al.,* 1987b, 1991), dialyzed into Tris–KCl buffer, and adsorbed to a glass coverslip. Taxol-stabilized microtubules and ATP (1 mM final concentration) were added, and the preparation was viewed by DIC video microscopy. During the 35-second interval (recorded 3 hours after preparing the coverslip), seven microtubules cross the field; one immobile microtuble may be seen at the lower right of the field. The average gliding velocity of these microtubules was 0.97 μm/s. Bar-2 μm. Reprinted, with the publisher's permission, from Paschal *et al.* (1987b).

Comments. Fresh preparations of cytoplasmic dynein assayed in Tris–KCl buffer will support microtubule gliding at an average velocity of up to 1.25 μm/s, a rate within the range for retrograde fast axonal transport (Grafstein and Forman, 1980). Bovine brain kinesin-driven motility is slower, inducing microtubules to glide at about 0.4 μm/s (Vale *et al.*, 1985b).

Cytoplasmic dynein and kinesin have distinct nucleotide and pharmacological profiles as well. Cytoplasmic dynein-driven motility requires MgATP and is inhibited by 10 mM sodium vanadate and 0.1 mM N-ethylmaleimide (Paschal and Vallee, 1987). In contrast, kinesin-driven motility is supported by MgATP or MgGTP, requires 100 mM sodium vanadate for complete inhibition, and is insensitive to 2 mM N-ethylmaleimide (Vale *et al.*, 1985c; Porter *et al.*, 1987).

The microtubule gliding assay has also been used to characterize the motility properties of axonemal dyneins, which induce very rapid gliding, consistent with the rates of intermicrotubule sliding in the axoneme. Flagellar outer-arm dynein from the sea urchin *Strongylocentrotus purpuratus* translocates microtubules at 3.5 μm/s (Paschal *et al.*, 1987a). Ciliary 14 S and 22 S dyneins from the protozoan *Tetrahymena thermophila* induce microtubules to glide 4.2 and 8 μm/s, respectively (Vale and Toyoshima, 1989).

In the best dynein or kinesin preparations, more than 90% of the microtubules can be seen to glide. Gliding activity can persist for many hours on the coverslip. In one cytoplasmic dynein preparation, gliding was still observed after 6 hours, though the rate was reduced by ~25%, apparently because of to ATP depletion.

A number of factors influence the activity of the motor proteins in the microtubule gliding assay. MAPs have been found to interfere with both cytoplasmic dynein- and kinesin-mediated gliding (Paschal *et al.*, 1989; von Massow *et al.*, 1989). This effect has been judged to result from competition for the same tubulin binding site in the case of cytoplasmic dynein (Paschal *et al.*, 1989; though see Rodionov *et al.*, 1990). It can also reflect cross-linking of the microtubule to the coverslip by the MAP (von Massow *et al.*, 1989). The inhibitory effect of MAPs seems to be one explanation for the low motility activity observed in the early stages of motor protein purification (Paschal *et al.*, 1987b).

Ionic strength is also an important variable in the gliding assay. This is probably due to the ionic nature of the interaction of the motor protein with microtubules, which is reflected in the extreme dependence of both cytoplasmic dynein and kinesin ATPase activities on ionic strength (Shpetner *et al.*, 1988; Wagner *et al.*, 1989).

The freshness of the motor protein preparation is another important variable in the motility assay. An extreme example of this was seen with sea urchin sperm axonemal dynein preparations (Paschal *et al.*, 1987a), which lose motility activity over a period of hours, while retaining ATPase activity for several days. In the case of cytoplasmic dynein, uniform motility of microtubules is observed for the first 24–48 hours following the sucrose gradient step of purification. After this period, microtubules are seen to adhere nonproductively to the coverslip. In many cases, only a portion of the microtubule becomes fixed in place, causing the remainder of the microtubule to engage in bending and whiplashing movements. The nonproductive adherence of microtubules to the coverslip is thought to reflect the formation of rigor complexes between a subfraction of the cytoplasmic dynein molecules and the microtubules. This is also thought to occur in assays of actin gliding on myosin. In this case it has been reported that the inactive myosin molecules can be selectively removed by cosedimentation with actin in the presence of ATP (Kron *et al.*, 1991).

Finally, it should be said that the translocating activities of cytoplasmic dynein and kinesin seem, in general, to be more sensitive to solution and pharmacological variables than are the ATPase activities (Cohn *et al.*, 1987; Paschal and Vallee, 1987; Shpetner *et al.*, 1988).

III. Polarity of Force Production

Chlamydomonas flagella are ideally suited for use in the polarity assay because the distal (plus) end of the axoneme has a tendency to fray during isolation, creating a fork that is readily visible in the light microscope. For this reason, the *Chlamydomonas* axoneme has been used as the standard for all microtubule polarity determinations, and the frayed and compact ends have been defined as plus and minus, respectively (Allen and Borisy, 1974). The direction of force production in the gliding assay is deduced by substituting the axonemes for the taxol-stabilized microtubules and noting whether the frayed end is leading or trailing. A surface coated with the retrograde motor cytoplasmic dynein causes axonemes to glide with their frayed end leading, whereas the anterograde motor kinesin causes axonemes to glide with their frayed end trailing (Fig. 3).

In addition to the use of axonemes alone, we have assembled microtubules from purified brain tubulin subunits onto the ends of the axoneme (Paschal and Vallee, 1987). Brain tubulin assembles more rapidly onto the frayed plus ends of the axoneme (Allen and Borisy, 1974), providing a further means for confirming the identification of the axonemal ends.

1. *Chlamydomonas* is grown by standard methods, and the flagella are released by dibucaine treatment, demembranated, and subjected to high salt

---

Fig. 3 Direction of force production determined using *Chlamydomonas* flagellar axonemes. Purified cytoplasmic dynein or kinesin was adsorbed to a glass coverslip; axonemes and ATP were added and the preparation was viewed by DIC video microscopy. Left: Axonemes glide on cytoplasmic dynein with their frayed (plus) ends leading, indicating that force produced against the coverslip is directed toward the microtubule minus ends (retrograde motor). Right: Axonemes glide on kinesin with their frayed ends trailing, indicating that force produced against the coverslip is directed toward the microtubule plus ends (anterograde motor). The *Chlamydomonas* axonemes were kindly provided by Dr. Steve King and Dr. George Witman of the Worcester Foundation for Experimental Biology. Bar-2 μm. Reprinted, with the publisher's permission, from Paschal and Vallee (1987).

Cytoplasmic Dynein   ## Kinesin

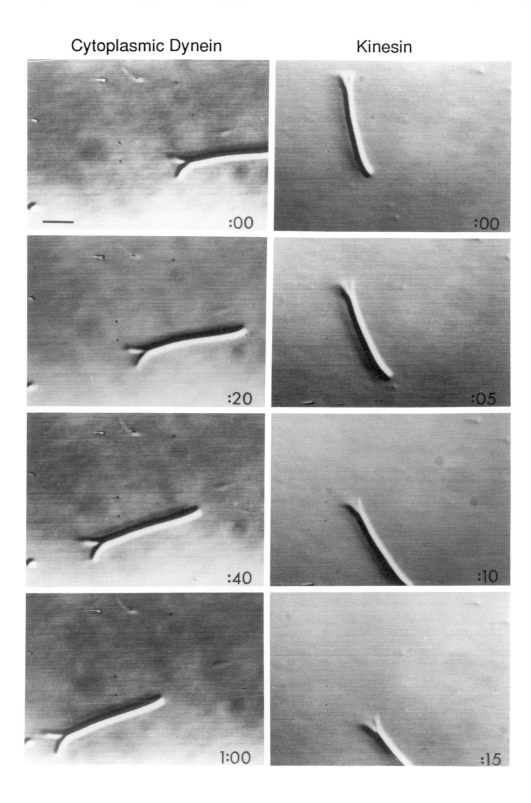

extraction (Witman, 1986). The axonemes are stored in Tris–KCl buffer containing 50% glycerol at $-20°C$ and are stable for at least 6 months. The glycerol is removed by dilution of the axonemes in 10 vol of Tris–KCl buffer, sedimentation at 12,000g for 10 minutes, and resuspension of the pellet in 0.1 vol of Tris–KCl buffer.

2. The axonemes are diluted into the Tris–KCl buffer to a concentration of approximately 10^7/ml, and 1 μl of the suspension is substituted for the taxol microtubules in the gliding assay described above. Although image processing is not essential for detection of the frayed end, which should be visible even with normal phase contrast optics, it does improve the image quality considerably.

3. To assemble microtubules onto the ends of the axonemes, the axonemes are incubated with purified tubulin subunits (1–2 mg/ml) and 1 mM GTP at 37°C for 5 minutes. Before the decorated axonemes are added to the assay, tubulin is applied to the coverslip to ensure that its final concentration in the motility assay is 1–2 mg/ml to prevent microtubule depolarization.

The direction of force production is evident from several aspects of the behavior of the axoneme/brain tubulin complexes. First, some of the complexes, despite their substantial size, glide well, and the direction of force production is deduced from both the morphology of the axoneme and the length of the microtubules grown onto the ends. Many of the complexes remain stationary. In these cases, microtubules at one end can be seen to writhe (Paschal and Vallee, 1987) in a manner comparable to that seen for individual microtubules in aging motor protein preparations (see above). This behavior reflects the thwarted efforts of individual microtubules to move against the immotile axoneme. Microtubules assembled at the opposite end of the same axoneme are invariably straight. This is the expected result if these microtubules are attempting to crawl away from the stationary axoneme.

Comments.   It may be possible to use buffers other than Tris–KCl in the preparation of the axonemes; however, 1,4-piperazinediethanesulfonic acid (Pipes) buffer should be avoided because it seems to destabilize the axonemes.

We routinely use axonemes that have been exposed to high salt to extract the outer dynein arms. This was initially done to minimize the chance that dynein would leach out of the axonemes and contribute to microtubule gliding activity during the motility assay, and, in fact, we have never seen gliding of axonemes in controls lacking cytoplasmic dynein or kinesin. It is uncertain, however, whether salt extraction is required.

We note that *Chlamydomonas* axonemes translocate more slowly (0.2 μm/s) than taxol microtubules on coverslips coated with cytoplasmic dynein (up to 1.25 μm/min). This seems to be the result of the off-axial interaction of the outer doublet microtubules with the coverslip within the frayed region of the axoneme. In fact, when using axonemes bearing brain microtubules on their ends, we often observed the axonemes to be pulled apart by these off-axial forces. This seemed to be caused by an adherence of the individual brain microtubules to the coverslip greater than that seen for the axonemal microtubules alone.

The growth of brain tubulin onto the ends of the axonemes provides a redundant means for checking the direction of force production. In the case of the *Chlamydomonas* axonemes, therefore, this is not routinely necessary, but with other axonemal preparations, it is required. For example, sea urchin sperm tail axonemes have also been used in polarity assays (Pryer *et al.*, 1986; Porter *et al.*, 1987; Lye *et al.*, 1987; B. M. Paschal and R. B. Vallee, unpublished results). These structures do not show intrinsic evidence of polarity because they are very long and break into smaller fragments during isolation. The individual pieces glide relatively well, but microtubules must be assembled onto the ends for the polarity of force production to be determined. This approach can be problematic because plus end-nucleated microtubules display a more frequent transition from growing to shrinking than minus end-assembled microtubules (Walker *et al.*, 1988). Under some conditions this produces longer microtubules at the minus end of the axonemes, the reverse of the expected condition, which can lead to errors in defining the direction of force production.

In general, we observe that at sufficiently large size, translocation of microtubule-containing structures becomes less efficient, especially with preparations of motor protein with less than optimal activity. Thus, gliding of structures of the size of axonemes and whole fixed sperm (Yang *et al.*, 1990) tends to be more difficult to reproduce than gliding of individual microtubules.

References

Allen, C. A., and Borisy, G. G. (1974). *J. Mol. Biol.* **90**, 381–402.

Allen, R. D., Allen, N. S., and Travis, J. L. (1981). *Cell Motil.* **1**, 291–302.

Allen, R. D., Weiss, D. G., Hayden, J. H., Brown, D. T., Fujiwake, H., and Simpson, M. (1985). *J. Cell Biol.* **100**, 1736–52.

Cohn, S. A., Ingold, A. L., and Scholey, J. M. (1987). *Nature (London)* **328**, 160–163.

Grafstein, B., and Forman, D. S. (1980). *Physiol. Rev.* **60**, 1167–1283.

Hyman, A. A., and Mitchison, T. J. (1991). *Nature (London)* **351**, 206–211.

Inoué, S. (1981). *J. Cell Biol.* **89**, 346–356.

Kron, S. J., Toyoshima, Y. Y., Uyeda, T. Q., and Spudich, J. A. (1991). *In* "Methods in Enzymology" (R. B. Vallee, ed.), Vol. 196, pp. 399–416. Academic Press, San Diego.

Lin, S. X. H., and Collins, C. A. (1992). *J. Cell Sci.* **101**, 125–137.

Lye, R. J., Porter, M. E., Scholey, J. M., and McIntosh, J. R. (1987). *Cell (Cambridge, Mass.)* **51**, 309–318.

Paschal, B. M., and Vallee, R. B. (1987). *Nature (London)* **330**, 181–183.

Paschal, B. M., King, S. M., Moss, A. G., Collins, C. A., Vallee, R. B., and Witman, G. B. (1987a). *Nature (London)* **330**, 672–674.

Paschal, B. M., Shpetner, H. S., and Vallee, R. B. (1987b). *J. Cell Biol.* **105**, 1273–1282.

Paschal, B. M., Obar, R. A., and Vallee, R. B. (1989). *Nature (London)* **341**, 569–572.

Paschal, B. M., Shpetner, H. S., and Vallee, R. B. (1991). *In* "Methods in Enzymology" (R. B. Vallee, ed.), Vol. 196, pp. 181–191. Academic Press, San Diego.

Pfarr, C. M., Coue, M., Grissom, P. M., Hays, T. S., Porter, M. E., and McIntosh, J. R. (1990). *Nature (London)* **345**, 263–265.

Porter, M. E., Scholey, J. M., Stemple, D. L., Vigers, G. P. A., Vale, R. D., Sheetz, M. P., and McIntosh, J. R. (1987). *J. Biol. Chem.* **262**, 2794–2802.

Pryer, N. K., Wadsworth, P., and Salmon, E. D. (1986). *Cell Motil. Cytoskel.* **6**, 537–548.

Rodionov, V. I., Gyoeva, A. S., Kashina, A. S., Kuznetsov, S. A., and Gelfand, V. I. (1990). *J. Biol. Chem.* **265,** 5702–5707.

Shpetner, H. S., Paschal, B. M., and Vallee, R. B. (1988). *J. Cell Biol.* **107,** 1001–1009.

Vale, R. D., and Toyoshima, Y. Y. (1989). *J. Cell Biol.* **108,** 2327–2334.

Vale, R. D., Schnapp, B. J., Reese, T. S., and Sheetz, M. P. (1985a). *Cell (Cambridge, Mass.)* **40,** 559–569.

Vale, R. D., Reese, T. S., and Sheetz, M. P. (1985b). *Cell (Cambridge, Mass.)* **42,** 39–50.

Vale, R. D., Schnapp, B. J., Mitchison, T., Steuer, E., Reese, T. S., and Sheetz, M. P. (1985c). *Cell (Cambridge, Mass.)* **43,** 623–632.

Vallee, R. B. (1986). *In* "Methods in Enzymology" (R. B. Vallee, ed.), Vol. 134, pp. 89–104. Academic Press, New York.

von Massow, A., Mandelkow, E. -M., and Mandelkow, E. (1989). *Cell Motil. Cytoskel.* **14,** 562–571.

Wagner, M. C., Pfister, K. K., Bloom, G. S., and Brady, S. T. (1989). *Cell Motil. Cytoskel.* **12,** 195–215.

Walker, R. A., O'Brien, E. T., Pryer, N. K., Sobbeiro, M., Voter, W. A., Erickson, H. P., and Salmon, E. D. (1988). *J. Cell Biol.* **107,** 1437–1448.

Witman, G. B. *In* "Methods in Enzymology" (R. B. Vallee, ed.), Vol. 134, pp. 280–290. Academic Press, New York.

Yang, J. T., Saxton, W. M., Stewart, R. J., Raff, E. C., and Goldstein, L. S. B. (1990). *Science* **249,** 42–47.

CHAPTER 5

Analyzing Microtubule Motors in Real Time

S. A. Cohn,[*] W. M. Saxton,[†] R. J. Lye,[‡] and J. M. Scholey[§]

[*] Department of Biological Sciences
DePaul University
Chicago, Illinois 60614

[†] Department of Biology
Indiana University
Bloomington, Indiana 47405

[‡] Department of Genetics
Washington University School of Medicine
St. Louis, Missouri 63110

[§] Section of Molecular and Cell Biology
University of California, Davis
Davis, California 95616

I. Introduction
II. Configuration of the Computer–Video Microscope Setup
    A. Optics
    B. Computer
    C. Superimposition of Video Signals
    D. The Velocity Program
III. The Microtubule Motility Assay
    A. Screening for Motor Activity
    B. Modification for Low Motor Protein Density
    C. Kinetics of Nucleotide-Dependent Motility and Its Inhibition
    D. Screening for Antibody-Dependent Inhibition
IV. Overview of Typical Results Obtained with the Real-Time Motility Assay
    A. Kinesin
    B. ncd Protein, a Kinesin-like Protein from *Drosophila*
    C. Cytoplasmic Dynein
    D. Flagellar Dynein

V. Conclusions
    Addendum
    References

I. Introduction

Video-enhanced microscopy has become one of the major techniques used for the identification, isolation, and characterization of intracellular movement and associated motor proteins (Allen and Weiss, 1985; Allen *et al.*, 1982, 1985; Lye *et al.*, 1987; Vale, 1987; Vale *et al.*, 1985a,b, 1986; Vallee and Shpetner, 1990); however, many of these video microscope systems have either proved to be relatively costly or require a good deal of time to accomplish quantitative analysis of the observed movements (such as measuring the movements on the screen during playback, marking the movements on the screen using acetates, or using separate x/y digitizers on prerecorded observations). Realizing the need for a rapid and quantitative motility assay, we developed a relatively inexpensive computer-assisted program that can analyze the movements displayed by video images in real time (Cohn *et al.*, 1987).

Superimpositions of x/y coordinate cursors onto video images to obtain positions and distances of microtubule (MT)-based movements had been used previously (e.g., Pryer *et al.*, 1986), and we decided to exploit the expanding capabilities of newer video-friendly minicomputers to develop a similar setup which could calculate the velocities in real time. Our setup involves directly superimposing the image of a mouse-controlled computer cursor over the microscope image, using the hand-operated mouse to move the computer cursor to desired points in the video image, and signaling start/stop times to the computer via the mouse button. The computer is then used to perform real-time calculations on the distances and times, resulting in quantitative velocity measurements. This chapter describes the basics of our computer-assisted video microscope setup and outlines some of the results we have obtained using the system.

II. Configuration of the Computer–Video Microscope Setup

The setup we have used (Fig. 1) consists basically of a high-quality differential interference contrast (DIC) light microscope connected to a high-sensitivity, high-resolution video camera with externally controllable gain and black level. The output of the video camera is mixed with the output of a computer, resulting in a superimposed video image of both the light microscope sample and the computer screen. By control of the position of the computer cursor with the "mouse," the position and velocity of moving microtubules can be measured in real time at relatively low cost (Fig. 1).

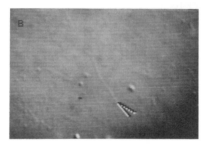 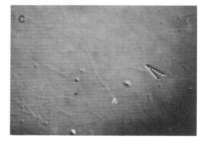

Fig. 1 The computer–video microscope setup and its superimposed video output. (A) The computer–video microscope setup described in the chapter (Cohn *et al.*, 1987, 1989) consists of a Zeiss standard microscope attached to a Dage 68 video camera with built-in video mixer, connected to an Amiga 1000 computer. The lower photographs illustrate the use of the system in measuring microtubules. (B) The computer cursor is moved over the superimposed image of a field of microtubules to the desired end of a microtubule. (C) The mouse button is then pressed, causing the computer to mark the position of that microtubule end. After the desired period of time has elapsed, the computer cursor can be moved to the new position of the same end of the moving microtubule and the mouse button again pressed. The computer then calculates the actual distance between the two points and the time elapsed between the mouse signals and displays the resulting velocity. Typically, the velocity of 10 to 25 microtubules is measured per coverslip, after which the computer can display a velocity histogram and calculate the mean velocity and standard deviation.

A. Optics

On the basis of the earlier microscope setup of Porter *et al.* (1987), we also found that with the proper optics and video enhancement, it is possible to clearly observe individual microtubules under the light microscope without the need for additional digital image enhancement (such as background subtraction). Digital enhancement does improve image quality and J.M.S. now routinely incorporates an Argus 10-HAMAMATSU system into the computer/video setup using a GenLock device. Without such digital enhancement, however, the correct adjustment of illumination and optics becomes more critical. The system we have used was set up using a Zeiss microscope containing a short light path from the illuminator to the sample (Zeiss standard or Zeiss axioskop) equipped with standard DIC optics and an additional set of magnification lenses (via slider or optavar system). The microscope is then fitted with a port to direct the light to a Newvicon Dage 68 video camera equipped with an external gain and black level control. The image resulting from this setup can clearly provide for detection of individual microtubules provided several steps are taken to ensure the maximum incident light on the subject and the maximal contrast:

1. Fit the microscope with the brightest light source available (typically we use a 100-W mercury arc burner). Lower-light-level illuminators often result in too low a contrast to be detected without digital enhancement.

2. Set the light to critical or near-critical illumination to maximize the light intensity on the MT sample.

3. Adjust the DIC slider to be close to extinction.

4. Set the gain on the camera to be at or near full amplification. (Sometimes, extra gain can be accomplished by setting the external box on auto gain and raising the black level accordingly.)

5. Set the black level on the camera relatively high (but not usually on maximum).

6. Occasionally, the condenser must be adjusted to slightly off-center to maximize and even out the brighter areas in the video field.

B. Computer

The computer used for this setup is a Commodore Amiga (1000 or 2000). These computers were chosen for their versatility, relatively low cost, and built-in-abilities for video interfacing. The Amiga computers all contain a built-in video output which can be connected to any of various video interface devices (such as a video mixer or GenLock). In addition, Amiga computers are built for ease of menu-driven programming and graphics, allowing the relatively rapid development of programs that could calculate positions and velocities and be controlled using the mouse and menus.

The computer can be set up as follows:

1. Connect the video output of the Amiga computer into the input port of the video camera (if using a modified camera) or into the Genlock Device.

2. Load and run the velocity program. As the computer program is run independently of the video microscope settings, care must be taken to ensure that the calibration settings used in the velocity program correspond to the magnification settings currently in use on the microscope.

3. Use the computer mouse to move the cursor to the desired locations on the screen.

4. Operate the velocity program as desired using the menu selections and signaling the starting and stopping points with the mouse button.

C. Superimposition of Video Signals

The measurement of moving objects observed in the video image is accomplished by superimposing the image of the computer screen directly over the video output of the camera. This can be accomplished in one of two ways:

1. *Installation of a video mixer within the camera.* In this type of setup, the video camera is installed with a video mixer chip (installed by DAGE), which can allow the input of an external video signal (in our case the video output of the computer) and use the external signal to time the video signal of the camera. The two signals are thus directly synchronized and are then superimposed. The resulting output from the camera is a combined image of the two signals.

2. *Synchronization of the computer/video outputs using an external "Gen-lock" device.* In this setup the video signals from both the computer and the video camera are fed into a "GenLock" device, which synchronizes the two signals and generates a combined video output. In general, Genlock devices provide output signals that are of higher quality than the internal video mixer, as well as a degree of control over the amount of superimposition.

In comparison, the in-camera video mixer provides ease of use, although no controls over the relative mixing of the two signals are available, and the line resolution of the output is limited by the resolution of the incoming computer. In contrast, the line resolution of the Genlocked output image is of standard broadcast quality. In either case, the same Amiga computer can be used.

For in-camera mixing, the interface devices are installed by connecting the video output of the camera (which should contain a mixed signal when the connected computer is on) to the monitor. This should result in a superimposed camera/computer image with a balanced (about 50/50) proportion of video and computer images. Because the relative proportion of video and computer signals cannot be altered with this method, care must be taken so that the background color of the computer screen does not minimize the quality or contrast of the superimposed "mixed" signal.

For GenLock mixing, the interface devices are installed as follows. The video output of the camera is connected to the video-in connection on the GenLock device. The video-out or "monitor-out" port of the computer is connected to the "computer-in" port of the GenLock device. The "video-out" port of the GenLock is connected to the monitor. The video image on the monitor should now be one of the superimposed camera and microscope images. The controls on the GenLock device are used to adjust the relative levels of the computer screen and the microscope signals on the superimposed image. The background color of the computer screen can also be altered if necessary to affect the quality of the mixed signal.

D. The Velocity Program

The computer program is designed to calculate the position of the cursor and the time intervals between clicks of the mouse. Moving the computer mouse moves the cursor on the computer screen. As the computer and video images are superimposed on one another, the cursor can be moved coordinately with the position of a moving object (Cohn and Schreuers, 1993; Cohn *et al.*, 1987, 1989) (see Fig. 1,B and C). For studies of microtubule motors, the cursor can be placed at the position of the microtubule end currently under scrutiny.

Functionally, a microtubule motility measurement is performed as follows. The sample to be tested is placed under the microscope and the DIC sliders and condenser are adjusted to maximize contrast and detection of MTs. The computer program is then started and the image of the computer screen is superimposed over the microscope image. The computer cursor is moved, using the computer's mouse, to an end of a MT. The button on the mouse is pressed, signaling the computer to record the position of the cursor and the start time of the event. The cursor is then used to follow the end of the MT by moving the mouse accordingly. At the desired later time (usually 5–15 seconds), the button on the mouse is pressed again, signaling the computer to record the second cursor position and the end time for the event. The computer program, using a previous calibration factor, calculates the actual distance represented by the start and end positions of the cursor and calculates the average velocity for that MT. The velocity of a number of MTs can be recorded this way and displayed as an average of all the MTs measured or as a histogram of the individual velocities (Fig. 2).

An updated version of the program, written by one of us (S.A.C.) in conjunction with Dr. W. Schreuers, is available from Dr. Cohn. The current version is applicable to measurement of a wide variety of cellular movements, and includes the ability to measure distances, velocities, angles, and continuous interval velocities, as well as to draw tracings of the movement on the computer screen, display a histogram of results, and determine the number of orientation and direction changes during a continuous set of movements (Cohn and Schreuers, 1993). The program also contains a routine to calibrate the distances observed under the video camera against a stage micrometer.

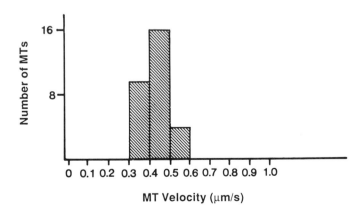

Fig. 2 Example of a histogram of microtubule velocities for a single set of microtubules. After a number of microtubules have been measured as in Fig. 1, the computer can display the data as a histogram, with the mean and standard deviation of the sample set. This histogram represents a sample set of kinesin-driven microtubules having an average velocity of about 0.43 μm/s.

The Microtubule Motility Assay

The assay that we perform in conjunction with the measurement program is basically a modification of the assay described previously by Vale and co-workers (1985a,b). It usually consists of incubation of the microtubule motor test solution on glass coverslips; addition and incubation of purified microtubules and the desired substrates, inhibitors, or reagents; and then observation of the resulting sample under the light microscope (Cohn *et al.*, 1987, 1989; Ingold *et al.*, 1988; Lye *et al.*, 1987; Porter *et al.*, 1987, Saxton *et al.*,1988). The microscopic sample can be quickly scanned to determine the area of adsorption and whether or not there is active motility. After the area of kinesin or sample adsorption is determined, microtubules are randomly selected and their velocities measured. After measurement of 10–30 microtubules, an average velocity is calculated and a histogram plotted (Fig. 2). Further details of the assay are described below.

A. Screening for Motor Activity

The assays to screen for activity were basically performed as follows. Fourteen to eighteen microliters of motor protein test solution was gently layered onto a glass coverslip using a Pipetman device. The solution was then incubated in a humidified petri plate at room temperature for 10–15 minutes to allow for the adsorption of kinesin (or other protein) onto the glass. At this point, 2 μl of 100–200 μg/ml MTs (polymerized from phosphocellulose-purified tubulin with 20 μM taxol and 10–100 μM GTP) along with 0–4 μl of nucleotide or buffer were added to the solution to bring the total volume to 20 μl. The MTs were ordinarily prepared from bovine brain tubulin. The mixture was allowed to incubate for another 2–5 minutes on the coverslip. The sample was then gently inverted onto a glass slide and sealed with VALAP (1:1:1 w:w:w Vaseline:lanolin:paraffin). The sample was then observed under 100× oil immersion DIC optics, with an oiled condenser lens for maximum light collection. Ten to twenty-five MTs were measured using the motility program, and mean velocities and standard deviations were calculated.

B. Modification for Low Motor Protein Density

In recent work (D. Cole, K. Hall, and J. M. Scholey) the assays have also been performed at low motor density, using the approach of Howard and co-workers (1989, and Chapter 10, this volume). These modified assays are performed as follows: 5 mg/ml casein (fresh, filtered, and clarified) is preadsorbed onto the coverslip, then removed prior to application of the MgATP/ motor protein solution, followed by addition of MTs as described previously.

In some experiments, a perfusion chamber has been used according to the design of Sale and co-workers (see Chapter 6, this volume). This modified assay is performed as follows: A perfusion chamber of approximately 10 μl is prepared by sealing a No. 1 glass coverslip over Parafilm spacers using VALAP. Samples (40 μl) of the perfusion solution are applied with a Pipetman device to one edge of the chamber. The solution is then pulled through the chamber using a piece of filter paper applied to the opposite side. For example, to assay sea urchin flagellar 21 S dynein, we perfuse five times (1-minute incubation each time) with 40 μl of 100 μg/ml dynein, then two times with 40 μl of motility buffer, followed by perfusion with 10 μl of 200 μg/ml taxol-stabilized MTs and 50 μl of motility buffer. The ends of the chamber are then sealed with VALAP and the slide is observed under the microscope as described previously. For the assay at low motor protein density, the coverslip is precoated by perfusion with 5 mg/ml casein.

C. Kinetics of Nucleotide-Dependent Motility and Its Inhibition

The nucleotide dependence of motor activity was determined by analyzing the kinetics of microtubule movement using samples treated with various concentrations of nucleotide substrate. To obtain good data points for Lineweaver–Burk or Dixon plot analysis, we typically made up a 100 mM stock solution of Mg–nucleotide and made working dilutions in sample buffer in the range of 10 μM to 10 mM. This could effectively produce final working concentrations on the coverslip of 1 μM to 10 mM. To eliminate any changes in ATP concentration caused by spontaneous hydrolysis, experiments using ATP concentrations of 100 μM or less were performed in the presence of an ATP-regenerating system consisting of 50 mM phosphocreatine and 10 U/ml creatine phosphokinase. These assays were run as follows.

A 15-μl sample of motor protein was layered onto a coverslip and allowed to incubate in a humidified chamber as described previously. After 5–10 minutes of incubation, the desired volume of nucleotide solution and buffer was added to obtain a final volume of 20 μl. (Typically, concentrations were chosen to allow 2–3 μl of nucleotide to be used.) For experiments investigating the inhibition of nucleotide-dependent movement, the desired reagent or inhibitor was added along with the nucleotide, incubated as described above, and the sample tested for motility. Samples were tested with final working concentrations of inhibitor in the range 10 μM to 10 mM.

D. Screening for Antibody-Dependent Inhibition

To screen for the inhibition of motile activity by antibodies, a two-stage incubation was performed. First, samples of motor protein (approximately 14 μl) were applied to the coverslip and allowed to incubate as described above. Then, after 10 minutes, the desired sample of antibodies was added to the solution, and the mixture allowed to further incubate for 10 minutes. Antibodies were typically added to final working concentrations of 1–100 μg/ml. Some hybridoma culture supernatants contained very low concentrations of antibody, so that large volumes greater than 5–10 μl were added to the initial sample of motor protein. In these cases much of the resulting mixture had to be removed after incubation, and replaced by smaller volumes of buffer. Controls for these latter experiments were performed by rinsing the initial samples with large volumes of buffer alone. MTs and nucleotide were then added to the mixture, after which the coverslip was sealed to a slide and the MT motility analyzed as described above.

IV. Overview of Typical Results Obtained with the Real-Time Motility Assay

This motility assay has allowed us to quantitatively measure the dose-dependent effects of nucleotides and inhibitors on several motor proteins (including kinesin, ncd protein, and cytoplasmic and flagellar dyneins) in real time, requiring each experiment to take an amount of time only minimally greater than would be required for a qualitative screening of the samples. By this method, quantitative characteristics, including some kinetic constants, can be quickly determined for the motor proteins and their association with substrates and potential inhibitors (e.g., Table I). These properties of *in vitro* microtubule motility can then be compared with other activities of the motor proteins such as ATPase and MT binding properties, to result in a more complete understanding of the structural and biochemical characteristics of the mechanoenzymes (Cohn 1990). A brief summary of some of the results we have obtained using the real-time motility assay on samples of kinesin, the kinesin like-motor ncd protein, cytoplasmic dynein, and flagellar dynein follows.

A. Kinesin

1. Kinesin Activity

Most of our studies have employed kinesin purified from sea urchin eggs and, to a lesser extent, *Drosophila* embryos. The described system was first used to characterize the dose-dependent inhibition of sea urchin egg kinesin-based MT motility. Active kinesin samples tend to result in large number of microtubules well adhered to the surface of the coverslip. The microtubules move smoothly and evenly, with an average velocity of about 0.5 μm/s. [Note that this velocity is at least somewhat independent of the source of the microtubules, as microtubules prepared from maize endosperm tubulin were transported by sea urchin egg kinesin with velocities similar to that observed using bovine brain microtubules (C. Troxell, unpublished observations)].

Using the computer analysis system we could rapidly measure the average velocity and velocity distributions of kinesin-driven microtubule samples, both with substrate alone and in the presence of various inhibitors,

Table I
Kinetic Parameters Measured Using the Computer–Assisted Video Microscope

Protein	Kinetic parameters	Inhibition
Sea urchin egg kinesin (Cohn *et al.*, 1989)	MgATP $K_m = 63 \pm 34\ \mu M$ $V_{max} = 0.56 \pm 0.10\ \mu\mathrm{m/s}$ MgGTP $K_m = 1.9 \pm 0.8\ \mathrm{m}M$ $V_{max} = 0.43 \pm 0.08\ \mu\mathrm{m/s}$	Competitive MgATPγS $K_i = 14 \pm 3\ \mu M$ MgADP $K_i = 156 \pm 52\ \mu M$ Not competitive AMPPNP (near competitive at low concentrations of MgATP with $K_i \approx 2\ \mu M$) Vanadate (increased the proportion of immotile MTs, required a vanadate:MgATP concentration ratio greater than about 1 : 10) EDTA and *PPP*$_i$ (Mg chelators)
Drosophila embryo kinesin (Saxton *et al.*, 1988)	MgATP $K_m = 44\ \mu M$ $V_{max} = 0.9\ \mu\mathrm{m/s}$	AMPPNP, ATPγS (complete inhibition by 2.5 mM inhibitor against 2.5 mM MgATP Apyrase, EDTA, Mg-free ATP (details of kinetics not determined)
Caenorhabditis elegans Cytoplasmic dynein (Lye *et al.*, 1987)	MgATP $V_{max} \geq 2\ \mu\mathrm{m/s}$	Vanadate (inhibition at <5 μM in the presence of 5 mM MgATP) AMPPNP, ATPγS (details of kinetics not determined)
Drosophila ncd protein	MgATP $V_{max} = 0.10 \pm 0.01\ \mu\mathrm{m/s}$	
Sea urchin egg 21 S flagellar dynein	MgATP $K_m = 27\ \mu M$ $V_{max} = 7.5\ \mu\mathrm{m/s}$	

such as ethylenediaminetetraacetic acid (EDTA), inorganic tripolyphosphate (*PPP*$_i$), AMPPNP, and vanadate. The quantified inhibition of kinesin-driven movement could thereby be correlated with the dose-dependent inhibition of MT-stimulated ATPase activity (Cohn *et al.*, 1987).

In addition, we were able to rapidly quantify the kinesin-directed movement of microtubules under a number of nucleotide and inhibitor concentrations, which allowed us to determine that the kinesin-directed movement generated by both *Drosophila* embryo and sea urchin egg kinesin follows Michaelis–Menten kinetics with respect to MgATP (Cohn, 1990; Cohn *et al.,* 1989; Saxton *et al.,* 1988). Moreover, by using Lineweaver–Burke and Dixon plot analysis of the average microtubule velocities under several conditions (Fig. 3), we determined that MgADP and MgATPγS behave as classic competitive inhibitors of kinesin-driven motility in the presence of MgATP (with K_i's of about 14 and 156 μM, respectively) and that the inhibition caused by AMPPNP is more complex (Cohn *et al.,* 1989).

As mentioned previously, the program is capable of performing histogram analysis of the microtubule measurements. Such analysis proved very useful in the analysis of vanadate inhibition, which indicated that vanadate is not a competitive inhibitor. Instead of progressively slowing down the microtubule gliding velocities at higher inhibitor concentrations (as do AMPPNP, ATPγS, and ADP), vanadate results in increasing numbers of immotile microtubules, with the remaining motile microtubules still moving at relatively control velocities. We interpreted this to mean that vanadate inhibits microtubule motility by forming inactive MT/kinesin complexes in a manner qualitatively quite different than that of either AMPPNP or the competitive inhibitors (Cohn *et al.,* 1987, 1989). The dose–response curves that we were able to generate with EDTA and PPP_i in the presence and absence of magnesium also helped us to determine that both of these inhibitors appear to act as chelating agents for magnesium.

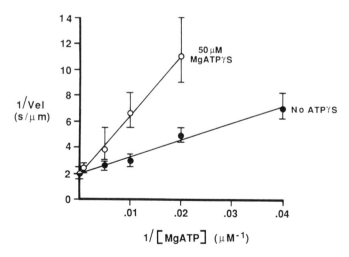

Fig. 3 Example of a Lineweaver–Burk plot of average MT velocities in the presence and absence of ATPγS. In this plot of kinesin-driven microtubule velocities, each point represents the average of several sample sets of microtubules measured at different MgATP concentrations. The linearity of the double-reciprocal plot is indicative of kinesin activity displaying Michaelis–Menten kinetics (in the absence of ATPγS, solid circles). The linear relationship shown by the velocities in the presence of ATPγS (open circles) indicates that ATPγS is a competitive inhibitor.

2. Screening for Antikinesin Antibodies

The assay was also used to screen for inhibitory activity of monoclonal antibodies raised against kinesin (Ingold *et al.*, 1988). Using the procedure described previously, we were able to screen seven mouse monoclonal antibodies that had tested positive against kinesin in both ELISA and Western blot analysis (Ingold *et al.*, 1988). Three of these antibodies (SUK 4, 6, and 7) caused dose-dependent inhibition of kinesin-driven microtubule motility. In addition, we were able to determine that the relative inhibitory ability of SUK 4 and 6 against *Drosophila* and squid kinesins corresponded directly with their relative ability to cross-react with these kinesins on Western blots. Neither of the two antibodies inhibited squid neuronal kinesin, whereas SUK 4 (which cross-reacts with *Drosophila* kinesin) could induce a decrease in the velocity of *Drosophila* kinesin-driven motility. Subsequent analysis (e.g., Scholey *et al.*, 1989) showed that the three function-blocking antibodies bind to the kinesin motor domains, where they presumably interfere with mechanochemical coupling.

B. ncd Protein, a Kinesin-like Protein from *Drosophila*

The real-time motility assay has been used (by J.M.S. and co-workers) to check the motility driven by the kinesin-like ncd protein purified from bacteria containing the pET-ncd expression vector (a gift of Dr. H. McDonald and Dr. L. Goldstein). We observed that in the presence of MgATP, the ncd protein moved microtubules over glass at about 0.10 ± 0.01 μm/s. More detailed studies of this protein are reported in Chapters 8 and 12 in this volume.

C. Cytoplasmic Dynein

Initially, analysis of *Caenorhabditis elegans* cytoplasmic dynein-driven motility was carried out by one of us (R.J.L.) using qualitative screenings. This was followed by quantitative measurements using a program designed by Derek Stemple and Dr. Guy Vigers that required frame-by-frame analysis from previously recorded video tape. Although this system enabled observation of the inhibition of dynein-driven motility by *N*-ethylmaleimide and Triton X-100, it was quite time consuming and did not allow for rapid quantitative analysis. The real-time assay described in this chapter allowed us to rapidly collect the information for dose-dependent inhibition, adjusting the concentrations as required, without having to replay prerecorded video tapes for determination of effect.

By use of the real-time assay, the activity of *C. elegans* cytoplasmic dynein was analyzed, indicating a V_{max} about three to four times faster than those of the kinesins that were measured. A direct dose-dependent inhibition of the motility by vanadate was also observed, at concentrations of vanadate that were much lower than those required to inhibit urchin kinesin motility (Lye *et al.*, 1987). Even at the faster speeds generated by the dynein, we were able to accurately analyze the motility by adjusting the Optavar magnification lenses to a lower setting, thus giving us a wider field of view and a longer distance over which to track the microtubules. The condenser diaphragm was also closed slightly more than with the kinesin to generate a slightly greater depth-of-field, as the dynein-driven microtubules are not always uniformly in contact with the glass coverslip as was the case with the kinesins we used.

D. Flagellar Dynein

The microtubule motility driven by 21 S sea urchin flagellar dynein has been analyzed in perfusion chambers, essentially according to the procedures of Sale and co-workers as described in this volume (see Chapter 6). Lineweaver–Burk plots of the kinetics of dynein-driven microtubule movements were linear, with a K_m of 27 μM for MgATP and a V_{max} of 7.5 $\mu m/s$, in the presence of an ATP-regenerating system (as in Cohn *et al.*, 1989). No motility was observed when the dynein was tested in the presence of 10 mM concentrations of either MgCTP, MgITP, MgUTP, MgTTP, or MgGTP.

V. Conclusions

The real-time system for analysis of microtubule motility we have described has proved to be an extremely useful and potentially versatile system for a number of applications. So far it has been used only to characterize sea urchin egg kinesin-driven motility in detail (Cohn *et al.*, 1987, 1989; Ingold *et al.*, 1988), but the preliminary results obtained with other microtubule motors described here suggest that it may be applicable to the detailed characterization of a number of mechanoenzymes, where the velocity of movement is constant and not much greater than a few micrometers per second. In addition, it has been used recently for measurements of cellular growth and chemotaxis (Cho *et al.*, 1991) and for analysis of diatom cell motility (Cohn and Disparti, 1993). The cost, ease of operation, and power of real-time analysis make this system a useful tool in the measurement and characterization of motile events.

Addendum

The computer system currently employed by S.A.C. uses an Amiga 2000 computer, GenLock device, and Zeiss axioskop connected to a DAGE 68 Newvicon camera, whereas the version used by J.M.S. uses an Amiga 1000 computer, in-camera video mixer, a DAGE 68 Newvicon camera, and a Zeiss standard microscope. The most recent version of the program is capable of measuring point-to-point velocities as described, as well as measuring distances, mouse-directed interval velocities, and cursor tracing with interval marks on the screen. Information on the current system setup or program is available from S.A.C.

References

Allen, R. D., and Weiss, D. G. (1985). An experimental analysis of the mechanisms of fast axonal transport in the squid giant axon. *In* "Cell Motility: Mechanism and Regulation" (H. Ishikawa, S. Hatano, and H. Sato, eds.), pp. 327–333. Univ. of Tokyo Press, Tokyo.

Allen, R. D., Metuzals, J., Tasaki, I., Brady, S. T., and Gilbert, S. P. (1982). Fast axonal transport in squid giant axon. *Science* **218**, 1127–1129.

Allen, R. D, Weiss, D. G., Hayden, J. H., Brown, D. T., Fujiwake, H., and Simpson, M. (1985). Gliding movement of and bidirectional transport along single native microtubules from squid axoplasm: Evidence for an active role of microtubules in cytoplasmic transport. *J. Cell Biol.* **100**, 1736–1752.

Cho, C.-W., Harold, F. M., and Schreurs, W. J. (1991). Electric and ionic dimensions of apical growth in achlya hyphae. *Exp. Mycol.* **15**, 34–43.

Cohn, S. A. (1990). The mechanochemistry of kinesin: A review. *Mol. Chem. Neuropathol.* **12**, 83–94.

Cohn, S. A., and Disparti, N. C. (1993). Environmental factors influencing diatom cell motility. (In preparation).

Cohn, S. A., and Schreuers, W. (1993). A computer program designed to assist real time analysis of movements observed under the video microscope. (In preparation).

Cohn, S. A., Ingold, A. L., and Scholey, J. M. (1987). Correlation between the ATPase and microtubule translocating activities of sea urchin egg kinesin. *Nature (London)* **328**, 160–163.

Cohn, S. A., Ingold, A. L., and Scholey, J. M. (1989). Quantitative analysis of sea urchin egg kinesin-driven microtubule motility. *J. Biol. Chem.* **264**, 4290–4297.

Howard, J., Hudspeth, A. J., and Vale, R. D. (1989). Movement of microtubules by single kinesin microtubules. *Nature (London)* **342**, 154–158.

Ingold, A. L., Cohn, S. A., and Scholey, J. M. (1988). Inhibition of kinesin-driven microtubule motility by monoclonal antibodies to kinesin heavy chains. *J. Cell Biol.* **107**, 2657–2667.

Lye, R. J., Porter, M. E., Scholey, J. M., and McIntosh, J. R. (1987). Identification of a microtubule-based cytoplasmic motor in the nematode C. elegans. *Cell (Cambridge, Mass.)* **51**, 309–318.

Porter, M. E., Scholey, J. M., Stemple, D. L., Vigers, G. P. A., Vale, R. D., Sheetz, M. P., and McIntosh, J. R. (1987). Characterization of microtubule movement induced by sea urchin egg kinesin. *J. Biol. Chem.* **262**, 2794–2802.

Pryer, N. K., Wadsworth, P., and Salmon, E. D. (1986). Polarized microtubule gliding and particle saltations produced by soluble factors from sea urchin eggs and embryos. *Cell Motil. Cytoskel.* **6**, 537–548.

Saxton, W. M., Porter, M. E., Cohn, S. A., Scholey, J. M., Raff, E. C., and McIntosh, J. R. (1988). Drosophila kinesin: Characterization of microtubule motility and ATPase. *Proc. Natl. Acad. Sci. U.S.A.* **85**, 1109–1113.

Scholey, J. M., Heuser, J., Yang, J. T., and Goldstein, L. S. B. (1989). Identification of globular mechanochemical heads of kinesin. *Nature (London)* **338**, 355–357.

Vale, R. D. (1987). Intracellular transport using microtubule-based motors. *Annu. Rev. Cell Biol.* **3**, 347–378.

Vale, R. D., Schnapp, B. J., Reese, T. S., and Sheetz, M. P. (1985a). Organelle, bead, and microtubule translocations promoted by soluble factors from the squid giant axon. *Cell (Cambridge, Mass.)* **40**, 559–569.

Vale, R. D., Reese, T. S., and Sheetz, M. P. (1985b). Identification of a novel force-generating protein, kinesin, involved in microtubule-based motility. *Cell (Cambridge, Mass.)* **42**, 39–50.

Vale, R. D., Scholey, J. M., and Sheetz, M. P. (1986). Kinesin: Possible biological roles for a new microtubule motor. *Trends in Biochem. Sci.* **11**, 464–468.

Vallee, R. B., and Shpetner, H. S. (1990). Motor proteins of cytoplasmic microtubules. *Annu. Rev. Biochem.* **59**, 909–932.

CHAPTER 6

Assays of Axonemal Dynein–Driven Motility

Winfield S. Sale, Laura A. Fox, and Elizabeth F. Smith

Department of Anatomy and Cell Biology
Emory University School of Medicine
Atlanta, Georgia 30322

I. Introduction
II. Dark-Field Video Microscopy
III. Procedures and Analysis of ATP-Induced Microtubule Sliding in Flagellar Axonemes
    A. Methods of ATP-Induced Microtubule Sliding in *Chlamydomonas* Flagellar Axonemes
    B. Method of Analysis of ATP-Induced Microtubule Sliding in Demembranated Sea Urchin Sperm Flagella
IV. The Second Assay: Dynein-Driven Microtubule Translocation Using Purified Proteins
V. Summary: Future
    References

I. Introduction

The purpose of this chapter is to provide a guide to microscopic assays of axonemal dynein function and illustrate novel observations derived from these assays. We describe two general types of functional motility assays each of which makes use of dark-field light microscopy, which was reintroduced by Ian Gibbons (Summers and Gibbons, 1971), to directly observe active microtubule sliding in isolated axonemes. Included in this chapter are a detailed description of the microscopic setup, step-by-step procedures for each experimental approach, and data derived from these techniques.

The first assay is video microscopic analysis of ATP-induced microtubule sliding in protease-treated, isolated axonemes (Summers and Gibbons, 1971). Dark-field microscopy coupled to video cameras provides a convenient means of analyzing patterns of microtubule sliding (Sale, 1986), the direction of force generation (Fox and Sale, 1987), and microtubule sliding velocity (Okagaki and Kamiya, 1986; Fox and Sale, 1987; Kamiya *et al.*, 1989; Sale *et al.*, 1989; Smith and Sale, 1992). This assay has the added value that dynein-driven motility can be quantitatively studied in otherwise paralyzed axonemes derived from flagella of useful mutant strains of *Chlamydomonas* (Witman *et al.*, 1978; Kamiya *et al.*, 1989; Smith and Sale, 1992).

The second approach is an *in vitro* microtubule translocation assay in which axonemal dynein fractions are first applied to a glass slide, microtubules are added in the presence of an ATP-containing motility buffer, and movement is viewed directly and recorded by video tape for subsequent analysis (Paschal *et al.*, 1987a; Vale and Toyoshima, 1988). Such assays were first described for analysis of microtubule-based motors in axoplasm (Allen *et al.*, 1985) and played a major role in the discovery of kinesin (Vale *et al.*, 1985) and cytoplasmic dynein (Paschal *et al.*, 1987b; Lye *et al.*, 1987; Schnapp and Reese, 1989). These functional assays have also led to the identification of functional domains in axonemal dyneins, verification of polarity of force generation, discovery of dynein's ability to produce torque as well and translocation, and discovery of a "weak-binding" state between dynein and the microtubule (Sale and Fox, 1988; Vale and Toyoshima, 1988; Vale *et al.*, 1990; Smith and Sale, 1991). We begin with description of the dark-field video microscope.

II. Dark-Field Video Microscopy

All of the assays described below make use of dark-field microscopy. The key elements include an intense light source, a dark-field oil immersion condenser with high numerical aperture, a high-quality 40× objective lens, and a sensitive video camera mounted on a trinocular head. An upright microscope is used for experiments requiring perfusion of samples in perfusion chambers described below. The inverted microscope is used for most other experiments.

For upright microscopy we use a standard Zeiss microscope body equipped with a 100-W mercury lamp, heat and interference filters, and the Zeiss ultra dark-field oil immersion condenser (NA 1.2/1.4). The sample is contained in a simple perfusion chamber constructed from a glass slide and coverslip separated by two parallel strips of double-sided tape. The chamber contains a total volume of 8 μl. Observations are made with the condenser adjusted to critical illumination. For most studies we use the 40x Zeiss Plan Neofluar oil immersion lens (NA 0.9, Zeiss 46-17-25), and the image is directed to the target of the DAGE 66 SIT camera, mounted on a trinocular head through either a 10x or a 10–20x zoom eyepiece. Recordings are directed through a time–date generator and made on $\frac{1}{2}$-in. video tape using the Panasonic NV-8950 or AG-1960 recorders.

Each of these recorders has an excellent pause mode which is required for velocity and motion analysis. The final magnification on the 13-in. screen of a black and white monitor is about 2800x.

For inverted microscopy the Zeiss Axiovert 35 is equipped with a 100-W mercury lamp and uses the same combination of heat and interference filters between the lamp and the Zeiss ultracondenser. The condenser height is usually adjusted for critical illumination. For the inverted scope we use the 40x Zeiss Plan Apochromatic oil immersion lens, with iris diaphragm, NA 1.0 (Zeiss 44-07-56), and the image is also projected onto the SIT camera target through an eyepiece mounted in the trinocular head.

Each configuration, upright or inverted, permits high-resolution detection of appropriately diluted purified single microtubules or doublet microtubules required for each assay. Any amount of excess light scatter, caused by dirty or defective slides or coverglasses, excess microtubules, or contaminating particulate material, will greatly diminish the ability to visualize single microtubules. We have found that many brands of precleaned slides and coverslips can be used directly without additional cleaning; however, suitability of the glass will vary greatly between different lot numbers irrespective of the source of the glass. Without washing the coverslips, we have discovered that most sources of axonemal dynein will not support microtubule binding or translocation on the coverslip. Therefore, all of our observations are made on the slide surface of perfusion chambers.

III. Procedures and Analysis of ATP-Induced Microtubule Sliding in Flagellar Axonemes

This section is divided into procedures used to analyze microtubule sliding in axonemes derived from *Chlamydomonas* flagella and microtubule sliding in demembranated sea urchin sperm tails.

A. Methods of ATP-Induced Microtubule Sliding in *Chlamydomonas* Flagellar Axonemes

In general we have adopted the methods of Okagaki and Kamiya (1986). Flagella are isolated from either gametic or vegetative *Chlamydomonas* cells by the dibucaine procedure as described by Witman (1986). Isolated flagella are stored on ice at an approximate protein concentration of 1 mg/ml in a buffer composed of

10 mM Hepes, pH 7.4

5 mM MgSO$_4$

1 mM dithiothreitol (DTT)

0.5 mM ethylenediaminetetraacetic acid (EDTA)

30 mM NaCl

0.1 mM phenylmethylsulfonyl fluoride (PMSF)

0.6 TIU aprotinin

For microtubule sliding experiments flagella are sonicated briefly with a Kontes cell disrupter at a power of about 6 for 30 seconds, yielding transversely fractured flagella about one-third the original length. Flagellar segments are then demembranated by adding Nonidet P-40 (10% stock solution) to a final concentration of 0.5%. For these experiments the upright microscope is used to facilitate perfusion using the glass chambers described above. Axonemal fragments are added directly to the glass perfusion chamber from one end using a micropipet. Perfusion is facilitated by careful adsorption from the opposite end of the chamber using narrow strips of filter paper. Nonsticking axonemal fragments are then washed away in the same buffer but without protease inhibitors. Sliding disintegration is induced by perfusing the axonemes with the same buffer containing 1 mM ATP and 1.5 μg/ml Nagarse (Type XXVII Protease, Sigma Chemical Co., St. Louis, MO), and axonemal fragments are observed on the video monitor. Microtubule sliding and disintegration of axonemes take place over the next minute while all motile events are recorded. Several examples, 4–6, of microtubule sliding can be recorded from each slide; however, as reported by others (Kurimoto and Kamiya, 1991), we have found that prolonged illumination of a region containing axonemes tends to inhibit ATP-induced sliding in the region of illumination. Thus, in practice, it is helpful to work rapidly and may be necessary to move to a newly illuminated position after perfusion.

Figure 1 is a dark-field image of axonemes before and after sliding. Analysis of sliding was restricted to axonemes in which sliding between pairs of microtubules or microtubule groups could be distinguished. For our experiments velocity was measured manually, combining Δt from the time–date-generated image superimposed on the video tape with Δd, the distance of microtubule translocation ($v = \Delta d/\Delta t$). Useful measurements within the range 0.2–25 μm/s have been made. As reported by Okagaki and Kamiya (1986) we find that Nagarse is the most reliable protease for producing uniform sliding in a large fraction of axonemes. Further, as reported (Kurimoto and Kamiya, 1991), we find that microtubule sliding velocity, in 1 mM MgATP, is dependent on the composition of the reactivation buffer. For example, in the 30 mM NaCl-containing buffer described above axonemes from wild-type *Chlamydomonas* cells slide at a reduced velocity when compared with velocity in a buffer made with potassium acetate [30 mM 4-(2-hydroxyethyl)-1-piperazineethanesulfonic acid (HEPES), pH 7.4, 5 mM MgSO$_4$, 1 mM DTT, 1 mM ethylene glycol bis(β-aminoethyl ether)N,N'-tetraacetic acid (EGTA), 50 mM potassium acetate, 0.5% polyethethylene glycol, and 1 mM ATP]. On the basis of a variety of data, reduced sliding velocity appears to be a consequence of Cl$^-$ inhibition of dynein (Gibbons *et al.*, 1985).

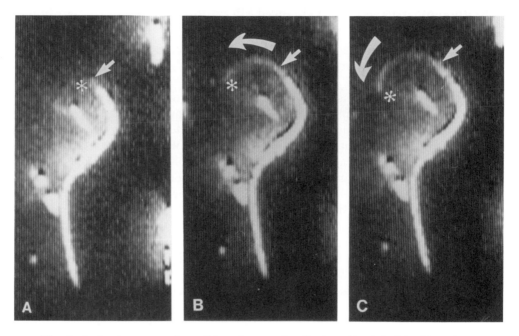

Fig. 1 Isolated axonemes from *Chlamydomonas* flagella viewed by dark-field video microscopy. The three frames selected illustrate a typical pattern of microtubule sliding, beginning at the small arrow, moving in the direction of the curved arrow, and ending at the asterisk.

This assay has been useful in analyzing microtubule sliding velocities of flagella from mutant *Chlamydomonas* cells. This analysis includes not only mutants defective in specific dynein components but also mutants defective in other axonemal structures. For example, results derived from this assay have begun to determine the relative contribution of the outer and inner dynein arms as well as individual inner-arm subforms to microtubule sliding velocities, and have also indicated that the outer and inner dynein arms may interact (Kamiya *et al.*, 1989; Kurimoto and Kamiya, 1991). In addition, analysis of sliding velocities of axonemes missing the central pair or radial spokes has indicated a possible role for these structures in the regulation of dynein (Smith and Sale, 1992a).

Our laboratory has also used this assay to evaluate function in reconstituted axonemes. We have successfully added isolated inner arms to either mutant axonemes missing a subset of inner-arm components or extracted axonemes missing all inner arms, and used the sliding disintegration assay as a method of assessing functional recovery (Smith and Sale, 1992). Undoubtedly, this functional assay when combined with *Chalmydomonmas* genetics and a reconstitution approach will continue to provide many clues to answering questions of flagellar motility.

B. Method of Analysis of ATP-Induced Microtubule Sliding in Demembranated Sea Urchin Sperm Flagella

Although fragments of isolated axonemes from sea urchin sperm tails were the first experimental specimens used for protease–ATP-induced microtubule sliding (e.g., Summers and Gibbons, 1971; Yano and Miki-Noumura, 1980; Sale and Gibbons, 1980), another useful approach is analysis of dynein-driven microtubule sliding in axonemes of intact, demembranated flagella (Sale, 1986). Conditions of demembranation have been defined in which flagella undergo a precise, ATP-induced sliding disruption, not necessarily requiring added proteases (Gibbons and Gibbons, 1980; Sale, 1986). Detailed methods are described next, but the form of sliding is illustrated in Figs. 2 and 3. By use of the methods of demembration and reactivation described, the demembranated cells assume a quiescent state with a characteristic, ATP-dependent basal, principal bend. Such axonemes are fragile and undergo microtubule sliding disruption as illustrated in Figs. 2 and 3 in which a subset of microtubules slides in the proximal direction, forming a loop. Sliding is easily visualized and analyzed by dark-field microscopy (see Sale, 1986).

To prepare cells for these experiments, sea urchin sperm (*Lytechinus pictus*) are shed by standard methods and stored undiluted on ice until used. Demembranation and induction of microtubule sliding are performed as described by Sale (1986). Briefly, 25 μl of undiluted cells is transferred to a demembranating buffer containing:

10 mM tris(hydroxymethyl)aminomethane (Tris) base (pH 8.15)

0.15 M potassium acetate

2.0 mM EGTA

1.0 mM MgSO$_4$

1.0 mM DTT

0.05% (w/v) Triton X-100

After 30 seconds, 25 μl of the demembranated cells is diluted into 2.5 ml "reactivation" buffer containing:

10 mM Tris base (pH 8.15)

0.15 M potassium acetate

2.0 mM MgSO$_4$

1.0 mM ATP

1.0 mM DTT

Fig. 2 Selected frames from dark-field video recording of microtubules sliding in demembranated "quiescent" sea urchin sperm. Sliding is restricted to shear between two sets of microtubules (see Fig. 3). Sliding can easily be monitored by dark-field microscopy as the distal end of microtubules of subset 2 (see Fig. 3) moves in the proximal direction (small arrow) in concert with development of the loop of microtubules.

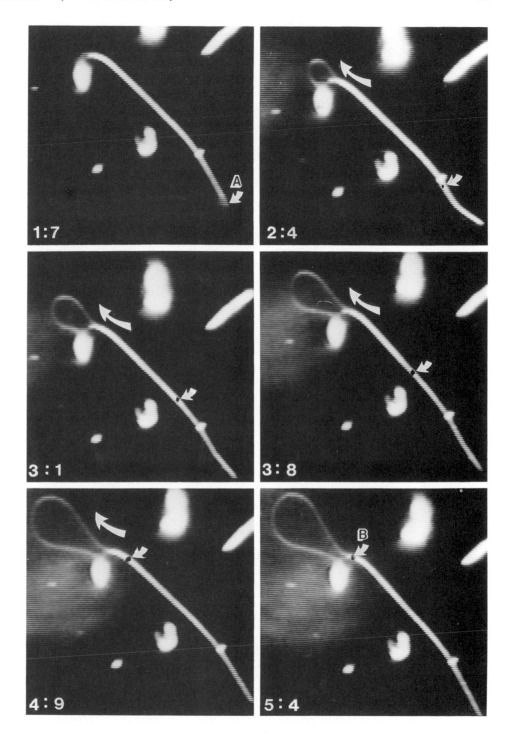

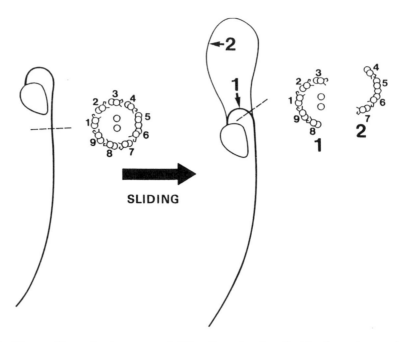

Fig. 3 Diagram illustrating the pattern of sliding disruption described for demembranated, quiescent sea urchin sperm tail axonemes (Fig. 2). Dynein-driven movement is characterized as sliding between two subsets of microtubules, with group 2 sliding proximally, forming a loop (see Sale, 1986).

0.4 m*M* CaCl$_2$

0.1 m*M* EGTA

These conditions result in quiescent sperm with an extreme basal principal bend, which spontaneously undergoes sliding disintegration as illustrated in Figs. 2 and 3. Alternatively the same pattern of microtubule sliding can be stimulated using proteases (Fox and Sale, 1987). For these experiments the upright microscope is used, and in some cases the 100-W halogen light source was used in place of the 100-W mercury burner.

The velocity of microtubule sliding is measured by two related methods. In the first method, the times of initiation and termination of sliding are determined from the time generator superimposed on the tape. The total distance traversed (e.g., point *a*, first frame to point *b*, last frame, Fig. 2) is measured and velocity calculated by $v = \Delta d/\Delta t$ and expressed as μm/s. In some cases the size of the microtubule loop is measured to determine Δd. For the second method, sliding distance is measured at time intervals (0.1- to 0.4-second intervals) for the entire sliding event, and distances are plotted as a function of time (Fig. 4). As described before (Yano and Miki-Noumura, 1980; Okagaki and Kamiya, 1986;

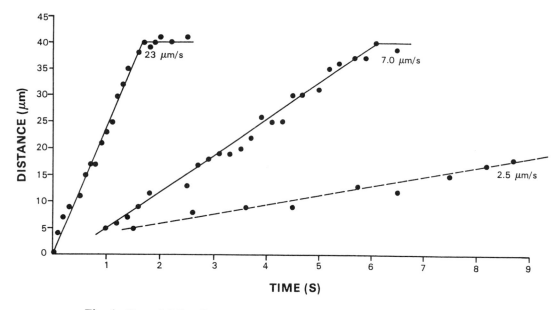

Fig. 4 Plots of sliding distance (small curved arrows, Fig. 2) against time for microtubule sliding in demembranated sea urchin sperm cells illustrated in Fig. 2. Plotted are three different examples of axonemal microtubules induced to slide at three different velocities. Notably, over a 10-fold difference (2.5–23 μm/s), velocity ($\Delta d/\Delta t$) is constant for each axoneme until movement stops.

Oiwa and Takahashi, 1988), sliding velocity (the slope of the lines in Fig. 4) remains constant throughout the distant traversed and, thus, is independent of the degree of microtubule overlap. This result suggests that for an unloaded set of microtubules, velocity of dynein-driven microtubule sliding is independent of the number of dynein arms. Therefore, velocity must be limited by some inherent rate-limiting step in the force-generating cycle (Oiwa and Takahashi, 1988).

IV. The Second Assay: Dynein-Driven Microtubule Translocation Using Purified Proteins

In vitro motility assays, as described above, have provided discovery of novel molecular motors, analysis of the functional properties of different axonemal dyneins, and investigation of the mechanisms of regulation of dyneins. In principle, microtubules are applied to a dynein-coated surface, and in the presence of ATP, microtubules are translocated across that surface (Fig. 5). The experimental requirements include production of purified, taxol-stabilized microtubules, a source of isolated axonemal dyneins, and the means to visualize and record dynein-driven movement.

For our studies we have used either bovine or porcine brain as a source

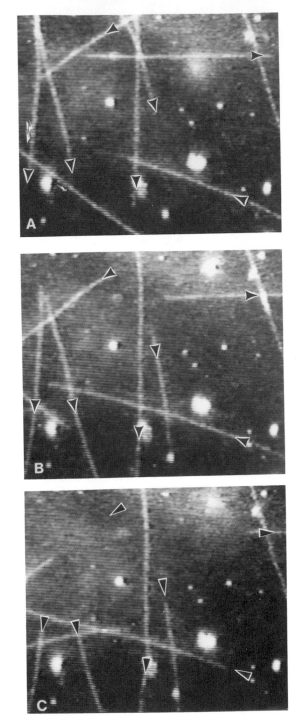

Fig. 5 Typical dark-field video image of single microtubules absorbed to and moving across a dynein-coated slide. Three frames were selected to illustrate the movement of each of several microtubules (arrowheads).

of tubulin. Tubulin is purified by a two-step process including cycles of temperature-dependent assembly and disassembly followed by purification from microtubule-associated proteins (MAPs) using O-diethylaminoethyl (DEAE) ion-exchange chromatography (Sale and Satir, 1987). Twice-cycled tubulin in a polymerization buffer (PM) containing 0.1 M 1,4-piperazine-diethanesulfonic acid (Pipes), pH 6.85, 2 mM EGTA, 1 mM MgSO$_4$, and 0.2 mM GTP is diluted 1 : 1 in PM containing 500 mM KCl for a final salt concentration of 250 mM KCl. The protein is loaded onto a 2 × 10-cm column of DEAE-Sephacel, equilibrated in the same buffer containing 250 mM KCl. Proteins are eluted by a two-step buffer wash: first, 250 mM KCl in PM to release MAPs, and second, 500 mM KCl in PM to elute tubulin. Fractions are immediately analyzed by UV spectrophotometry and by mini-gel electrophoresis to determine peak tubulin fractions and degree of purification. Peak fractions, which usually contain only small amounts of contaminating high-molecular-weight MAPs, are pooled, concentrated in an Amicon ultrafiltration cell on an Amicon Diaflo YM30 filter to about 5–8 mg/ml protein, and subsequently desalted on a Bio-Rad desalting column in polymerization buffer containing 0.5 mM GTP. The assembly-competent, purified tubulin is then stored as frozen drops in a liquid nitrogen storage container. As required, microtubules are assembled at 2–3 mg/ml protein by incubation at 37°C for 10 minutes in PM buffer and finally stabilized by dilution with polymerization buffer containing 20 μM taxol. Taxol-stabilized microtubules are then sedimented for 5 minutes in a Beckman Airfuge (80,000g), and the microtubule pellet is gently resuspended in a motility buffer containing 20 μM taxol, 10 mM Tris, pH 7.3, 2 mM MgSO$_4$, 1 mM EGTA, 0.1 M potassium acetate, 1 mM DTT, and 1 mM ATP. In some cases microtubules were assembled from the plus ends of axonemal doublet microtubule fragments as described by Vale and Toyoshima (1988). The process involves mixing equal parts of N-ethylmaleimide (NEM)-treated tubulin and tubulin with a small amount of doublet microtubule pieces at 37°C. Such axonemal–microtubule constructs offer a polarity marker (the minus end is distinguishable by dark-field microscopy) and a potential to analyze microtubule rotation as well as translocation (Vale and Toyoshima, 1988).

For many studies axonemal dynein subforms can be solubilized using high-ionic-strength buffers and then partially purified by sucrose gradient centrifugation as large, monodisperse particles. For the purpose of this chapter, we focus discussion on the outer-arm dynein from sea urchin sperm tail axonemes (Tang et al., 1982; Sale et al., 1985); however, the assay has been applied to the study of axonemal dynein subforms from several sources including the dyneins from the inner row of arms (Kagami et al., 1990; Smith and Sale, 1991).

The outer arm from sea urchin is selectively solubilized and purified as a functionally intact, 21 S multimeric ATPase with a mass of 1.3 MDa and at least nine polypeptide subunits (Gibbons and Fronk, 1979). Each dynein is a complex of the α and β heavy chains (Tang et al., 1982), and electron microscopy has shown that it consists of two structurally distinct, large globular head domains joined by flexible stems and smaller globular subunits (Sale et al., 1985). The isolated outer arm can be dissociated and fractionated into subunits (Fig. 6) (Tang *et al.*, 1982; Sale *et al.*, 1985), and the functional activities of individual subunits examined by a variety of approaches including the microtubule translocation assay described here.

Following sucrose gradient purification, dynein-containing fractions were generally adjusted to about 0.1 mg/ml protein either by dilution into the motility buffer described above or by concentration in Centricon 30 cells (Amicon Corp., Danvers, MA) and then appropriate dilution to 0.1 mg/ml protein using motility buffer. As described (Vale and Toyoshima, 1988; Sale and Fox, 1988) we found that there is a threshold of about 0.08 mg/ml dynein for successful microtubule binding and motility. In developing these assays, we also discovered that, when present at subthreshold concentrations, dynein could be concentrated on the slide by successive incubations of 5- to 7-μl aliquots in the perfusion chamber for 1 minute each. Our interpretation is that the glass surface must be thoroughly coated with dynein for efficient microtubule binding. Thus, for dynein samples adjusted to approximately 0.1 mg/ml in motility buffer, our standard experimental protocol was as follows: dynein fractions were applied to the perfusion chamber in two successive 6-μl aliquots for a 1-minute incubation period each (20°C). After incubation, the chamber is perfused with 50 μl motility buffer to rinse away nonadherent protein. A 6-μl aliquot of taxol-stabilized microtubules (~0.2 mg/ml) suspended in motility buffer containing 1 mM ATP was applied, and nonadherent microtubules were rinsed away by a further rinse with 50 μl motility buffer. It is important to note that, for reasons not yet clarified, microtu-

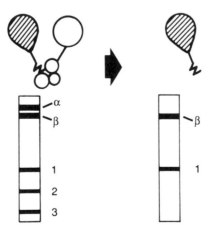

Fig. 6 Diagram of corresponding structure and polypeptide composition of the outer dynein arm from sea urchin sperm (see Sale *et al.*, 1985). The β heavy chain/intermediate chain subunit (β/IC1) can be isolated by dialysis of the outer-arm dynein (21 S) in a low-ionic-strength buffer followed by separation on sucrose gradients made up in the low-ionic-strength buffer (Tang *et al.*, 1982; Sale *et al.*, 1985).

bules will not stick to sea urchin sperm tail dynein in the absence of ATP (Moss *et al.*, 1992). Therefore, for these experiments microtubules were always added in the presence of 1 m*M* ATP. For experiments requiring lower concentrations of ATP (Figs. 7 and 8), an enzymatic, ATP-regenerating system composed of 20 U/ml creatine kinase and 2 m*M* phosphocreatine is added to the motility buffer (see Vale and Toyoshima, 1988).

Most observations were performed using the inverted microscope coupled to the SIT 66 camera; however, microtubule movement could be observed equally well using the upright microscope. The intact 21 S outer dynein arm generally induced rapid, unidirectional translocation of microtubules (Paschal *et al.*, 1987a; Sale and Fox, 1988; Vale *et al.*, 1989). In the absence of dynein, microtubules do not bind to the surface, and at subthreshold concentrations of dynein, microtubules either do not bind or the microtubules bind weakly by one end. Movement of longer microtubules (5–50 μm) was continuous, whereas movement of short microtubules (0.5–3 μm) (which could be produced by passage of a mixture of long microtubules once or twice through a 26-gauge needle attached to a tuberculin syringe) was often intermittent. As described before (Vale and Toyoshima, 1988), however, velocity of moving microtubules was independent of their length (Fig. 7).

The isolated β heavy chain/intermediate chain one (β/IC1) subunit of the outer dynein arm also induces rapid, unidirectional movement of microtubules (Sale

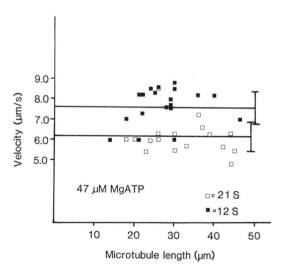

Fig. 7 Plot of microtubule translocation velocity in 47 μM MgATP as a function of microtubule length for both the 21 S outer arm and the 12 S β/IC1 subunit. As described before, velocity is independent of microtubule length (see Paschal *et al.*, 1987a; Vale and Toyoshima, 1988).

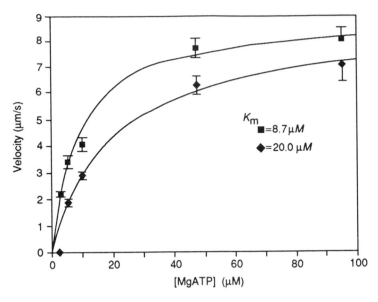

Fig. 8 Plots of microtubule translocation velocity versus [MgATP] for 21 S outer-arm dynein (diamonds) and the β/IC1 subunit (squares). The data were fit with the Michaelis equation and yield V_{max} and K_m for each dynein. To perform these experiments the motility buffer contained an ATP-regenerating system (20 U/ml creatine kinase and 2 mM phosphocreatine).

and Fox, 1988; Vale *et al.*, 1989). In this case, motility was more rapid than that of the intact outer arm from which the subunit was derived (see Fig. 8) (Sale and Fox, 1988). In Fig. 8 velocity was plotted as a function of [MgATP]. From such plots we have been able to calculate V_{max} and K_m for the dynein fractions.

The advantage of using sea urchin sperm axonemes as a dynein source is that large quantities of outer-arm dynein can be obtained, and methods for dissecting the outer dynein arms can be combined with this assay to determine outer-arm functional domains. This, however, is not the case for *Chlamydomonas* flagellar axonemes. In isolating dyneins from *Chlamydomonas* axonemes it is important to note that, unlike sea urchin sperm, high-ionic-strength buffers remove both the inner and outer dynein arms. To obtain a pure preparation of *Chlamydomonas* dynein, that is, inner arms or subsets of inner arms, the dyneins are obtained from mutants and subsequently purified by either sucrose density gradient centrifugation (Smith and Sale, 1991) or FPLC (cf. Kagami *et al.*, 1990). Therefore, by using various dynein-defective *Chlamydomonas* mutants as a dynein source, the investigator gains access to the inner dynein arms and specific subsets of inner-dynein-arm subforms.

V. Summary: Future

We have described two *in vitro* assays used to investigate the mechanism and regulation of flagellar dyneins. In the sliding disintegration assay dynein's ability to translocate microtubules is assessed *in situ*. Therefore, this assay has the advantage that all axonemal components that might regulate dynein's ability to induce microtubule translocation are present. The second assay has the advantage that the number of components are pared down to a minimum, permitting careful control of factors that might regulate dynein (see Hamasaki *et al.,* 1991). For either assay, well-characterized mutant *Chlamydomonas* cells that are defective in certain axonemal components will continue to be useful. Further, for both assays variation in buffer conditions can be used to investigate the mechanism of dynein regulation. Finally, these functional assays together with improved spatial and temporal resolution and the means of measuring force (cf. Oiwa and Takahashi, 1988; Gelles *et al.,* 1988) provide new opportunities for understanding the mechanism of axonemal dynein's crossbridge cycle.

References

Allen, R. D., Weiss, D. G., Hayden, J. H., Brown, D. T., Fujiwake, H., and Simpson, M. (1985). Gliding movement and bidirectional organelle transport along single native microtubules from squid axoplasm: Evidence for an active role of microtubules in cytoplasmic transport. *J. Cell Biol.* **100,** 1736–1752.

Fox, L. A., and Sale, W. S. (1987). Direction of force generated by the inner row of dynein arms on flagellar microtubules. *J. Cell Biol.* **105,** 1781–1987.

Gelles, J., Scnapp, B., and Sheetz, M. (1988). Tracking kinesin—driven movements with nanometer—scale precision. *Nature (London)* **331,** 450–453.

Gibbons, B. H., and Gibbons, I. R. (1980). Calcium induced quiescence in reactivated sea urchin sperm. *J. Cell Biol.* **84,** 13–27.

Gibbons, B. H., Tang, W.-J., and Gibbons, I. R. (1985). Organic anions stabilize the reactivated motility of sperm flagella and the latency of dynein I. *J. Cell Biol.* **101,** 1281–1287.

Gibbons, I. R., and Fronk, E. (1979). A latent ATPase form of dynein I from sea urchin sperm flagella. *J. Cell Biol.* **254,** 187–196.

Hamasaki, T., Barkalow, K., Richmond, J., and Satir, P. (1991). cAMP-stimulated phosphorylation of an axonemal polypeptide that co-purifies with 22S dynein arm regulates microtubule translocation velocity and swimming speed in *Paramecium. Proc. Natl. Acad. Sci. U.S.A.* **88,** 7918–7922.

Kagami, O., Takada, S., and Kamiya, R. (1990). Microtubule translocation caused by three subspecies of inner-arm dynein from *Chlamydomonas* flagella. *FEBS Lett.* **264**(2), 179–182.

Kamiya, R., Kurimoto, E., Sakakibara, H., and Okagaki, T. (1989). A genetic approach to the function of inner and outer arm dynein. *In* "Cell Movement" (F. D. Warner, P. Satir, and I. R. Gibbons, eds.) Vol. 1. pp. 209–218. Liss, New York.

Kamiya, R., Kurimoto, E., and Muto, E. (1991). Two types of *Chlamydomonas* flagellar mutants missing different components of inner-arm dynein. *J. Cell Biol.* **112,** 441–447.

Kurimoto, E., and Kamiya, R. (1991). Microtubule sliding in flagellar axonemes of *Chlamydomonas* mutants missing inner and outer arm dynein: Velocity measurements on new types of mutants with an improved method. *Cell Motil. Cytoskel.* **19,** 275–281.

Lye, R. J., Porter, M. E., Scholey, J. M., and McIntosh, J. R. (1987). Identification of a microtubule-based cytoplasmic motor in the nematode *Caenorhabditis elegans*. *Cell (Cambridge, Mass.)* **51,** 305–318.

Moss, A. G., Sale, W., Fox, L., and Witman, G. (1992). The α subunit of sea urchin sperm outer arm dynein mediates structural and rigor binding to microtubules. *J. Cell Biol.* **118,** 1189–1200.

Oiwa, K., and Takahashi, K. (1988). The force–velocity relationships for microtubule sliding in demembranated sperm flagella of the sea urchin. *Cell Struct. Funct.* **113,** 193–205.

Okagaki, T., and Kamiya, R. (1986). Microtubule sliding in mutant *Chlamydomonas* axonemes devoid of outer or inner dynein arms. *J. Cell Biol.* **103,** 1895–1902.

Paschal, B. M., King, S., Moss, A., Collins, C., Vallee, R., and Witman, G. (1987a). Isolated flagellar outer arm dynein translocates brain microtubules *in vitro*. *Nature (London)* **330,** 672–674.

Paschal, B. M., Shpetner, H. S., and Vallee, R. B. (1987b). MAPIC is a microtubule activated ATPase that translocates microtubules *in vitro* and has dynein like properties. *J. Cell Biol.* **105,** 1273–1282.

Sale, W. S. (1986). The axonemal axis and Ca-induced asymmetry of active microtubule sliding in sea urchin sperm tails. *J. Cell Biol.* **102,** 2042–2052.

Sale, W. S., and Fox, L. A. (1988). Isolated β-heavy chain subunit of dynein translocates microtubules *in vitro*. *J. Cell Biol.* **107,** 1793–1798.

Sale, W. S., and Gibbons, I. R. (1980). Study of the mechanism of vanadate inhibition of the dynein cross-bridge cycle in sea urchin sperm flagella. *J. Cell Biol.* **82,** 291–298.

Sale, W. S., and Satir, P. (1977). Direction of active sliding of microtubules in *Tetrahymena* cilia. *Proc. Natl. Acad. Sci. U.S.A.* **74,** 2045–2059.

Sale, W. S., Goodenough, U., and Heuser, J. E. (1985). The substructure of isolated and *in situ* outer dynein arms of sea urchin sperm flagella. *J. Cell Biol.* **101,** 1400–1412.

Sale, W. S., Fox, L. A., and Milgram, S. A. (1989). Composition and organization of the inner row of dynein arms. *In* "Cell Movement" (F. D. Warner, P. Satir, and I. R. Gibbons, eds.) Vol. 1, pp. 89–102. Liss, New York.

Schnapp, B. J., and Reese, T. R. (1989). Dynein is the motor for retrograde axonal transport of organelles. *Proc. Natl. Acad. Sci. U.S.A.* **86,** 1548–1552.

Smith, E. F., and Sale, W. S. (1991). Microtubule binding and translocation by inner dynein arm subtype I1. *Cell Motil. Cytoskel.* **18,** 258–268.

Smith, E. F., and Sale, W. S. (1992). Structural and functional reconstitution of inner dynein arms in *Chlamydomonas* flagellar axonemes. *J. Cell Biol.* **117,** 573–582.

Smith, E. F., and Sale, W. S. (1992a). Regulation of dyneins driven microtubule sliding by the radial spokes in flagella. *Science* **257,** 1557–1559.

Summers, K., and Gibbons, I. R. (1971). Adenosine triphosphate-induced sliding of tubules in trypsin treated flagella of sea urchin sperm. *Proc. Natl. Acad. Sci. U.S.A.* **68,** 3092–3096.

Tang, W.-J. Y., Bell, C. W., Sale, W. S., and Gibbons, I. R. (1982). Structure of the dynein-1 outer arm in sea urchin sperm flagella. I. Analysis by separation of subunits. *J. Biol. Chem.* **257,** 568–575.

Vale, R. D., and Toyoshima, Y. Y. (1988). Rotation and translocation of microtubules *in vitro* induced by dyneins from *Tetrahymena* cilia. *Cell (Cambridge, Mass.)* **52,** 459–469.

Vale, R. D., and Toyoshima, Y. Y. (1989). Microtubule translocation properties of intact and proteolytically digested dyneins from *Tetrahymena* cilia. *J. Cell Biol.* **108,** 2327–2334.

Vale, R. D., Reese, T. S., and Sheetz, M. P. (1985). Identification of a novel force-generating protein, kinesin, involved in microtubule-based motility. *Cell (Cambridge, Mass.)* **42,** 39–50.

Vale, R. D., Soll, D. R., and Gibbons, I. R. (1989). One dimensional diffusion of microtubules bound to flagellar dynein. *Cell (Cambridge, Mass.)* **59,** 915–925.

Witman, G. B. (1986). Isolation of *Chlamydomonas* flagella and flagellar axonemes. *In* "Methods in Enzymology" (R. B. Vallee, ed.), vol. 134, pp. 280–290. Academic Press, Orlando, FL.

Witman, G. B., Plummer, J., and Sander, G. (1978). *Chlamydomonas* flagellar mutants lacking radial spokes and central tubules. *J. Cell Biol.* **76,** 729–747.

Yano, Y., and Miki-Noumura, T. (1980). Sliding velocity between outer doublet microtubules of sea urchin sperm axonemes. *J. Cell Sci.* **44,** 169–196.

CHAPTER 7

Preparation of Marked Microtubules for the Assay of the Polarity of Microtubule-Based Motors by Fluorescence Microscopy

Jonathon Howard† and Anthony A. Hyman*

† Department of Physiology and Biophysics
University of Washington
Seattle, Washington 98195

* Department of Pharmacology
University of California, San Francisco
San Francisco, California 94143

  I. Introduction
 II. Preparation of Labeled Microtubules
III. Reagents
 IV. Preparation of the Bright Microtubule Seeds
  V. Preparation of the Polarity-Marked Microtubules
 VI. Visualization of Marked Microtubules in Motor Polarity Assays
     References

I. Introduction

Microtubules are polymers of the heterodimeric protein tubulin. Because all the dimers have the same orientation in the lattice, and because the dimers are not mirror symmetric, the microtubule polymer has an inherent polarity. This intrinsic polarity has two major functional implications. First, the growth rates at the different ends of the microtubules are different: growth at the so-called plus end is about three times faster than that at the minus end. The ends are thus functionally different. Second, different microtubule-based motor proteins move in opposite directions along the microtubule: kinesin moves toward the plus end, and dynein and ncd move toward the minus end. Thus, the cylindrical surface of the microtubule has a different texture in the two directions, a texture that the motor proteins can detect. As knowledge of the polarity of movement of a motor protein is essential for interpreting its biological function, it is essential to have a quick and reliable method for determining the polarity of a newly discovered motor. In this chapter we describe how to assemble microtubules with obvious polarity when viewed by fluorescence microscopy; the direction of movement of a motor protein with respect to such polarity-marked microtubules indicates the motor's polarity.

II. Preparation of Labeled Microtubules

Briefly, the strategy for making polarity-marked microtubules is shown in Fig. 1. It takes advantage of the properties of tubulin polymerization. Microtubule nucleation will not occur below a certain critical concentration, but new microtubule growth will occur from the preexisting ends of microtubules down to tubulin concentrations below that critical concentration. Short microtubules are constructed by polymerizing highly rhodamine-labeled tubulin at concentrations that support the nucleation of microtubules. The brightly labeled microtubules formed are then diluted into lightly labeled tubulin, at concentrations of tubulin just below the critical concentration for nucleation. Almost all polymerization of the dimly labeled tubulin now takes place from the ends of the preexisting brightly labeled microtubules. Because plus ends grow faster than minus ends, the longer dimly labeled rhodamine segment can be distinguished as the plus end from the shorter dimly labeled segment at the minus end. The microtubules are then stabilized in taxol.

III. Reagents

Tubulin is purified according to Weingarten et al. (1974) with the modifications described (Mitchison and Kirschner, 1984). The tubulin is labeled with rhodamine according to Hyman et al. (1991). The buffer used for microtubule polymerization is BRB80 [80 mM 1,4-piperazinediethanesulfonic acid (Pipes), 1 mM GTP, 1 mM MgCl$_2$, pH 6.8, with KOH].

IV. Preparation of the Bright Microtubule Seeds

Short brightly labeled microtubules can be prepared in three ways. We have noticed that rhodamine-labeled tubulin polymerizes poorly with GTP if the stochiometry of labeling is over 1. In cases where the labeling stochiometry is

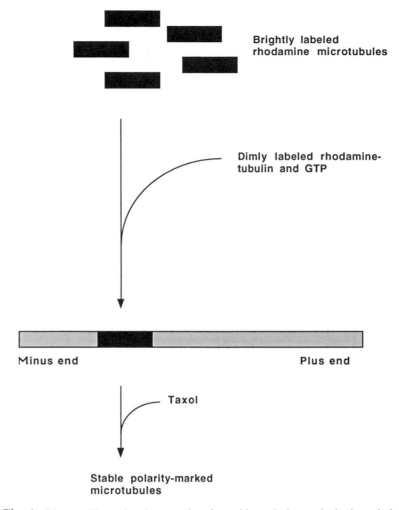

**Brightly labeled
rhodamine microtubules**

**Dimly labeled rhodamine-
tubulin and GTP**

Minus end   **Plus end**

Taxol

**Stable polarity-marked
microtubules**

Fig. 1 Diagram illustrating the procedure for making polarity-marked microtubules.

over 1, the rhodamine-tubulin should be diluted 1:1 with unlabeled tubulin. For convenience, assembly mixtures can be stored as frozen aliquots at −70°C and warmed at 37°C for polymerization. Prior to freezing all the components are mixed on ice and spun at 30 psi in the airfuge at 4°C for 5 minutes to remove any aggregates. The supernatant is fast-frozen in 5 μl aliquots in liquid nitrogen and stored at −70°C. When required, the aliquot can be placed at 37°C to poly- merize.

1. Stable seeds can be made by polymerizing tubulin with a nonhydrolyzable

analog of GTP, GMPCPP (Hyman *et al.*, 1992). Unfortunately, a commercial source of GMPCPP is not available and it must be synthesized (Hyman *et al.*, 1992). To do this 4 mg/ml rhodamine tubulin is polymerized with 150 μM GMPCPP in BRB80 to 37°C for 15 minutes.

2. Short GTP-microtubules can be formed by polymerizing tubulin in the presence of glycerol. Rhodamine-labeled tubulin (4 mg/ml) is placed in BRB80 augmented with 4 mM MgCl$_2$ and glycerol to 40% (v/v) at 37°C for 30 minutes.

3. Microtubule seeds polymerized with GTP can be stabilized by chemical crosslinking using the method of Koshland *et al.* (1988). This technique has the advantage of creating a large number of seeds which can be kept at room temperature up to 2 months. The following protocol schematized in Fig. 2 is derived from Koshland *et al.* (1988). Microtubules whose ends have been crosslinked with EGS cannot support further polymerization from their ends; however, if they are broken into small microtubules by shearing, the resulting ends of these short microtubules can now polymerize. It is therefore essential that the microtubules grow extremely long before crosslinking so that after breakage, there is a high concentration of newly exposed ends. To achieve this, short seeded microtubules are used to nucleate the growth of microtubules (step B of Fig. 2). Under these conditions, because of the properties of microtubules growing under dynamic instability conditions (Mitchison and Kirschner, 1984), some of the microtubules will grow extremely long.

1. To make glycerol seeds, to 100 μl tubulin 4 mg/ml, add 50 μl glycerol, 1.5 μl 100 mM GTP, and 0.5 μl 1 M MgCl$_2$. Incubate at 37°C for 50 minutes. Shear five times with a 22- to 30-gauge needle.

2. Dilute the glycerol seeds into rhodamine–tubulin at 20 μM. On ice, mix 100 μl tubulin 4 mg/ml, 10 μl rhodamine–tubulin 40 mg/ml, 90 μl BRB80, and 2 μl 100 mM GTP. Warm to 37°C for 1 minute; then add 1/100th volume of seeds. Incubate at 37°C for 60 minutes.

3. Crosslink by adding 1/10th volume of 15 mM EGS (Pierce 21565) in dimethylsulfoxide (DMSO) (Aldrich Sure Seal). Incubate at 37°C for 60 minutes.

4. Quench by adding 2 ml BRB80 containing 50% sucrose, 10 mM potassium glutamate, and 0.1% β-mercaptoethanol. Incubate at 37°C for 60 minutes.

5. Shear four times with a 22- to 30-gauge needle.

6. Clean the seeds by diluting the seed mixture 1:1 with BRB80 and layering, on a step gradient, 2 ml seeds, 1 ml 40% sucrose in BRB80 plus 0.1% β-mercaptoethanol, and 2 ml 75% sucrose in BRB80 plus 0.1% β-mercaptoethanol. Spin 40 minutes 40,000 rpm in a SW 50.1 rotor. Collect seeds at the 40%/75% interface.

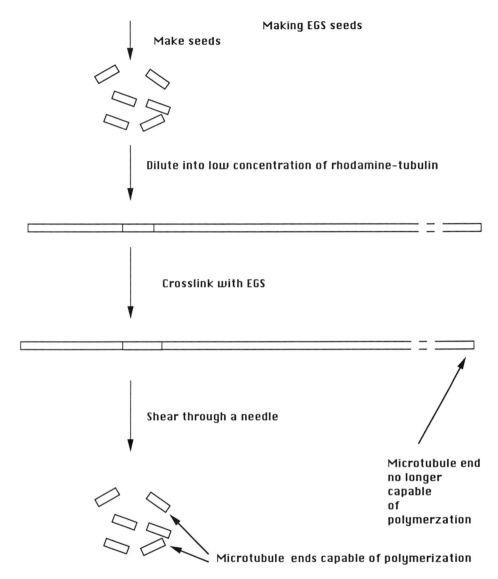

Fig. 2 Diagram illustrating the procedure for preparing EGS crosslinked seeds.

================= **V. Preparation of the Polarity-Marked Microtubules**

The highly labeled seeds are then diluted into a mixture containing a 10:1 molar ratio of unlabeled to rhodamine-labeled tubulin. The polarity is easily identified, as the slower-growing minus ends are shorter than the plus ends. This disparity in plus- and minus-end lengths can be increased by including *N*-ethylmaleimide (NEM)-treated tubulin, which prevents minus-end growth (Hyman *et al.*, 1991), in the assembly mixture. NEM–tubulin is prepared according to Hyman *et al.* (1991) at a molar ratio of 0.7:1.0. This difference can be seen by comparing Fig. 3 with Fig. 4. In Fig. 3, the microtubule has obvious minus-end and plus-end segments. In Fig. 4, the minus-end growth of the microtubules has been suppressed using NEM–tubulin, and the seed can be seen marking the minus end of the microtubules.

1. Prepare a dimly labeled assembly mixture containing unlabeled tubulin at 1.5 mg/ml and rhodamine–tubulin at 0.15 mg/ml with 1 m*M* GTP in BRB80. Warm to 37°C for 1 minute.

2. After 1 minute, dilute the rhodamine seeds into the tubulin aliquot at a 1:10 (v/v) ratio into the prewarmed assembly mixture. Allow the microtubules to polymerize until they reach a desired length, generally about 10 μm.

3. Stabilize the microtubules by adding BRB80 augmented with 10 μ*M* taxol prewarmed to 37°C, taking care to avoid shear.

4. To remove any remaining free rhodamine–tubulin monomer, pellet the taxol-stabilized microtubules in the airfuge at 20 psi for 3 minutes and resuspend in BRB80 augmented with 10 μ*M* taxol.

VI. Visualization of Marked Microtubules in Motor Polarity Assays

Bleaching of fluorescently labeled microtubules poses a major problem with long periods of illumination. In addition, microtubules will break on observation with sufficiently high illumination (Vigers *et al.*, 1988). To prevent photobleaching we use the oxygen-scavenging system developed for the observation of single actin filaments (Kishino and Yanigida, 1988). In the motility buffer to be used, 0.1 mg/ml catalase, 0.03 mg/ml glucose oxidase, 10 m*M* glucose, and 0.1% β-mercaptoethanol are added. With this system, microtubules have been successfully recorded after 10 minutes of illumination with an unattenuated 100-W mercury arc lamp. Microtubule breakage is very rare. Nevertheless, for extended recording, we recommend the use of shuttered light source and a sensitive camera such as a silicon-intensified-target camera.

This technique has been used to follow the polarity of movement of a number of different isolated motors (Belmont, *et al.*, 1990; Hyman and Mitchison, 1991;

Fig. 3 A microtubule moving by kinesin-based motility. Kinesin is adsorbed onto the surface. Images were taken at 4-second intervals. The velocity of movement is 1.0 μm/s. In such preparations, about 50% of microtubules have obvious bright seeds. The remaining microtubules are a mixture of very short seeds, broken microtubules, and microtubules that arise by spontaneous nucleation. Bar - 10 μm.

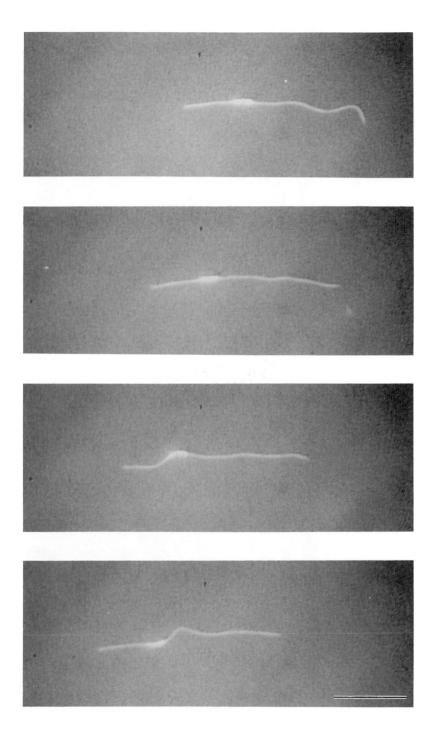

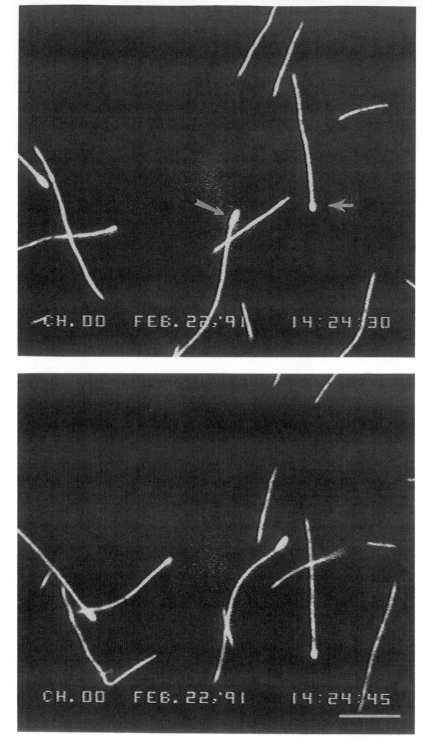

Fig. 4 Polarity-marked microtubules prepared in the presence of *N*-ethylmaleimide–tubulin. The brightly labeled seeds are marked with arrows. Note the absence of a minus-end segment. Images were taken at 15-second intervals. Bar - 10 μm.

McDonald and Goldstein, 1990; Sawin *et al.*, 1992; Walker *et al.*, 1990). An example of movement driven by kinesin is shown in Fig. 3.

Acknowledgments

The authors are grateful to Tim Mitchison and David Drechsel for many helpful discussions and advice. A.A.H. is a Lucille P. Markey scholar and this work was supported in part by a grant from the Lucille P. Markey charitable trust.

References

Belmont, L., Hyman, A. A., Sawin, K. S., and Mitchison, T. J. (1990). Real-time visualization of cell cycle dependent changes in microtubule dynamics in cytoplasmic extracts. *Cell (Cambridge, Mass.)* **62,** 579–589.

Hyman, A. A., and Mitchison, T. J. (1991). Two different microtubule-based motor activities with opposite polarites in kinetochores. *Nature (London)* **351,** 206-211.

Hyman, A. A., Drexel, D., Kellog, D., Salser, S., Sawin, K., Steffen, P., Wordeman, L., and Mitchison, T. J. (1991). Preparation of modified tubulins. *In* "Methods in Enzymology" (R. B. Vallee, ed.), Vol. 196, pp. 478-485. Academic Press, San Diego.

Hyman, A. A., Salser, S., Drechsel, D., Unwin, N., and Mitchison, T. J. (1992). The role of GTP hydrolysis in microtubule dynamics: Information from a slowly hydrolyseable analogue GMPCPP. *Mol. Biol. Cell* **3,** 1155–1167.

Kishino, A., and Yanigida, T. (1988). Force measurements by manipulation of a single actin filament. *Nature (London)* **334,** 74–76.

Koshland, D., Mitchison, T. J., and Kirschner, M. W. (1988). Chromosome movement driven by microtubule depolymerization in vitro. *Nature (London)* **311,** 499–504.

McDonald H. B., and Goldstein, L. S. B. (1990). Identification and characterization of a gene encoding a kinesin-like protein in *Drosophila. Cell (Cambridge, Mass.)* **61,** 991–1000.

Mitchison, T. J., and Kirschner, T. J. (1984). Dynamic instability of microtubule growth. *Nature (London)* **312,** 237–242.

Sawin, K. E., LeGuellec, K., Philippe, M., and Mitchison, T. J. (1992). Mitotic spindle pole organization by a plus-end directed microtubule motor. *Nature (London)* **359,** 540–543.

Vigers, G. P. A., Coue, M., and McIntosh, J. R. (1988). Fluorescent microtubules break up under illumination. *J. Cell Biol.* **107,** 1011–1024.

Walker, R. A., Salmon, E. D., and Endow, S. A. (1990). The *Drosophila* claret segregation protein is a minus-end directed motor molecule [see comments]. *Nature (London)* **347,** 780–782.

Weingarten, M. D., Suter, M. M., Littman, D. R., and Kirschner, M. W. (1974). Properties of depolymerization products of microtubules from mammalian brain. *Biochemistry* **13,** 5520–5529.

CHAPTER 8

Expression of Microtubule Motor Proteins in Bacteria for Characterization in *in Vitro* Motility Assays

Rashmi Chandra and Sharyn A. Endow

Department of Microbiology
Duke University Medical Center
Durham, North Carolina 27710

I. Introduction
II. Construction of Plasmids for Expression in Bacteria
III. Vectors Used for Expression of Motor Proteins
    A. pET Expression System
    B. pGEX Expression System
IV. Expression of Proteins in *Escherichia coli*
V. Partial Purification of Proteins from Bacteria for *in Vitro* Motility Assays
VI. Brief Description of Motility Assays for Testing the Functional Properties of the Expressed Proteins
VII. Microtubule Binding Assays
VIII. Measurement of ATPase Activity as a Determinant of Motor Protein Function
IX. Problems Encountered with Prokaryotic Expression Systems and Some Solutions
    References

I. Introduction

Expression of foreign proteins in bacteria has become a very powerful technique for obtaining large quantities of protein. Proteins that are expressed in minute amounts *in vivo* and therefore require very large amounts of starting material and enormous effort to purify by conventional methods can now be expressed in bacteria. Various protein domains can also be expressed to determine their function.

METHODS IN CELL BIOLOGY, VOL. 39

Several microtubule motor proteins, including kinesin heavy chain (Yang *et al.*, 1989) and ncd (nonclaret disjunctional) (Walker *et al.*, 1990a, McDonald *et al.*, 1990) have been successfully expressed in bacteria for analysis of functional properties. This chapter describes some of the bacterial expression systems in use, with emphasis on those that have been used to express microtubule motor proteins. We also describe protocols used in this laboratory to obtain active motor proteins from bacteria, assays that can be used for biochemical characterization of motor proteins, and a brief description of *in vitro* motility assays.

II. Construction of Plasmids for Expression in Bacteria

Some of the frequently used prokaryotic expression systems are described in Table I. The DNA sequence encoding a protein (or a segment of it) is ligated into a unique restriction site in the expression vector with the initiating methionine in the correct reading frame. Standard cloning techniques (Sambrook *et al.*, 1989) are used to transform the ligated DNA into bacterial cells, select transformants carrying the DNA insert in the correct orientation, and transfer constructs into appropriate host cells for protein induction. The DNA fragment for insertion into a vector can be made in one of two ways.

In the first method, primers corresponding to the ends of the region of interest are synthesized and the DNA is amplified by polymerase chain reaction (PCR). The PCR product can then be inserted into the plasmid by ligation into a blunt-end or Klenow-repaired restriction site. Alternatively, the same restriction site as the one into which the PCR product will be cloned can be added at the 5' end of each primer; the PCR product can then be digested with the restriction enzyme and ligated to the vector. A few bases (three to five) beyond the 5' end of the restriction site should be added to the primer for efficient digestion by the restriction enzyme. The PCR product should be gel-purified before ligation in order to remove template DNA, incorrect products, and PCR primers. Orientation of the insert in the vector can be determined by digesting the construct with restriction enzymes that cut at a site in the insert positioned asymmetrically with respect to a site in the vector. This excises a fragment that differs in size depending on the orientation of the insert. DNA sequence analysis should be carried out on clones constructed using PCR-amplified DNA to ensure that no bases were changed in the sequence. Single base changes, especially in the active or functional site, can drastically alter the properties of the expressed protein. Minipreps (Kraft *et al.*, 1988) of chloramphenicol-amplified plasmid DNA (Sambrook *et al.*, 1989) can be sequenced using the dideoxy chain termination method (Sanger *et al.*, 1977).

In the second method, restriction sites in the foreign DNA can be used to excise fragments from available clones and insert them in the expression vector. Overdigestion of DNA should be avoided by using purified DNA, following the manufacturers' guidelines for DNA/enzyme ratios, and setting up test restric-

Table I
Features of Commonly Used Expression Vectors

Vector	Fusion/nonfusion	Promoter	Host cells	Advantages	Disadvantages	Reference
pET (plasmid for expression by T7 RNA polymerase)	Nonfusion or fusion protein of 2–13 residues	T7 promoter, IPTG inducible	BL21 (DE3) BB21(DE3)pLysS	High levels of expression Nonfusion protein	Protein purification by conventional methods	Studier *et al.* (1990)
pET-His	Fusion protein with 6 histidine residues, which can be removed by proteolysis	As for pET	As for pET	Affinity purification over a Ni column	Cannot be purified over the affinity column under denaturing conditions	Marketed by Novagen
pGEX	Fusion with N-terminal 26-kDa glutathione S-transferase (GST); the GST portion can be removed by proteolysis	tac promoter, IPTG inducible	E. coli strains, e.g., DH5α	Generally, proteins are soluble Can be purified by a single affinity chromatography step	As for pET-His Sometimes the protease cleavage site between GST and the foreign protein is inaccessible as a result of folding of the fusion protein, and the GST moiety cannot be removed	Smith and Johnson (1988)
pRIT-2T protein A	Protein A fusion protein, can be purified over IgG–Sepharose column	Thermoinducible	E. coli strains	Affinity purification	Protein A sequence cannot be removed	Marketed by Pharmacia

tion enzyme digestions to determine the minimal amount of enyzme that will restrict the DNA. Junctions that cannot be recut by the restriction enzyme used to generate them can be sequenced to confirm that the insert is in the correct reading frame. Orientation of inserted DNA is determined as described above. DNA insertions of restriction fragments do not require complete sequence analysis because base changes or losses are rare if care is taken with the DNA restriction enzyme digestions. Frequently, convenient restriction sites may not be present in the sequence and constructs must be made using PCR.

For pET constructs, ligated DNAs are usually transformed initially into a nonexpression host (e.g., DH5α) and then into a BL21 expression strain, as lower frequencies of transformation are obtained with BL21 host cells.

III. Vectors Used for Expression of Motor Proteins

The pET (Studier *et al.*, 1990) and pGEX (Smith and Johnson, 1988) bacterial expression systems have been used successfully for the expression of microtubule motor proteins.

A. pET Expression System

The pET system is useful for expressing either a nonfusion protein or a fusion protein with only 10 to 13 additional amino acids at the N terminus of the expressed protein. Expression of proteins cloned in pET vectors is under the control of a T7 promoter (Rosenberg *et al.*, 1987). Plasmids carrying the DNA insert can be expressed in BL21(DE3) or BL21(DE3)pLysS strains of *Escherichia coli*. In both of these strains, expression of T7 RNA polymerase (which controls the T7 promoter) is under the control of the isopropyl-β-D-thiogalactoside (IPTG)-inducible *lac*UV5 promoter. Protein expression is induced by adding IPTG to a logarithmically growing culture of bacteria. Basal levels of T7 RNA polymerase promote some transcription of the target gene. This may retard cell growth if the protein product of the cloned gene is toxic to bacteria. In such cases it is useful to express the protein in BL21(DE3)pLysS cells, which contain an additional plasmid that expresses low levels of T7 lysozyme. T7 lysozyme is an inhibitor of T7 RNA polymerase and therefore reduces basal levels of expression of the foreign protein, so that proteins are expressed only after addition of IPTG. There is a lag period after addition of IPTG during which the protein of interest is not expressed, because the cellular T7 lysozyme first must be titrated out by the expression of T7 RNA polymerase. The presence of lysozyme also facilitates lysis of bacterial cells during protein purification (Studier *et al.*, 1990). Additional pET vectors (pET-10 and pET-11) are now available for expressing extremely toxic proteins in bacteria. These vectors have a *lac* operator in close proximity to the T7 promoter (T7 *lac* promoter) and carry a copy of the *lacI* gene. The additional *lac* repressor available from the *lacI* gene on the plasmid represses both the T7 *lac* promoter and the chromosomal gene for T7 RNA polymerase, providing almost complete inhibition of T7 RNA polymerase expression. On addition of the inducer IPTG, repression is released and expression of T7 RNA polymerase drives protein synthesis (Dubendorff and Studier, 1991).

B. pGEX Expression System

The pGEX system allows the expression of a fusion protein with a 26-kDa N-terminal glutathione *S*-transferase (GST) domain. An advantage of the pGEX system is that the expressed fusion protein can be easily purified by binding to glutathione-linked agarose beads via the GST portion of the protein. Unbound bacterial proteins remain in the supernatant after low-speed centrifugation of the beads. The GST moiety can be removed by proteolyzing the fusion protein with thrombin or factor Xa, which cleave at engineered sites between GST and the foreign polypeptide. If the proteolysis is done with the fusion protein bound to the glutathione–agarose, the protein of interest fractionates to the supernatant and can be obtained in relatively pure form. Sometimes, however, because of folding of the recombinant protein, the protease site is inaccessible and the GST portion cannot be removed. Proteins expressed from genes cloned in the pGEX vectors can be expressed in DH5α or BL21 strains of *E. coli*.

IV. Expression of Proteins in *Escherichia coli*

Bacteria are routinely grown in LB (10 g bactotryptone, 5 g yeast extract, 10 g NaCl per liter of medium supplemented with sterile glucose to 0.1% w/v after autoclaving) or M9ZB [1 liter of medium is made by mixing 860 ml of ZB (10 g *N*-Z-amine A, 5 g NaCl), 100 ml 10× M9 salts (1 g NH_4Cl, 3 g KH_2PO_4, 6 g Na_2HPO_4 per 100 ml), 40 ml 10% glucose, and 1ml 1 M $MgSO_4$; the solutions are autoclaved separately and mixed prior to use] containing the selective antibiotic [100 μg/ml ampicillin for plasmids in DH5α, BL21, and BL21(DE3) strains; 100 μg/ml ampicillin + 25 μg/ml chloramphenicol for BL21(DE3)pLysS cells]. An overnight culture grown in LB or M9ZB supplemented with 100 μg/ml ampicillin or ampicillin + chloramphenicol is diluted 1 : 100 in fresh medium the following morning and grown with shaking at 37°C to OD_{550} = 0.5–0.7. The cultures are then induced by adding IPTG to 0.1–0.4 mM. The cultures can be supplemented with 100 μg/ml ampicillin at this point to inhibit growth of cells that may have lost the plasmid. After induction, the bacterial cells are harvested by centrifugation at 4000g and the cells washed once in 10 mM $NaPO_4$, pH 7.4, 100 mM NaCl. Cell pellets are weighed, frozen in liquid nitrogen, and stored frozen at −70°C.

Expression at different temperatures, such as 18, 24, and 37°C, can be tried to determine conditions under which the protein is most soluble. The time of induction should also be optimized based on protein stability and level of expression. Protein expression can be monitored by inducing small (25-ml) cultures at different temperatures. Two aliquots of 1 ml are withdrawn at 1, 2, 3, and 4 hours and the cells are pelleted in a microfuge. Cells in one set of tubes are suspended in 80 μl of 1× sodium dodecyl sulfate–polyacrylamide gel electrophoresis (SDS–PAGE) sample buffer (Laemmli, 1970) to determine the expression level with time of the induced protein. The amount of protein in the soluble and insoluble fractions is monitored by suspending the cells in the second set of tubes in 100 μl of 10 mM Tris–HCl, pH 8.0, 50 mM ethylenediaminetetraacetic acid (EDTA) containing 100 μg/ml lysozyme and protease inhibitors [1 μg/ml pepstatin A, 1 μg/ml leupeptin, 2 μg/ml aprotinin, 2 μg/ml $N\alpha$-p-tosyl-L-arginine methyl ester (TAME), 1 mM phenylmethylsulfonyl fluoride (PMSF)]. The cell suspension is incubated for 30 minutes at 37°C and the tube is centrifuged for 5 minutes at 12,000g in a microfuge. The supernatant is transferred to another tube containing 20 μl of 5× sample buffer, and 120 μl of 1× SDS–PAGE sample buffer is added to the pellet. All samples are boiled for 5 minutes and spun 5 minutes in a microfuge, and 10 μl is loaded on a SDS–polyacrylamide gel (Laemmli, 1970). The expressed protein can usually be detected in gels stained with Coomassie blue. Antibodies that cross-react with the protein of interest can be used to monitor expression and to determine the amount of expressed protein present in the soluble and insoluble fractions.

V. Partial Purification of Proteins from Bacteria for *in Vitro* Motility Assays

Motility properties of microtubule motors expressed in bacteria can be analyzed using partially purified proteins, as bacteria lack microtubules and microtubule motor proteins, and bacterial proteins are not known to affect binding of microtubules to motor proteins or microtubule movement in *in vitro* assays. Motor proteins may, however, require purification for determination of biochemical properties such as ATPase or GTPase activity, protein association state or for structural studies.

Frozen bacterial cell pellets are thawed on ice and resuspended (3.5 ml/g cells) in cold buffer [20 mM Pipes, pH 6.9, 1 mM MgCl$_2$, 1 mM EGTA, 1 mM dithiothreitol (DTT)] containing protease inhibitors (as above). Other buffer and salt conditions may be used depending on the nature of the protein (Cull and McHenry, 1990). Cells are lysed by one of three methods: sonication, French press, or lysis by freeze–thawing.

1. Sonication is commonly used although there is a possibility of protein denaturation because of heating or foaming of the protein. Sonication should therefore be performed carefully. Heating of the sample can be prevented by

sonicating on ice, and foaming of proteins can be avoided by maintaining the microprobe in the center of the tube. Sonication is done for 30 seconds at a setting of 4 (Yang *et al.*, 1989) or 10 to 15 times for 3 seconds each.

2. In the French press method, pressure is applied mechanically to a metal chamber containing the bacterial cell suspension, causing cell lysis. Some workers believe this to be the method of choice because the possibility of protein denaturation is less than with sonication and expressed proteins associated with inclusion bodies may become partially soluble. The cell suspension is passed twice through the precooled cell for efficient lysis. However, the French press may not be available in some laboratories.

3. We have routinely been using lysis by freeze–thawing in our protein preparations. Briefly, lysozyme is added to the cell suspension to 0.1 mg/ml together with protease inhibitors (as above) and kept for 30 minutes on ice. The cells are lysed by two to three cycles of freezing for 5 to 10 minutes in liquid nitrogen and then thawing in a 37°C water bath with mixing so that the temperature of the cell lysate does not rise above 0°C. Lysis is monitored by the viscosity of the solution. The lysate becomes very viscous on release of bacterial chromosomal DNA. The bacterial DNA is digested by adding DNase I (Sigma) to 40 μg/ml and $MgCl_2$ to 10 mM, and supplemented with protease inhibitors. After 20 to 30 minutes on ice, the viscosity of the suspension decreases greatly. The presence of salt (0.3–1.0 M NaCl, NaBr, KCl, or KI) in the lysis buffer helps to solubilize proteins that are loosely associated with inclusion bodies; however, high salt inhibits DNase I, and cells lysed in >0.2 M salt should be passed through a 20-gauge needle two or three times after freeze–thawing to shear the DNA. The presence of salt (>0.3 M) in the lysis buffer inactivated some of our expressed motor proteins; therefore, the effect of salt on protein activity should be monitored. After DNase treatment, the lysate is centrifuged for 20 minutes at 25,000 g. The pellet contains the bacterial membranes, cellular debris, and inclusion bodies. The supernatant contains the soluble expressed protein and several bacterial proteins.

If the protein has been expressed using the pGEX vector, the supernatant is recentrifuged at 40,000g for 30 minutes to pellet any residual particulate matter. Soluble protein is then adsorbed to the affinity resin and purified by washing and eluting the resin using standard protocols (Smith and Johnson, 1988). Briefly, the lysate is incubated with glutathione–agarose (Sigma) or glutathione–Sepharose beads (Pharmacia) for 10–30 minutes at 24°C. The beads are pelleted by centrifuging at 1000g for 2 minutes. The supernatant is removed and the beads are washed two or three times with 10 mM $NaPO_4$, pH 7.4, 100 mM NaCl. Triton X-100 (1%) can be added to the washing buffer to remove nonspecfically bound proteins from the beads; however, Triton X-100 may denature some motor proteins, and its effects should be monitored. It is advantageous to cleave the GST moiety while the fusion protein is bound to the beads. This eliminates the additional step of eluting the fusion protein from the beads by adding glutathione and then cleaving and removing the GST portion. The beads can then be washed with 5 mM glutathione, pH 7.0, 3 M NaCl and reused after equilibration in the binding buffer. Beads can be stored at 4°C and reused several times. In some cases, the GST–motor fusion proteins either did not bind efficiently to glutathione–agarose beads or could not be eluted in an active form after binding (Yang *et al.,* 1990; R.C. and S.A.E., unpublished observations). Successful purification of an active GST–motor protein using glutathione–agarose beads has, however, been reported by Sawin *et al.,* 1992.

Proteins expressed using the pET system require conventional methods for purification. If any of the expressed protein is soluble, purification should be attempted from this fraction, as isolation of proteins from inclusion bodies requires protein denaturation and some motor proteins may be irreversibly denatured. We have used the following methods for obtaining active ncd proteins: The expressed protein in the bacterial cell lysate can be concentrated by ammonium sulfate precipitation. A trial experiment can be carried out to determine the ammonium sulfate concentration at which the protein precipitates. An undercut and/or overcut can then be used to concentrate the protein of interest and eliminate bacterial proteins. The proteins are precipitated with ammonium sulfate for 30–60 minutes on ice, with occasional mixing, and recovered by centrifugation at 15,000g for 20 minutes. The precipitated protein is resuspended in buffer and dialyzed, and can be further purified using gel-filtration and/or ion-exchange chromatography. During the course of our experiments we have observed that the ncd protein constructs bind to S-Sepharose, whereas most bacterial proteins flow through (Chandra *et al.,* 1993). If the expressed protein binds S-Sepharose, chromatography on S-Sepharose can be an important purification step.

Unfortunately, a large number of foreign proteins expressed in bacteria are partitioned into inclusion bodies. Proteins can be recovered by solubilizing inclusion bodies in 6–8 M urea or 4–6 M guanidine–HCl (Marston and Hartley, 1990). The extract is then passed over a column (ion-exchange or gel-filtration) to purify the protein. Because inclusion bodies contain very large amounts of the expressed protein, it is relatively simple to purify proteins from this fraction; however, the activity of the protein should be assayed to ensure that functional properties are retained. Protein renaturation from urea can be attempted by dialyzing away the urea. Both rapid and slow dialysis of urea can be tried; some proteins renature faster than others. More information on protein refolding methods can be obtained from Jaenicke and Rudolph (1989) and Kohno *et al.* (1990).

Our attempts to purify ncd by solubilization of inclusion bodies in high salt (1 M NaBr, KI), 6 M urea, 3.5 M guanidine–HCl, and Sarkosyl detergent (Frankel *et al.,* 1990) have resulted in protein that did not bind microtubules to the glass surface in *in vitro* motility assays.

VI. Brief Description of Motility Assays for Testing the Functional Properties of the Expressed Proteins

Activity of the expressed motor protein can be assayed by adding microtubules and ATP to a small amount of protein on a glass coverslip and observing the movement of microtubules using video-enhanced differential interference contrast (VE-DIC) microscopy (Walker *et al.*, 1990b). Motility assays can be carried out using dialysed cell lysates enriched in motor proteins by ammonium sulfate precipitation; however, this protein fraction may contain bacterial ATPases which can alter the apparent ATP concentration optimum.

We make slides for video microscopy assays using two different methods. In the method used for routine assays, 5 μl of protein, 1.6 μl taxol-stabilized microtubules (1 mg/ml), 0.9 μl 50 mM MgATP (1:1 dilution of 100 mM MgCl$_2$ with 100 mM ATP in 100 mM Pipes, pH 6.9), and 1.5 μl 6× salt solution (70 mM Pipes, pH 6.9, 480 mM NaCl, 30 mM MgCl$_2$, 0.7 mM MgSO$_4$, 1.4 mM EGTA, 4.8 mM DTT) are added sequentially to a clean glass coverslip and mixed, and the coverslip is inverted onto a glass microscope slide. The edges are sealed with VALAP (1 : 1 : 1 mix of vaseline : lanolin : paraffin, heated to form a uniform mixture). Coverslips are cleaned by sonication for 15–30 minutes in a mild solution of Alconox laboratory detergent, followed by three washes in distilled water (10 minutes each with sonication) and one wash in 95% ethanol (10 minutes with sonication). The coverslips are stored under ethanol in a covered container.

Flow chambers are used to measure the activity of the motor protein under different assay conditions. Flow chambers can be made by putting two strips of double-stick tape approximately 3 mm apart on a glass slide (Gliksman *et al.*, 1992). A clean coverslip is then pressed down onto the tape. Excess tape that extends beyond the coverslip is removed by scoring with a scalpel and removing it from the slide with a pair of forceps. Protein (5–7 μl) is added to the flow chamber and allowed to bind to the glass surface for 1 minute. The unbound protein can be washed away by adding 50 μl of buffer to one end of the chamber and absorbing it with a Kimwipe at the other end. A mixture of 1 mg/ml taxol-stabilized microtubules (1.6 μl), 50 mM Mg-ATP (0.9 μl), and 6× salts (1.5 μl) in 9 μl (5 μl water) is then made to flow into the chamber. Precoating the flow chamber with tubulin and cytochrome c (Howard *et al.*, 1989) or casein (Romberg and Vale, 1993) reduces denaturation of the motor proteins on the glass. With the use of a flow chamber, solutions can be made to flow sequentially into the chamber and the binding of protein to microtubules can be assayed under different conditions.

Other chapters in this volume describe the measurement of microtubule velocity and directionality of microtubule translocation on protein-coated glass surfaces.

VII. Microtubule Binding Assays

In vitro assays of binding of motor proteins to microtubules in the presence of ATP or ATP analogs help to confirm the results of protein–microtubule interactions observed in motility assays. These assays are described in several publications (Vale *et al.*, 1985, Ingold *et al.*, 1988; Yang *et al.*, 1989; Chandra *et al.*, 1993). Briefly, 20–50 μg/ml microtubule motor protein is incubated with 1 mg/ml taxol-stabilized microtubules in the presence of 5 mM 5′-adenylylimidodiphosphate (AMP-PNP) for 30 minutes at 24°C. Binding of the protein to microtubules can also be analyzed in the presence of GTP, ADP, ATP, ATP + vanadate, and no added nucleotide. Controls of microtubules alone and protein alone are usually run at the same time. The tubes are centrifuged for 20 minutes at 100,000g at 20°C. The supernatant is carefully removed and the pellet and supernatant are analyzed by SDS–PAGE (Laemmli, 1970). The amount of protein in the pellet and supernatant can be quantitated by gel scanning densitometry of Coomassie blue-stained gels.

VIII. Measurement of ATPase Activity as a Determinant of Motor Protein Function

Measurement of the ATPase activity of expressed motor proteins is a useful criterion for correct folding of the protein. A simplified protocol is given below.

Reaction mixtures (300 μl) containing 20 μg/ml pure protein + 4 mM MgATP (5000–35,000 cpm/nmol ATP) $\pm$ 1 mg/ml taxol-stabilized microtubules in 15 mM imidazole, pH 7.0, 2 mM MgCl$_2$, 1 mM EGTA, 1 mM DTT are incubated at 24°C (Cohn *et al.*, 1987, Wagner *et al.*, 1989, Chandra *et al.*, 1993). Aliquots (40 μl) are withdrawn at 0, 2, 5, 10, 20, 40, and 60 minutes into microfuge tubes containing 0.76 ml of 5% activated charcoal (Norit A, Aldrich) suspended in 50 mM NaH$_2$PO$_4$ (Higashijima *et al.*, 1987). The activated charcoal binds unhydrolyzed ATP but not free P_i, which is left in the supernatant. The charcoal mixtures are kept on ice for 15 minutes, tubes are centrifuged for 5 minutes at 15,600g, and the supernatants are collected and recentrifuged to remove charcoal particles. Samples (150 and 300 μl) of the final supernatants are taken for scintillation counting; 2- and 4-μl samples of the reaction mixture are counted in scintillation fluid to determine ATP specific activity.

The specific activity of ATP in the assay is calculated by dividing the total counts per minute (cpm) in 1 μl of the assay mixture by the nanomoles of ATP in 1 μl. The cpm for each time point (obtained after scintillation counting) are averaged, and the total cpm in 0.8 ml of the activated charcoal supernatant calculated. Counts at 0 minutes are subtracted to eliminate background cpm. Values are then divided by the specific activity of ATP to determine the nanomoles of P_i released at each time point. The slope of the line from a plot of nanomoles of P_i released versus time divided by the amount of protein in the

40-μl aliquot sampled gives the ATPase specific activity in nanomoles per minute per milligram. Alternatively, the ATPase specific activity can be expressed in k_{cat} (reciprocal seconds). The k_{cat} values are obtained by dividing the amount of P_i released in nanomoles per second by the nanomoles of protein present in the 40-μl aliquots.

IX. Problems Encountered with Prokaryotic Expression Systems and Some Solutions

The most common problem encountered with bacterial expression is that the expressed protein partitions as insoluble aggregates in inclusion bodies and must be denatured for the protein to be purified. Use of lower temperatures during induction, induction for shorter periods, or growth of cells in a fermenter to achieve better aeration and temperature control can be tried to overcome this problem. Expressing a fusion protein from pGEX or another vector carrying an affinity purification tag may be useful, as small amounts of soluble protein can be purified by affinity chromatography. Growing large cultures (1–5 liters) can result in sufficient quantities of protein for *in vitro* motility assays. The level of expressed protein has also been reported to vary depending on the bacterial host used for expression (Frorath *et al.*, 1992). We have compared protein expression in DH5α, BL21, LE392, and TG1 cells for several pGEX/ncd constructs. Protein expression was lower in DH5α cells compared with the other host strains. The amount of protein in the soluble fraction also varied with the host cell for some constructs.

The rules that determine protein solubility on expression in bacterial cells are not understood. Proteins that differ by only a small number of amino acids can partition differently to the soluble and insoluble cellular fractions. If time permits, several constructs can be made and tested for expression. The N terminus of major products of bacterial protease degradation can be sequenced to determine the domains of protein that can exist in the cell as stable proteins, and the sequence information can be used to direct plasmid construction for protein expression.

Acknowledgments

This work was supported by USPHS Grants GM-31279 and GM-46225 to S.A.E. We thank Dr. E. D. Salmon for instruction in VE-DIC microscopy and microtubule motility assays, and Dr. P. Casey for suggestions regarding the ATPase assays.

References

Chandra, R., Salmon, E. D., Erickson, H. P., Lockhart, A., and Endow, S. A. (1993). Structural and functional domains of the *Drosophila* ncd microtubule motor protein. *J. Biol. Chem.* **268,** 9005–9013.

Cohn, S. A., Ingold, A. L., and Scholey, J. M. (1987). Correlation between theATPase and microtubule translocating activities of sea urchin egg kinesin. *Nature* (*London*) **328,** 160–163.

Cull, M., and McHenry, C. S. (1990). Preparation of extracts from prokaryotes. *In* "Methods in Enzymology" (M. Deutscher, ed.), Vol. 182, pp. 147–153. Academic Press, San Diego.

Dubendorff, J. W., and Studier, F. W. (1991). Controlling basal expression in an inducible T7 expression system by blocking the target T7 promoter with *lac* repressor. *J. Mol. Biol.* **219,** 45–59.

Frankel, S., Condeelis, J., and Leinwald, L. (1990). Expression of actin in *Escherichia coli* Aggregation, solubilization, and functional analysis. *J. Biol. Chem.* **265,** 17980–17987.

Frorath, B., Abney, C. C., Berthold, H., Scanarini, M., and Northemann, W. (1992). Production of recombinant rat interleukin-6 in *Escherichia coli* using a novel highly efficient expression vector pGEX-3T. *BioTechniques* **12,** 558–563.

Gliksman, N. R., Parsons, S. F., and Salmon, E. D. (1992). Okadaic acid induces interphase to mitotic-like microtubule dynamic instability by inactivating rescue. *J. Cell Biol.* **119,** 1271–1276.

Higashijima, T., Ferguson, K. M., Smigel, M. D., and Gilman, A. G. (1987). The effect of GTP and Mg^{2+} on the GTPase activity and the fluorescent properties of G_o*. *J. Biol. Chem.* **262,** 757–761.

Howard, J., Hudspeth, A. J., and Vale, R. D. (1989). Movement of microtubules by single kinesin molecules. *Nature* (*London*) **342,** 154–158.

Ingold, A. L., Cohn, S. A., and Scholey, J. M. (1988). Inhibition of kinesin-driven microtubule motility by monoclonal antibodies to kinesin heavy chains. *J. Cell Biol.* **107,** 2657–2667.

Jaenicke, R., and Rudolph, R. (1989). Folding proteins. *In* "Protein Structure: A Practical Approach" (T. E. Creighton, ed.), pp. 191–223. IRL Press, Oxford, England.

Kohno, T., Carmichael, D. F., Sommer, A., and Thompson, R. C. (1990). Refolding of recombinant proteins. *In* "Methods in Enzymology" (D. Goeddel *et al.,* eds.), Vol. 185, pp. 187–195. Academic Press, San Diego.

Kraft, R., Tardoff, J., Krauter, K. S., and Leinwald, L. A. (1988). Using mini-prep plasmid DNA for sequencing double-stranded templates with Sequenase. *BioTechniques* **6,** 544–546.

Laemmli, U. (1970). Cleavage of structural proteins during the assembly of the bacteriophage T4. *Nature* (*London*) **227,** 680–685.

Marston, F. A. O., and Hartley, D. L. (1990). Solubilization of protein aggregates. *In* "Methods in Enzymology" (M. Deutscher, ed.), Vol. 182, pp. 264–276. Academic Press, San Diego.

McDonald, H. B., Stewart, R. J., and Goldstein, L. S. B. (1990). The kinesin-like *ncd* protein of *Drosophila* is a minus end-directed microtubule motor. *Cell* (*Cambridge, Mass.*) **63,** 1159–1165.

Romberg, L., and Vale R. D. (1993). Chemomechanical cycle of kinesin differs from that of myosin. *Nature* (*London*) **361,** 168–170.

Rosenberg, A. H., Lade, B. N., Chui, D.-S., Lin, S.-W., Dunn, J. J., and Studier, F. W. (1987). Vectors for selective expression of cloned DNAs by T7 RNA polymerase. *Gene* **56,** 125–135.

Sambrook, J., Fritsch, E. F., and Maniatis, T. (1989). "Molecular Cloning: A Laboratory Manual," 2nd ed. Cold Spring Harbor Lab, Cold Spring Harbor, New York.

Sanger, F., Nicklen, S., and Coulson, A. R. (1977). DNA sequencing with chain-terminating inhibitors. *Proc. Natl. Acad. Sci. U.S.A.* **74,** 5463–5467.

Sawin, K. E., LeGuellec, K., Philippe, M., and Mitchison, T. J. (1992). Mitotic spindle organization by a plus-end-directed microtubule motor. *Nature* (*London*) **359,** 540–543.

Smith, D. B., and Johnson, K. S. (1988). Single-step purification of poly-peptides expressed in *Escherichia coli* as fusions with glutathione *S*-transferase. *Gene* **67,** 31–40.

Studier, F. W., Rosenberg, A. H., Dunn, J. J., and Dubendorff, J. W. (1990). Use of T7 RNA polymerase to direct expression of cloned genes. *In* "Methods of Enzymology" (D. Goeddel *et al.,* eds.), Vol. 185, pp. 60–89. Academic Press, San Diego.

Vale, R. D., Reese, T. S., and Sheetz, M. P. (1985). Identification of a novel force-generating protein, kinesin, involved in microtubule-based motility. *Cell* (*Cambridge, Mass.*) **42,** 39–50.

Wagner, M. C., Pfister, K. K., Bloom, G. S., and Brady, S. T. (1989). Copurification of kinesin polypeptides with microtubule-stimulated Mg-ATPase activity and kinetic analysis of enzymatic properties. *Cell Motil. Cytoskel.* **12,** 195–215.

Walker, R. A., Salmon, E. D., and Endow, S. A. (1990a). The *Drosophila claret* segregation protein is a minus-end directed motor molecule. *Nature (London)* **347,** 780–782.

Walker, R. A., Gliksman, N. R., and Salmon, E. D. (1990b). Using video-enhanced differential interference contrast microscopy to analyze the assembly dynamics of individual microtubules in real time. *In* "Optical Microscopy for Biology" (B. Herman and K. Jacobson eds.), pp. 395–407. Wiley-Liss, New York.

Yang, J. T., Laymon, R. A., and Goldstein, L. S. B. (1989). A three-domain structure of kinesin heavy chain revealed by DNA sequence and microtubule binding analyses. *Cell (Cambridge, Mass.)* **56,** 879–889.

Yang, J. T., Saxton, W. M., Stewart, R. J., Raff, E. C., and Goldstein, L. S. B. (1990). Evidence that the head of kinesin is sufficient for force generation and motility *in vitro*. *Science* **249,** 42–47.

CHAPTER 9

Tracking Nanometer Movements of Single Motor Molecules

Michael P. Sheetz and Scot C. Kuo

Department of Cell Biology
Duke University Medical Center
Durham, North Carolina 27710

I. Introduction
II. Movements by Single Motor Molecules
III. Microtubule Movements on Motor-Coated Glass
    A. Purification and Storage of Motor Proteins for Motility
    B. Bead Movements on Microtubules
    C. Recording Nanometer-Level Movements
    D. Video-Enhanced DIC Microscopy of Nanometer Movements
    E. Analysis
IV. Conclusion
    References

I. Introduction

To understand the mechanism of motor movements it is necessary to understand how individual molecules move with sufficient spatial precision. Until recently, motility of motor proteins, such as muscle myosin, required an ensemble of motor molecules interacting with the same filament. In the last few years, the microtubule motor kinesin was demonstrated to work as single molecules (Howard *et al.*, 1989; Block *et al.*, 1990). Similar approaches with cytoplasmic dynein are showing promising results (Lopez and Sheetz, 1992). In these approaches, limiting dilutions of motor protein are adsorbed to an anionic surface, either acid-washed glass or carboxylate-modified polystyrene microspheres. The addition of carrier protein, such as bovine serum albumin, cytochrome c, and casein, increases the range of active motor dilutions at least 50-fold during surface adsorption. Although specific activation of the motor protein cannot be excluded, the number of carrier proteins that preserve motor activity suggests that surface denaturation of the adsorbed motor protein is a major cause of reduced motor activity. As described in the later protocols, a number of carrier protein protocols have been published (Howard *et al.*, 1989; Block *et al.*, 1990), but our best activity for both kinesin and cytoplasmic dynein from either chicken or squid is obtained using casein alone.

A variety of different technologies have been used to follow particle movements with high precision. A complete description of these technologies, including the necessary microscopy, is beyond the scope of this chapter, but the general principles are described. Approaches based on video are most convenient; however, improved temporal resolution and spatial sensitivity are achieved by custom photodiode detection systems with custom illumination.

II. Movements by Single Motor Molecules

In general, the greatest obstacle to performing studies of single motor molecules is proving that the motility is indeed produced by single and not multiple molecules. As most motility assays often show less than a tenth of the motor molecules as active, arguments based on adsorbed motor density are ineffective and statistical arguments must be used. In the case of microtubules gliding on motor-coated glass, however, translocation while pivoting around a nodal point is a strong indicator of motility by a single motor molecule. The absence of nodal translocation clearly indicates multiple attachments to the glass. The presence of nodal translocation still requires confirmation by careful statistics to prove translocation by a single molecule. In the case of bead movements, again, statistical arguments have been used to show that the movements observed are driven by single motors (Block *et al.*, 1990). The use of the optical tweezers in those assays is important to make an unbiased sample of the population of beads.

III. Microtubule Movements on Motor-Coated Glass

A. Purification and Storage of Motor Proteins for Motility

Many procedures have been published for the purification of either kinesin or cytoplasmic dynein (see Schroer *et al.*, 1989; Kuznetsov *et al.*, 1988; Vale *et al.*, 1985; Paschal *et al.*, 1987). In general, the motor activity is best within the first day, but the addition of reducing agents and storage on ice have preserved detectable activity for weeks. It is important to titrate the activity each time an experiment is done to be in a single motor range.

1. Kinesin

Coating the glass surface with other proteins that block the denaturation of kinesin has greatly improved the motility of kinesin attached to glass surfaces. Originally, a sequential coating of dimeric tubulin and cytochrome c was found to preserve motility at dramatically lower concentrations of kinesin, exhibiting linear dependence (Hill coefficient = 1) of probability of surface attachment and movement of microtubules with the concentration of kinesin (Howard *et al.*, 1989). This protocol (see Table I, bottom) has worked for many other kinesins, but even better results have been obtained with α-casein (see Table I). In some cases there is still a threshold phenomenon in which motility will not occur unless there is a minimum concentration of motor. This appears to be dependent on the age of the motor, and the freshest preparations of motor have no detect-

Table I
Microtubule Motility on Motor-Coated Glass[a]

Time	Volume	Solution
Acid-washed coverslips		
≥1 h		10–20% Hydrochloric or nitric acid
≥20 min		Rinsing in a stream of filtered, distilled water
Spin dry or rinse in 50% ethyl alcohol and air-dry		
		Optional Adsorption of Fiducial Markers
5 min	15 μl	1:20,000 dilution in PEM of 2% 150-nm latex microspheres
2 min	30 μl	Motor dilution in CP (typically 1:5000 or greater)
Rinse	30 μl	CP
2–15 min	30 μl	Microtubules in CP with nucleotides and 5 μM taxol
Observe under microscope		
		Buffers and Solutions
PEM 100 mM		Pipes (KOH), pH 6.8
1 mM		EGTA
1 mM		MgSO$_4$
CP, 150 μg/ml		α-casein in PEM
		Alternate Published Compositions of Carrier Protein CP
10 mg/ml BSA in 80 mM Pipes, 50 mM KCl, 0.5 mM DTT, 1 mM MgCl$_2$, 1 mM EGTA, pH 6.8 (Block *et al.*, 1990)		
50 μg/ml casein and 50 μg/ml cytochrome C in 80 mM Pipes, 50 mM KCl, 0.5 mM DTT, 1 mM MgCl$_2$, 1 mM EGTA, pH 6.8 (Block *et al.*, 1990)		
50 μg/ml cytochrome C in 80 mM Pipes (KOH), 1 mM EGTA, 2 mM MgCl$_2$, pH 6.85 (Howard *et al.*, 1989)		
		Published protocols
1 min		Blocking protein: CP (Block *et al.*, 1990) or tubulin (Howard *et al.*, 1989)
2 min		Motor dilution in CP
2–15 min		Microtubules in CP with 1–2 mM ATP and 5 μM (10 μg/ml?) taxol

[a] Pipes, 1,4-piperazinediethanesulfonic acid; EGTA, ethylene glycol bis (β-aminoethyl ether) N,N'-tetraacetic acid; DTT, dithiothreitol

able threshold. Treatment of the glass, the concentration of the surface blockers, and the motor source also affect the linearity of motor activity.

2. Cytoplasmic Dynein

Cytoplasmic dynein-dependent motility also benefits from the use of blocking proteins. Although cytochrome c inhibits motility, coating with dimeric tubulin alone maintains cytoplasmic dynein motility on dilution. As for kinesin, the use of casein as the blocking protein provides the most sensitive assay for motility. Unlike kinesin, cytoplasmic dynein shows a consistent threshold effect requiring a minimal motor density for motility. Above this threshold, the probability of microtubules attaching to the surface and moving was monotonic with increasing motor concentration.

3. Attachment of Marker Particles to Microtubules

Current microscope systems do not detect microtubules with sufficient contrast to allow high-precision measurements of position; however, marker particles on the surface of microtubules provide sufficient contrast. Microtubules containing biotinylated tubulin heterodimers have been described elsewhere (Hyman et al., 1991) and streptavidin-coated colloidal gold is commercially available. For 0.5-μm latex particles, however, we had to develop a sandwich technique to construct streptavidin-coated particles. Prior to particle tagging of biotinylated microtubules, all biotinylated dimeric tubulin must be removed by centrifugation or washing in a flow cell, even when microtubules are stabilized by taxol. The bond between bead and microtubule is stronger than multiple kinesin motors (>10 pN per bond), allowing us to characterize the force of a single kinesin molecule. With a standard video-enhanced differential interference contrast (DIC) microscope system it is possible to resolve movements of latex particles with 1- to 5-nm precision.

B. Bead Movements on Microtubules

Very small (200–400 nm) glass beads have been coated with the same blocking proteins as coverslips, which has produced motility of a fraction of beads with less than one kinesin per particle (Block et al., 1990). A variation of this protocol is described in Table I for the microtubule gliding assay.

A protocol for coating anionic latex beads with motors after adsorption with a surface blocker is provided in Table II. In general the criterion for single-motor-dependent movement is that the percentage of beads capable of movement is linearly dependent on the concentration of motor on the bead surface. As previously described, optical tweezers can provide an unbiased sampling of motor-coated particles and show that the frequency of bead movement follows poissonian statistics (Block et al., 1990). At a limiting dilution of motors, if 20% of the particles attach and move on microtubules, a Poisson distribution predicts

Table II
Latex Bead Movement by Single Motors

Time	Volume	Solution
Attachment of Microtubules to Coverslip		
1 h	50 μl	Antitubulin antibody (1:800 dilution in PMEE′), should be dried at 37°C on coverslip surface
1 min	50 μl	SB to wash lanes of coverslip sandwich (parallel lines of silicone grease between coverslips form the sandwich)
10 min	30 μl	Diluted microtubules (~5 ml of stock in 1 ml of SB)
1 min	50 μl	Rinsing buffer (1:1 of H$_2$O:PMEE′ with 1 mM ATP, 9.6 μM taxol, 240 mM NaCl, 220 μg/ml α-casein)
Addition of motor-coated beads		
<1 m	30 μl	Motor/bead mixture
Seal and observe under microscope		
Preparation of Kinesin/Bead Mixture		
5 m	4 μl	1:1 of H$_2$O:PMEE′ with 5 mM ATP, 48 μM taxol, and 400 mM NaCl
	20 μl	125 μg/ml of α-casein in PMEE′
	2 μl	Kinesin (diluted in CP)
	6 μl	143-nm carboxylated beads (2% stock diluted 1:100 in H$_2$O)
after vortexing in plastic tube		
1 m	2 μl	2.2 mg/ml of α-casein in PMEE′
Preparation of Stock Microtubules		
30 m	20 μl	Freshly thawed PC tubulin (1.5 mg/ml)
	5 μl	Polymerization buffer (PMEE′ with 5 mM GTP and 100 μM taxol)

Optional step to remove monomeric tubulin: Load 20 μl of polymerized tubulin onto 100 μl of 20% sucrose in PMEE′ and spin in airfuge (30 psi) for 5 min. Suspend pellet in 20 μl of stabilizing buffer (SB).

		Buffers and Solutions
PMEE′	35 mM	PIPES[a] (KOH), pH 7.2
	5 mM	EGTA
	1 mM	EDTA
	5 mM	MgCl$_2$
SB		
		PMEE′
	1 mM	GTP
	20 μM	Taxol

[a] See Table I for explanation of abbreviations.

that 1.75% of the particles (<9% of the motile beads) have more than one motor molecule. A statistical argument shows that the majority of the movements are driven by single motors.

C. Recording Nanometer-Level Movements

To record the position of particles moving by single motors requires a major commitment of resources, either to a specialized system or to a high-quality video-enhanced DIC microscope. The lack of image formation in the photodiode systems is compensated by their greater temporal and spatial resolution. The limitations of the photodiode systems have been reported for hair cell bundle measurements (about 0.01 nm at 1 millisecond) (Denk *et al.*, 1989). In contrast, the video-enhanced DIC systems have lower resolution (about 0.5 nm at 30 milliseconds but variable scan cameras can give 3 milliseconds) and give additional information about the microtubule position and allow the use of fiducial marker beads in the system. The major problem is to match the instrument to the question being asked because the acquisition of more data than is needed makes analysis cumbersome.

D. Video-Enhanced DIC Microscopy of Nanometer Movements

Earlier works have dealt with some of the important parameters in video-enhanced DIC recording of nanometer movements (Schnapp *et al.*, 1988; Gelles *et al.*, 1988; Kuo *et al.*, 1991). Resolution of the position measurement is affected by the size of the recorded field, the intensity of the light source, particle contrast, and the mode of recording. These measurements require stationary reference beads in the same field. Time resolution in standard video systems is limited to 30 milliseconds, and standard Newvicon cameras have a frame-to-frame carryover of 15–20%. Solid-state cameras (CCD or CID) have variable scan features that allow portions of the video field to be scanned as rapidly as 3 milliseconds; however, solid-state cameras require digital contrast enhancement to approach the performance of Newvicon cameras on low-contrast DIC images, such as microtubules.

The recording medium is important as it often limits the final resolution obtained. In comparing direct digital video recorders (a real-time disk system from Applied Memory Technologies that records up to 3000 digitized frames), optical memory disk recorders, and S-VHS tape recorders, we have found that the final noise contributed by the recording medium is a small factor in routine measurements. The noise contributed by the S-VHS tape players is as low as 0.04 pixel (typically 2 nm). Optical memory disk recorders, when frame-averaged to reduce playback noise, perform better and introduce only 0.01 pixel (typically 0.5 nm). As a result, it is only justifiable to use the more expensive media when the experimental system requires improved spatial precision.

E. Analysis

Several commercial analysis routines are available and most investigators share their software with other researchers; however, efficient analysis requires a hardware-dependent system, which usually involves a major commitment of money or time. Two modes of analysis have commonly been employed: a simple centroid calculation or a cross-correlation analysis followed by a centroid calculation of the cross-correlation peak. With either fluorescence microscopy of labeled particles (Gross and Webb, 1988) or bright-field microscopy of gold particles (DeBrabander *et al.*, 1988) the simple centroid calculation should give maximal precision. In general, the signal-to-noise ratio for particle detection is greatest for DIC microscopy, and analysis of only the dark or light portion of the particle image gives only half of the possible precision. To obtain the greatest possible precision we have used a cross-correlation analysis (Gelles *et al.*, 1988).

Stage drift is a major problem, particularly for high-precision or long-term measurements. Latex particles will bind strongly to the glass surface and can be used as reference particles to compensate for drift. Several problems with reference particles will actually increase the error of the final motion analyses. The simplest is that the particles, although stationary to the eye, are actually moving about their attachment site. This is often evident in the analysis and can be overcome by averaging the positions of several stationary particles. The other problem is inherent in the use of reference particles in that the noise of the position measurement of the reference particle is added to the noise of the position measurement of the moving particle. As the measurement noise is high frequency, it is possible to use a sliding-window temporal average of the position of the reference particle to diminish the error of position measurement.

IV. Conclusion

The detailed analysis of motor movements at the molecular level has great promise for aiding in the understanding of the basic mechanisms of motor function. The technology is now available to perform these analyses routinely at the nanometer level.

References

Block, S. M., Goldstein, L. S. B., and Schnapp, B. J. (1990). Bead movement by single kinesin molecules studied with optical tweezers. *Nature (London)* **348,** 348–352.

DeBrabander, M., Nuydens, R., Geerts, H., and Hopkins, C. R. (1988). Dynamic behavior of the transferrin receptor followed in living epidermoid carcinoma (A431) cells with nanovid microscopy. *Cell Motil. Cytoskel.* **9,** 30–47.

Denk, W., Webb, W. W., and Hudspeth, A. J. (1989). Mechanical properties of sensory hair bundles are reflected in their Brownian motion measured with a laser differential interferometer. *Proc. Natl. Acad. Sci. U.S.A.* **86,** 5371–5375.

Gelles, J., Schnapp, B. J., and Sheetz, M. P. (1988). Tracking kinesin-driven movements with nanometer precision. *Nature (London)* **331,** 450–453.

Gross, D. J. and Webb, W. W. (1988). Cell surface clustering and mobility of the liganded LDL receptor measured by digital fluorescence microscopy. *In* "Spectroscopic Membrane Probes" (L. M. Leow, ed.), Vol. 11, pp. 19–48. CRC Press Inc. Boca Raton, Florida.

Howard, J., Hudspeth, A. J., and Vale, R. D. (1989). Movement of microtubules by single kinesin molecules. *Nature (London)* **342,** 154–158.

Hyman, A., Drechsel, D., Kellogg, D., Salser, S., Sawin, K., Steffen, P., Wordeman, L., and Mitchison, T. (1991). Preparation of modified tubulins. *In* "Methods in Enzymology" (R. B. Vallee, ed.), Vol. 196, pp. 478–485. Academic Press, San Diego.

Kuo, S. C., Gelles, J., Steuer, E., and Sheetz, M. P. (1991). A model for kinesin movement from nanometer level movements of kinesin and cytoplasmic dynein and force measurements. *J. Cell Sci., Suppl.* **14,** 135–138.

Kuznetsov, S. A., Vaisberg, Y. A., Shanina, N. A., Magretova, N. N., Chernyak, V. Y., and Gelfand, V. I. (1988). The quaternary structure of bovine brain kinesin. *EMBO J.* **7,** 353–356.

Lopez, L. A., and Sheetz, M. P. (1992). Inhibition of dynein and kinesin motility by MAP2. *Cell Motil. Cytoskel.* (in press).

Paschal, B. M., Shpetner, H. S., and Vallee, R. B. (1987). MAP 1C is a microtubule-activated ATPase which translocates microtubules *in vitro* and has dynein-like properties. *J. Cell Biol.* **105,** 1273–1282.

Schnapp, B. J., Gelles, J., and Sheetz, M. P. (1988). Nanometer-scale measurements using video light microscopy. *Cell Motil. Cytoskel.* **10,** 47–53.

Schroer, T. A., Steuer, E. R., and Sheetz, M. P. (1989). Cytoplasmic dynein is a minus-end directed motor for membranous organelles. *Cell (Cambridge, Mass.)* **56,** 937–946.

Vale, R. D., Reese, T. S., and Sheetz, M. P. (1985). Identification of a novel force-generating protein, kinesin, involved in microtubule-based motility. *Cell (Cambridge, Mass.)* **42,** 39–50.

CHAPTER 10

Assay of Microtubule Movement Driven by Single Kinesin Molecules

Jonathon Howard, Alan J. Hunt, and Sung Baek

Department of Physiology and Biophysics
University of Washington
Seattle, Washington 98195

I. Introduction
II. Protein Preparation
    A. Tubulin
    B. Kinesin
III. Flow Cells and Microscopy
    A. Construction of a Flow Cell
    B. Solution Exchange in the Flow Cell
    C. Adsorbing Proteins to the Flow Cell Surfaces
    D. Choosing the Appropriate Spacer and Coverglass Thickness
    E. Cleaning Slides for Dark-Field Microscopy
IV. Microscopy and Motion Analysis
    A. Microscope and Video Equipment
    B. Measuring Motility
V. *In Vitro* Motility Assay
    A. Buffers and Stock Solutions
    B. Growing Microtubules
    C. Diluted Microtubule Solution for the Motility Assay
    D. Standard Upside-Down High-Density Assay
    E. Low-Density Upside-Down Assay
    F. Low-Density Bead Assays
VI. Evidence That Single Molecules Suffice for Microtubule-Based Motility
    References

METHODS IN CELL BIOLOGY, VOL. 39

I. Introduction

Low-density motility assays are being used in this and other laboratories to explore the mechanism of force generation by kinesin (Howard *et al.*, 1989; Block *et al.*, 1990; Kuo and Sheetz, 1991; Schnapp *et al.*, 1991). The assay confers advantages over the standard high-density assays because the movement of a microtubule by a single motor molecule can be observed. This is important for two reasons. First, uncertainty about the number of motors cooperating in the movement of one microtubule in a high-density assay prevents the extrapolation of a measured property, such as microtubule speed or total force acting on the microtubule, down to the single motor molecule level. Second, even if the number of motors moving a microtubule were known (but >1), it is still not possible to confidently extrapolate down to the single molecule level because the manner in which the motors interact with one another as they cooperate in the movement of one microtubule is not understood. The existence of such interactions has been proven unequivocally for the motor protein myosin (Warshaw *et al.*, 1990): when actin filaments are observed to move over surfaces coated with co-polymers of phosphorylated and unphosphorylated smooth muscle myosin or with copolymers of smooth muscle and skeletal myosin, the speed of movement depends nonlinearly on the mole fraction of motors. This indicates that motor proteins differ in a fundamental way from most other enzymes. For example, consider an ion pump: when several pump molecules reside in the one membrane, each pump operates independently of the others, and each feels the same transmembrane electric field. But when several motor proteins move a single filament, it is expected that the force generated by one will be felt by and possibly influence the other attached motors. Understanding how one motor influences another is central to testing the ''independent force generator hypothesis'' of Huxley (1980): Do two motors generate twice the force as one motor? The approach taken in this laboratory is first to understand force generation by a single motor, then to use this information to understand the interactions between several motors moving the one filament.

The assay is extremely sensitive: movement at densities as low as one motor molecule per square micrometer can be observed. This corresponds to a total of about 10^8 motor molecules in the entire observation chamber of $\sim$100 mm^2. By making the chamber 100 smaller, the activity of as few as 10^6 motor molecules might be detected and studied; such sensitivity may make it feasible to study motor proteins obtained from a single cell. But such sensitivity complicates the study of other potential, microtubule-based motor proteins; trace amounts of kinesin could easily confer motility to any protein fraction. Thus, to prove that a protein is a microtubule-based motor requires exclusion of the possibility of contamination by other motors such as kinesin.

II. Protein Preparation

Unless indicated, all chemicals are obtained from the Sigma Chemical Company (St. Louis, MO).

A. Tubulin

Tubulin is prepared from bovine brain by two and a half cycles of depolymerization and polymerization, followed by phosphocellulose chromatography to remove microtubule-associated proteins (Weingarten *et al.*, 1974). The purity is greater than 95%. Tubulin is then cycled again to remove any dead protein and to exchange buffers if required (Hyman *et al.*, 1991). Aliquots (10–100 μl) of tubulin at 5–10 mg ml^{-1} are quickly frozen in liquid nitrogen and stored for up to 6 months at −80°C.

B. Kinesin

Kinesin is prepared from bovine brain using a method similar to that of Wagner *et al.* (1991), except that endogenous microtubules are used for the initial affinity step (preparation developed by Fady Malik, University of California, San Francisco). Endogenous microtubules are polymerized as in the tubulin preparation from the first high-speed supernatant and stabilized with 10 μM taxol. Apyrase (grade V from potato) at 0.5 U ml^{-1} is added to deplete nucleotide tri- and diphosphates, and 20 μM AMP-PNP is added to further promote the binding of kinesin to the microtubules. The microtubules with kinesin attached are pelleted twice through a glycerol cushion, the kinesin is released with 9 mM ATP, and the microtubules are pelleted and discarded. The proteins in the supernatant are concentrated by ammonium sulfate precipitation and resuspension and loaded on a gel filtration column (Bio-Gel A5M, 100–200 mesh, Bio-Rad, Richmond, CA). The kinesin is reconcentrated over an ion-exchange column (S-Sepharose, Pharmacia, Piscataway, NJ) and eluted with high salt. Two brains typically yield ~250–500 μg at ~90% purity, with the chief contaminant being tubulin. Kinesin in 10% sucrose and 100 μM ATP is stored at −80°C in 10-μl aliquots, it may lose some activity over 6 months.

III. Flow Cells and Microscopy

The experiment chamber is called a flow cell because it permits the exchange of solutions by perfusion. The effects of ATP concentration, ionic strength, agonist or antagonist concentration, and protein modification on the surface can be studied using the one flow cell. The flow cell described here is the poor person's version of the Berg and Block (1984) cell.

A. Construction of a Flow Cell

The flow cell is constructed by making two ~1-mm-wide, ~25-mm-long lines of Apiezon M grease (VWR) (extruded from an 18-gauge needle) about 10 mm apart and parallel to the length of a $1 \times 25 \times 75$-mm microscope slide. Two ~1-mm^2 shards of broken coverglass in each line of grease serve as spacers for the overlying 18×18-mm coverglass, which is pressed down on top. The chamber has the dimensions 18 mm $\times$ ~5 mm $\times$ (cover glass spacer thickness). Approximate coverglass thicknesses are 200 μm for No. 2, 175 μm for No. 1½, 150 μm for No. 1, 100 μm for No. 0, and 75 μm for No. 00 (the latter were obtained from Dr. David Warsaw, University of Vermont). The chamber is filled by capillary action, and solutions are perfused by simultaneously presenting new solution via a pipet at one end and withdrawing fluid from the other end with a piece of absorbant filter paper. Perfusion of one chamber volume of solution is complete within a second or two (depending on the spacer thickness and the filter paper). Evaporation from the flow cell is slow; slides can be used for up to half an hour without humidifying the surrounding air. Evaporation can be prevented by sealing the chamber with immersion oil.

B. Solution Exchange in the Flow Cell

The dimensions of the flow cell makes it ideal for adsorbing proteins to its inner surfaces. The diffusion coefficient for a moderately large, prolate protein like kinesin (length 75 nm, average diameter ~2 nm) is about 20 μm^2 s^{-1} (Brennen and Winet, 1977; Hackney *et al.*, 1992). This means that it will diffuse the ~50-μm half-depth of a flow cell in ~60 seconds. If diffusion to the surface is the rate-limiting step, then we expect this protein to adsorb to the surface with a half-time of ~1 minute. This is observed. This half-time is short enough that a few minutes suffice for adsorption, but is long enough that there is little adsorption during the initial seconds it takes to get the solution into the cell. This means that at nonsaturating protein concentrations, the coating of the surfaces will be uniform along the length of the flow cell. If the cell were only a tenth the thickness, the diffusion time would decrease 100-fold, whereas the perfusion time would increase one 100-fold; this thin flow cell would not be uniformly coated.

On the other hand, the chamber is thin enough that diffusion ensures the exchange of molecules from the center to the surfaces of the chamber and vice versa (Berg and Block, 1984). During fluid perfusion, the velocity profile is roughly parabolic through the ~100-μm depth of the chamber (except near the greasy edges); the velocity is greatest at the center, and zero at the top and bottom two surfaces, where there exists an unstirred, stationary layer (Landau and Lifshitz, 1987). Exchange of solutes into and out of this layer, which is ~10 μm deep, must occur by diffusion. The average velocity is two-thirds that of the maximum, and is achieved at points ~21% of the distance across the flow cell (~20 μm). Assuming laminar flow and no diffusive mixing, we expect perfusion of one chamber volume into the flow cell will give complete flow-through of the central ~60% of the fluid and about half removal of the unstirred layers near the two surfaces. We thus expect that the perfusion of one chamber volume will remove about 80% of the initial fluid. This agrees with experiment; solutions can actually be recovered by using a second pipet instead of the filter paper. If two volumes are perfused, the unstirred layer will be only half as thick. Because the flow cell is thin, small molecules such as ATP (diffusion coefficient ~200 μm^2 s^{-1}) and larger molecules like kinesin will diffuse into or out of the unstirred layer, effectively reequilibrating throughout the flow cell in ~1 second and ~10 seconds, respectively. Thus each wash, separated by about 1 minute, can effectively exchange about 80% of the molecules in the fluid. The exception is at the edges of the flow cell near the grease. Even microtubules can diffuse down to the surface quite rapidly: it takes a 2-μm-long microtubule (diffusion coefficient ~1 μm^2 s^{-1}) about 1 minute to diffuse through an ~10-μm-thick boundary layer.

C. Adsorbing Proteins to the Flow Cell Surfaces

The amount of protein that adsorbs to the surfaces of the flow cell depends on the size and shape of the protein. A hypothetical, spherical protein of molecular weight M Daltons, has a volume $V \approx M/1200$ nm^3 and a projected area $A \approx (\frac{3}{4})^{2/3} \pi^{1/3} V^{2/3}$. If the protein adsorbs at random onto the surface, approximately half the area will be covered; the other half will be gaps too small for another protein to squeeze in. Thus, the binding capacity of the surface is $10^6/2A$ molecules μm^{-2} $\approx 3.2 \times 10^7/M^{2/3}$ molecules μm^{-2}. We therefore expect that for globular proteins such as bovine serum albumin and streptavidin, which both have molecular weights of ~65 kDa, the binding capacity should be ~20,000 μm^{-2}. This agrees reasonably well with measurement. For our hypothetical spherical protein, the binding capacity corresponds to $53 \cdot M^{1/3}$ pg mm^{-2}. Assuming that both surfaces are equally coated, and that the flow cell has a depth of 100 μm, this binding capacity corresponds to the adsorption of ~$M^{1/3}$ μg ml^{-1} protein from the solution perfused into the flow cell. For bovine serum albumin or streptavidin, this corresponds to ~40 μg ml^{-1}. For highly elongated molecules like kinesin (360 kDa, axial length 75 nm, axial ratio ~40), the binding capacity depends on the geometry of the adsorption. The two extreme cases are on-end binding (higher capacity) and lengthwise binding (lower capacity) and the capacity must be determined empirically. For kinesin the binding capacity to the glass is ~2000–5000 μm^{-2}, though most is probably inactive with respect to motility unless the glass is pretreated with other proteins.

D. Choosing the Appropriate Spacer and Coverglass Thickness

Choice of coverglasses for the spacers and for the top surface depends on the microscope optics to be used and whether the top (coverglass closest to the objective) or bottom (slide, closest to the condenser) surface is to be viewed. Microscope slides and coverglasses are made of different types of glass; they wet differently, and the motility assay works at lower kinesin density (about 10-fold) on the microscope slide surface.

1. *Looking for the top surface.* For epifluorescence, the coverglass surface has better contrast because there is less stray light from fluorescence in the solution or on the other surface. When the coverglass surface is being observed by fluorescence or any other technique, all objectives will be well corrected if a No. $1\frac{1}{2}$ coverglass is used on top.

2. *Looking at the bottom surface.* For transilluminated dark-field microscopy, the bottom surface is preferred because many high-numerical-aperture dark-field condensers (illuminating annular cone of light between numerical apertures 1.2 and 1.4) do not have a depth of field much greater than the 1-mm thickness of the microscope slide. Also, there is a strong evanescent light wave that penetrates only ~100 nm into the solution. The evanescent wave comes from the most oblique light rays, which undergo total internal reflection (the rays with numerical aperture between 1.33, the refractive index of water, and 1.4). The evanescent component nearly doubles the intensity of light scattered by microtubules at the slide surface. Indeed, by restricting the numerical aperture of the dark-field condenser to 1.33–1.4 by placing an appropriate mask in the back focal plane of the condenser to block the rays with numerical aperture between 1.2 and 1.33, light diffracted by objects in solution or on the other surface can be almost completely eliminated. This corresponds to total-internal-reflection dark-field microscopy (Spencer, 1982, p. 39).

The problem with looking at the bottom surface of the flow cell is that because the objective must look through the depth of the chamber, which is filled with solution of refractive index 1.33, different to air (1.0) or glass (~1.5) or immersion oil (1.515), all objectives will suffer spherical and other aberrations. In spherical aberration, the axial and oblique rays are not simultaneously in focus and the image is blurred. The higher the numerical aperture, the more serious the image degradation. Even a 40X/0.66NA (X stands for magnification, NA for numerical aperture) air objective is quite seriously compromised. This problem can be almost completely solved by making the optical depth (depth × refractive index) of the air/oil–coverglass–solution combination equal to that of the air/oil–No. $1\frac{1}{2}$ coverglass combination for which the objective was designed. For the 40X/0.66NA/0.55mm working distance/infinity-corrected Reichert air objective (Leica, Buffalo NY), a No. 00 coverglass spacer with No. 0 coverglass on top has almost no spherical aberration. For the Reichert 100X/1.3NA/ 0.15mm working distance/infinity-corrected objective (with iris diaphragm), good correction for numerical apertures up to ~1.0 is achieved with the No. 00 spacer, No. 0 coverglass, and a special immersion oil of refractive index 1.696 (R. P. Cargille Laboratories, Cedar Grove, NJ). For other objectives, a trial-and-error search will be facilitated by using the star test (Spencer, 1982) to assess the sign and magnitude of the spherical aberration.

E. Cleaning Slides for Dark-Field Microscopy

The advantage of dark-field microscopy is that unmodified microtubules can be seen without video or digital image processing (which must be used in conjunction with differential interference or phase-contrast microscopy). But the major disadvantage is that the technique is extremely susceptible to scratches, pits, etch marks, or dust on the surface. We use Fisher precleaned microscope slides, but order them in small batches (<5 gross) in case we get a dirty batch. The chamber side of the microscope slide and both surfaces of the top coverglass need to be cleaned of dust. We use the following technique: 2–4 μl of distilled water is placed on the glass surface, and a Ross Optical Lens Tissue (Thomas Scientific) is laid parallel to the surface onto the drop so that the water soaks into the lens paper, pulling it tight against the glass. The paper is then pulled parallel to the surface over the area to be cleaned. The surface tension should make the pull quite difficult and care must be taken lest the coverglass breaks. Finger grease must be kept away from the water and from the relevant areas of the glass and lens tissue. The first person who significantly improves on this method before 1995 will receive a fine bottle of red wine from the first author.

IV. Microscopy and Motion Analysis

A. Microscope and Video Equipment

This section describes briefly the microscope and video equipment we use for observing and recording the motion of microtubules in the assay. We use fluorescence microscopy for rhodamine-labeled microtubules (Hyman *et al.,* 1991) and dark-field microscopy for unlabeled microtubules. The Diastar upright microscope (Leica, Buffalo, NY) has a fixed stage and binocular tube. To focus the infinite-tube-length objective, only the nosepiece is moved. This provides good stability and the ability to bolt camera and detectors directly to the table, as can be done when using an inverted microscope. For both dark-field and fluorescence, 100-W Hg arc lamps are used with critical illumination. The magnified image is projected onto a silicon-intensified-target camera (Hamamatsu C2400-8, Bartels and Stout, Bellevue, WA). The video signal is passed through a time–date generator which marks every frame (Panasonic WJ-810, Burtek, Seattle, WA), recorded on VHS (Panasonic AG-7350, Proline, Seattle, WA), and displayed on a high-resolution black and white monitor (Sony PVM-122). Hard copy of video images from $\frac{1}{2}$-in. video tape are made using a video printer (Sony UP-5000) or an oscilloscope camera (Tekronic C-5C, Beaverton, OR).

B. Measuring Motility

The easiest way to analyze motility is by tracing the images directly off the video monitor onto acetate film. The principal disadvantage is that the magnification is different in the vertical and horizontal directions. To circumvent this problem we use a personal computer-based board (Measure, Mike Walsh Electronics, San Dimas, CA) which projects cross-hairs onto the monitor, the position of which is controlled by the mouse. The software, kindly supplied by Dr. S. Block (described in Sheetz *et al.*, 1986), is able to read the cursor position, correcting for unequal magnification. The board also reads the time directly from the Panasonic AG-7350 video recorder. By use of these features, computer-assisted measurement of microtubule positions and speeds is fast and accurate.

V. *In Vitro* Motility Assay

A. Buffers and Stock Solutions

BRB80 buffer [80 mM 1,4-piperazinediethanesulfonic acid (Pipes), free acid, 1 mM MgCl$_2$, 1 mM ethylene glycol bis (β-aminoethyl ether) N,N'-tetraacetic acid (EGTA), pH 6.85, with KOH] is filtered (0.2 μm), and 10-ml aliquots are stored at $-20°$C. Because of the large number of individual components and the difficulties encountered debugging the assays when they go wrong, we make 10- to 100-μl aliquots of the following stock solutions, store them at $-20°$C, and thaw a new aliquot at the start of each day: 25 mM GTP, 100 mM ATP, 100 mM MgCl$_2$ (!), 1 mM taxol in dimethylsulfoxide (Adrich Sure Seal, Milwaukee WI), dimethylsulfoxide, 2.5 mg ml^{-1} casein BRB80, 40 mg ml^{-1} glucose oxidase in BRB80, 8 mg ml^{-1} catalase in BRB80, and 2 M D-glucose. All solutions are filtered at 0.2 μm.

B. Growing Microtubules

We usually use microtubules polymerized in the following way: To freshly thawed tubulin at $\geq$1.5 mg ml^{-1} on ice, add extra MgCl$_2$ to 4 mM, GTP to 1 mM, and dimethylsulfoxide to 5% (v/v). Warm to 37°C for 20 minutes. To the warm microtubules add taxol to 10 μM and stir by passing the microtubules several times through a yellow-tip (20–200 μl) pipet. Store at room temperature. They last for several days for motility assays.

C. Diluted Microtubule Solution for the Motility Assay

To BRB80 at room temperature, add extra $MgCl_2$ to 1 mM, ATP to 1 mM, and taxol to 10 μM. Mix well, and then add ~0.1-1% (v/v) microtubule solution (above) and stir well by passing the solution several times through a yellow-tip pipet. If rhodamine-labeled microtubules are used, then the solution must be augmented with glucose oxidase to 0.2 mg ml^{-1}, catalase to 0.04 mg ml^{-1}, D-glucose to 20 mM, and 2-mercaptoethanol to 1% (v/v). This antifade mixture slows photobleaching to several minutes under the full intensity of the 100-W Hg arc lamp.

D. Standard Upside-Down High-Density Assay

Fill the flow cell with ~10 μl of kinesin at a concentration $\geq$50 μg ml^{-1} (50 μg ml^{-1} corresponds to a density of ~4000 molecules μm^{-2}, assuming a depth of 100 μm). Allow 2–3 minutes for adsorption. Perfuse with 20 μl of diluted-microtubule solution (above).

E. Low-Density Upside-Down Assay

Fill the flow cell with ~10 μl casein (2.5 mg ml^{-1} in BRB80) and wait 2 minutes. Perfuse with one flow cell volume (~10 μl) of kinesin (diluted in BRB80 augmented with 50 μg ml^{-1} casein) and wait 3 minutes. Perfuse with diluted-microtubule solution (above) augmented with 50 μg ml^{-1} casein. In the presence of dilute kinesin (<1 μg ml^{-1}), silanized microfuge tubes should be used.

F. Low-Density Bead Assays

The general procedure is as follows but the reader is referred to Block *et al.* (1990) for more experimental details. Microtubules are first adsorbed to a cover-glass. Silica beads are then preincubated with buffer containing carrier proteins, either 10 mg ml^{-1} bovine serum albumin or 50 μg ml^{-1} casein plus 50 μg ml^{-1} cytochrome c, to precoat the surface and therefore prevent kinesin inactivation on the surface. The beads are coated with kinesin from a solution containing dilute kinesin in addition to the carrier protein(s). The kinesin-coated beads are then presented to the fixed microtubules using an optical trap to hold the bead close to the microtubule. The optical trap increases the encounter rate by about 1 million times.

VI. Evidence That Single Molecules Suffice for Microtubule-Based Motility

The advantage of the so-called "low-density" assay is that movement occurs at kinesin densities as low as 1 μm^{-2}, some 3 to 4 orders of magnitude lower than the minimum density required in the standard assay in which the kinesin is adsorbed directly to the glass. Pretreatment of the glass microscope slide surface with casein prior to kinesin adsorption (Block *et al.*, 1990) increases by a factor of 5, the rate at which microtubules attach to and start moving across the glass surface compared with pretreatment with cytochrome c and tubulin (Howard *et al.*, 1989). This indicates that in the original low-density assay of Howard *et al.*, over 80% of the adsorbed kinesin molecules were nonfunctional. An additional advantage of the casein pretreatment over the cytochrome c/tubulin pretreatment is that the casein also works on the coverglass surface (though with only about 1/10th of its efficiency on the microscopeslide glass), whereas the latter pretreatment has little effect on the coverglass surface.

At kinesin densities ≤ 7 μm^{-2} (casein pretreatment, Romberg and Vale, 1993) or <20 μm^{-2} (cytochrome c and tubulin, Howard *et al.*, 1989), several lines of evidence suggest that the motion is by single kinesin molecules. First, microtubules moving across the surface pivot about a single fixed point on the surface. It appears that the microtubule is attached to only a single point on the surface, a point at which one functional motor is presumably located. Second, the microtubule appears to "thread through" this point of attachment, and when its trailing end reaches this point, the microtubule diffuses back into the solution. Third, dilution studies show that the number of functional motors on the surface is directly proportional to the amount of kinesin perfused into the chamber and, therefore, to the density of kinesin on the surface. This implies either that the functional motor is the species that exists in solution, the single kinesin molecule (an $\alpha_2\beta_2$ tetramer), or that aggregation takes place at the surface. This latter possibility, which would occur if a kinesin bound to the surface acted as a nucleation site for the adsorption of more kinesin molecules to form a motor complex, is most likely ruled out by the following kinetic argument. At the lowest densities, the kinesin concentration in the solution is <100 pM, and even if the association rate for the formation of these complexes were 10^7 M^{-1} s^{-1}, an upper limit for protein–protein association, the formation of such aggregates would occur with a time constant greater than 16 minutes, much longer than the time taken for the functional motors to adsorb to the surface in the assay.

Independent evidence that a single kinesin molecule suffices to form a functional motor comes from Block *et al.* (1990). By adsorbing kinesin to silica beads (see above), they showed that when a bead contained an average of only one kinesin molecule, it had a high probability (>0.5) of binding to and moving along the microtubule. The probability of the bead not binding to the microtubule varied with the average number of kinesin molecules per bead according to the Poisson distribution. These results are strong evidence that the presence of only one kinesin molecule on the bead is sufficient for motility.

Acknowledgments

This work was supported by a grant from the National Institutes of Health (AR-40593). Work was performed while J.H. was a Pew Scholar in the Biomedical Sciences, an Alfred P. Sloan Research Fellow, and a recipient of the FESN fellowship on Sensory Transduction.

References

Berg, H. C., and Block, S. M. (1984). A miniature flow cell designed for rapid exchange of media under high-power microscope objectives. *J. Gen. Bacteriol.* **130,** 2915–2920.

Block, S. M., Goldstein, L. S. B., and Schnapp, B. J. (1990). Bead movement by single kinesin molecules studied with optical tweezers. *Nature (London)* **348,** 348–352.

Brennen, C., and Winet, H. (1977). Fluid mechanics of propulsion by cilia and flagella. *Annu. Rev. Fluid Mech.* **9,** 339–398.

Hackney, D. D., Levitt, J. D., and Suhan, J. (1992). Kinesin undergoes a 9S to 6S conformational transition. *J. Biol. Chem.* **267,** 8696–8701.

Howard, J., Hudspeth, A. J., and Vale, R. D. (1989). Movement of microtubules by single kinesin molecules. *Nature (London)* **342,** 154–158.

Huxley, A. F. (1980). "Reflections on Muscle." Princeton Univ. Press, Princeton, NJ.

Hyman, A., Drechsel, D., Kellog, D., Salser, S., Sawin, K, Steffen, P., Wordeman, L., and Mitchison, T. (1991). Preparation of modified tubulins. *In* "Methods in Enzymology" (R. B. Vallee, ed.), Vol. 196, pp. 478–485. Academic Press, San Diego.

Kuo, S. C., and Sheetz, M. P. (1991). Measuring the isometric force of single kinesin molecules using the laser optical trap. *Biophys. J.* **59,** 568a.

Landau, L. D., and Lifshitz, E. M. (1987). "Fluid Mechanics," 2nd ed. Pergamon, Oxford.

Romberg, L., and Vale, R. D. (1993). Chem-mechanical cycle of kinesin differs from that of myosin. *Nature (London)* **361,** 168–170.

Schnapp, B. J., Block, S. M., Goldstein, L. S. B., Stewart, R. J., and Godek, C. P. (1991). Can a single kinesin head drive microtubule motility? *Biophys. J.* **59,** 567a.

Sheetz, M. F., Block, S. M., and Spudich, J. A. (1986). Myosin Movement *in Vitro:* A Quantitative Assay Using Oriented Actin Cables from Nitella. *In* "Methods in Enzymology" (R. B. Vallee, ed.), Vol. 134, pp. 531–544. Academic Press, San Diego.

Spencer, M. (1982). "Fundamentals of Light Microscopy." Cambridge Univ. Press, Cambridge.

Wagner, M. C., Pfister, K. K., Brady, S. T., and Bloom, G. S. (1991). Purification of kinesin from bovine brain and assay of microtubule-stimulated ATPase activity. *In* "Methods in Enzymology" (R. B. Vallee, ed.), Vol. 196, pp. 157–175. Academic Press, San Diego.

Warshaw, D. M., Derosiers, J. M., Work, S. S., and Trybus, K. M. (1990). Smooth muscle myosin cross-bridge interactions modulate actin filament sliding velocity in vitro. *J. Cell Biol.* **111,** 453–463.

Weingarten, M. D., Suter, M. M., Littman, D. R., and Kirschner, M. W. (1974). Properties of the depolymerization products of microtubules from mammalian brain. *Biochemistry* **13,** 5529–5537.

CHAPTER 11

In Vitro Motility Assays Using Microtubules Tethered to *Tetrahymena* Pellicles

Vivian A. Lombillo, Martine Coue,[1] and J. Richard McIntosh

Department of Molecular, Cellular and Developmental Biology
University of Colorado
Boulder, Colorado 80309

I. Introduction
II. Isolation of Assay Components
    A. Preparation of *Tetrahymena* Pellicles
    B. Purification of Bovine Brain Tubulin
    C. Isolation of Mitotic Chinese Hamster Ovary Chromosomes
    D. Further Characterization of Assay Components: Insignificant Levels of Contaminating ATP
    E. Preparation of Axonemes from *Chlamydomonas*
III. *In Vitro* Motility Assay I: Inducing Chromosome and Vesicle Motions with Microtubule Depolymerization
IV. *In Vitro* Motility Assay II: Testing Microtubule Sliding Activity of Motor Enzymes
V. Other Procedures
    A. Fluorescence Observations
    B. Optimal Fixation of Pellicle-Initiated Microtubules
VI. Additional Remarks
References

I. Introduction

In this chapter we describe a way to initiate and grow microtubules (MTs) from purified tubulin to form a tethered array of polymers with known polar

[1] Present address: Institut Jacques Monod, University of Paris, Paris, France.

orientation. This array can then be used to study both the binding and the movement of objects that interact with MTs under well-defined experimental conditions. We have used this system to study two forms of motility: (1) chromosome movement induced by MT depolymerization alone and (2) MT–MT sliding driven by conventional motor enzymes. The value of the system, in general terms, is that it permits an analysis of interactions between cytoplasmic objects and a MT array that resembles the MT complex grown from a centrosome *in vivo*. The specific advantage of this *in vitro* system is that motility occurs along anchored MTs of uniform polarity, and therefore the geometry of motion is defined. In addition, conditions can be varied easily during these assays to control the dynamic state of the tethered MTs.

The first assay we developed allows the study of movements of cellular objects driven by MT depolymerization (Coue *et al.*, 1991). In this assay, chromosomes and vesicles, which are bound to the tethered MTs, will move only when the MTs are induced to depolymerize. This motility occurs in the absence of ATP, with rates that are fast enough to resemble motions *in vivo* and with forces large enough to be physiologically significant. Our results with this assay resurrect an earlier idea that MT disassembly might serve to power chromosome movement during anaphase A (reviewed in Inoue, 1981; Koshland *et al.*, 1988). Although the *in vitro* movements are analogous in direction to anaphase A, it is not yet clear whether our system is truly a valid model for mitotic chromosome motions in cells. What our results with this assay do demonstrate unambiguously, however, is that MT depolymerization can exert sufficient force to move chromosomes and vesicles. One can use this model system to characterize this form of motility in detail and, more generally, to examine the role of MT disassembly during mitotic chromosome movement.

The second assay we perform using tethered MTs allows one to test whether a particular motor of interest has the ability to slide MTs past each other (Nislow *et al.*, 1992). In this particular assay, MTs grown off added axonemes are translocated by motors along the surface of the tethered MTs. Unlike the first assay described above, the sliding assay requires ATP and does *not* involve MT dynamics for translocation. The geometry and general mechanics of MT sliding in this *in vitro* assay are reminiscent of spindle elongation *in vivo*. This second assay may therefore be useful as a model for anaphase B motility.

The assays described in this chapter test motility in real time within a light microscope perfusion chamber using video-enhanced differential interference contrast (DIC) optics. The basic components of our experimental "spindle" are (1) basal bodies lodged within a *Tetrahymena* pellicle as the "centrosomes" and (2) MTs nucleated from these basal bodies with purified brain tubulin as the "spindle MTs." In the first assay, chromosomes and vesicles are added to the system; they bind to MTs tethered to the pellicle and move in toward the basal bodies when we induce MT depolymerization (either by tubulin dilution or by the addition of calcium). In the second assay, *Chlamydomonas* axonemes are added to the system, they bind to the tethered MTs when we add a motor protein and subsequently move when ATP is introduced into the chamber. Below we describe how to prepare the components to construct these assays (in Section II) and then outline the procedure we use for each assay (in Sections III and IV).

II. Isolation of Assay Components

A. Preparation of *Tetrahymena* Pellicles

1. Maintaining *Tetrahymena*

Pellicles are made using a mucus-free strain of *Tetrahymena thermophila*, SB255. Any mucus-free strain of *Tetrahymena* may work fine, though this particular one can be obtained from us by request. Stocks of cells are maintained in 2% proteose peptone (Oxoid Ltd.), 0.003% sequestrine (an iron supplement, Ciba–Geigy), and 0.005% each of streptomycin and penicillin (Gibco Labs) at room temperature in (100-ml) bottles with the lids slightly unscrewed to allow air exchange. These stocks should be changed when the density of cells is about 5×10^7; this density is reached about 10 days after 50 μl of stock cells is added to 100 ml of fresh medium. Long-term anaerobic stocks (that need only be changed every 6 months) can be stored in a sealed bottle at room temperature in Neff's medium (0.75% proteose peptone, 0.75% yeast extract, 80 mM glucose, 0.003% sequestrine, and both 1 mM MgSO$_4$ and 50 μM CaCl$_2$ added sterile after autoclaving).

2. Growing *Tetrahymena* to Density prior to Preparing Pellicles

To grow cells up for a pellicle preparation, transfer 1–2 ml of cells from the stock solution into a 500-ml capped Erlenmyer flask (Bellco) that contains 200 ml of the 2% proteose peptone medium described in Section II,A,1. Amounts can be scaled according to need; 200 ml of cells typically yields enough pellicles for approximately 100 assays. Shake cells at 200 rpm on an orbital shaker for about 1.5–2 days at room temperature until they reach mid-log phase ($\sim 5 \times 10^6$). At this density, the medium should look slightly opaque.

3. Preparing *Tetrahymena* Pellicles

In this procedure, *Tetrahymena* are harvested from the shaking cultures described in Section II,A,2, deciliated by mechanical shearing, and then partially extracted. This method yields pellicles with basal bodies that are competent to nucleate growth of MTs. We find that deciliation methods using chemicals like dibucaine diminish the nucleation efficiency of the basal bodies. The relative activity of the basal bodies should be tested at the end of each preparation by examining their ability to assemble MTs from phosphocellulose-purified tubulin. The entire prep, from cells to frozen samples, takes about 1.5 hours.

1. Grow *Tetrahymena* (SB255) to mid-log phase in 2% proteose peptone, 30 μg/ml sequestrine, and 50 μg/ml penicillin/streptomycin as described above.

2. Harvest cells in 50-ml conical tubes with a table-top centrifuge at 500g, 5 minutes, room temperature; 200 ml of starting material yields about 1.5 ml of pelleted cells.

3. Wash pelleted cells in fresh medium (about 15 ml) and centrifuge as above.

4. Resuspend washed pellet in 25 ml cold PME buffer + protease inhibitors (see recipes below) and centrifuge (500g, 3 minutes, 4°C).

5. Resuspend pellet in a smaller volume (about 10 ml) of PME + inhibitors and transfer sample to a 15-ml conical centrifuge tube. Centrifuge (500 g, 3 minutes, 4°C).

6. Resuspend cell pellet gently for about 30 seconds in (5 ml) PME (+ protease inhibitors) containing 0.25% Nonidet P-40 (Particle Data Labs, Ltd.) and then immediately deciliate the cell suspension by mechanical shearing using a Sorvall Omni-Mixer for three 10-second pulses at setting 4 (4°C).

7. Check the sample with DIC microscopy to be sure that all cilia have been sheared (see Fig. 1B). If some cilia remain continue shearing with the Omni-Mixer for an additional 10 seconds. *Note: Cells should not be in the lysis buffer longer than 10 minutes or they will begin to bleb and break apart.*

8. Add buffer to 15 ml to dilute the detergent slightly; then pellet these permeabilized, deciliated cells (hereafter called pellicles) in PME + inhibitors (without detergent) three times at 500g, 3 minutes, 4°C.

9. If the pellicles are still not completely deciliated by this stage, repeat the entire shearing/lysis step (step 6). If deciliation was successful, add 2–3 ml of the lysis buffer used in step 6 to further extract pellicles for an additional minute.

10. Rinse the pellicles in PME + inhibitors two or three times as in step 8.

11. The volume of the final pellet of pellicles should be about 300–500 μl from 200 ml of mid-log-phase cells. Resuspend cell pellet in a threefold volume of PME + inhibitors. This dilution will yield a concentration of pellicles appropriate to perform roughly 100 assays.

12. Divide the preparation into 40- to 50-μl aliquots, quick-freeze in liquid nitrogen, and store at −70°C. Thaw as needed; pellicles remain good for at least 4 months (but once thawed, an aliquot should not be frozen again).

13. To gauge the preservation of the basal bodies lodged within the pellicles, test their ability to nucleate tubulin polymerization by adding pellicles to a solution of 30 μM pure tubulin + 1 mM GTP in PME; mix well and incubate at 37°C for 15 minutes. Examine pellicle-initiated MTs with video-enhanced DIC microscopy (see Fig. 1D).

Buffer for Pellicle Preparation. PME buffer contains 80 mM piperazinediethane sulfonic acid (Pipes), pH 6.9, 2 mM MgCl$_2$, and 1 mM ethylene glycol

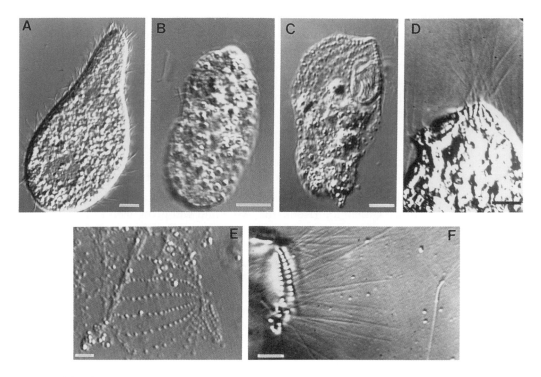

Fig. 1 Preparation of *Tetrahymena* particles. (A) A ciliated *Tetrahymena* cell harvested in step 2. (B) A deciliated, partially extracted cell after step 6. (C) A pellicle after the second extraction in step 9. (D) MT growth intiated by a pellicle in a solution of 25 μM tubulin. (E) If desired, fragments of pellicles that contain fewer basal bodies than intact pellicles may be prepared with this procedure by extracting the pellicles three or more times. (F) These portions of pellicles will nucleate a modest array of MTs that generally grow along the coverslip surface. Bar (A–D) = $10\,\mu$m, bar (E,F) = $5\,\mu$m.

bis(β-aminoethyl ether) N,N'-tetraacetic acid (EGTA). The protease inhibitors used in PME are 0.2 mg/ml soybean trypsin inhibitor, 10 μg/ml leupeptin, 1 μg/ml aprotinin, 1 μg/ml pepstatin A, 10 μg/ml *p*-tosyl-L-arginine methyl ester, 10 μg/ml benzamidine.

B. Purification of Bovine Brain Tubulin

These assays require phosphocellulose (PC)-purified bovine brain tubulin to form MTs from the pellicles. Briefly, we prepare bovine brain microtubule protein essentially using the method of Berkowitz *et al.* (1977) with the reassembly buffer described by Weisenberg (1972). This procedure involves two cycles of microtubule polymerization and depolymerization to yield a preparation of tubulin + microtubule-associated proteins (MAPs). Tubulin is separated from the MAPs on a PC column according to the method of Williams and Detrich (1979). Before the tubulin is used for *in vitro* assays, batches of PC-purified tubulin are cycled two more times exactly as described by Hyman *et al.* (1991). This recycling step yields tubulin that is extremely potent to poymerize (presumably because of the reduction of ''dead'' or nonpoymerizable tubulin dimer) and the activity of *in vitro* movements is greatly improved. This tubulin is good for months when stored at −70°C.

C. Isolation of Mitotic Chinese Hamster Ovary Chromosomes

Chinese hamster ovary (CHO) cells can be obtained from the American Type Culture Collection (Rockville, MD) and will grow well in Ham's F-12 medium (Gibco Labs) supplemented with 5% fetal calf serum. To isolate mitotic chromosomes from CHO cells, we have worked from a method described by Mitchison and Kirschner (1985), which is a modification of the Lewis and Laemmli (1982) isolation that uses polyamines. Although the final sample is composed primarily of chromosomes, both vesicles and nonmembranous particles also exist free in solution as well as on the chromosomes themselves (Fig. 2).

In this procedure, mitotic CHO cells are collected following vinblastine treatment by gently shaking the flasks to detach rounded mitotic cells into the medium. The cells are then lysed and homogenized in a digitonin-based lysis buffer. Chromosomes are released from the cells and can then be collected on a sucrose step gradient. The entire prep takes about 2.5 hours.

1. Add 10 μg/ml of vinblastine sulfate (Sigma) to each of 10 tissue culture flasks (150 cm^2, Corning) containing CHO cells at about 80% confluence. *Incubate cells in the drug for 10 hours at 37°C.*

2. Detach mitotic cells by gently shaking the flasks and swirling the medium across the surface. (Shaking the flasks too vigorously to try to increase the yield of mitotic cells causes interphase cells to detach also. This results in the contamination of interphase nuclei in the final chromosome fraction.)

3. Collect the medium from all the flasks into 50-ml conical centrifuge tubes.

4. Pellet cells (350g, 3 minutes, room temperature).

5. Aspirate supernatant, leaving about 0.5 ml in each tube. Resuspend pellets gently and combine them into one 15-ml conical centrifuge tube.

6. To the cell suspension, add 10 ml of (low-ionic-strength) swelling buffer at room temperature (see buffer recipes below). Incubate for 10 minutes at room temperature.

7. Centrifuge the suspension to pellet cells (350g, 3 minutes, 4°C).

8. Add 2–3 ml of lysis buffer to the pellet and gently resuspend for 30 seconds.

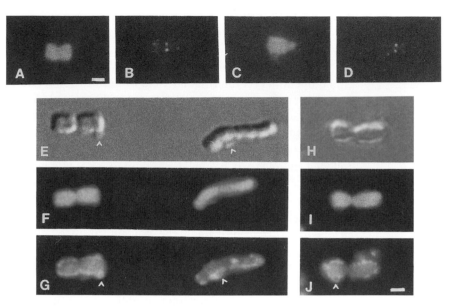

Fig. 2 Characterization of isolated mitotic chromosomes. Images A–D corroborate other reports that dynein remains associated with chromosomes during their isolation (Pfarr *et al.*, 1990). In A and C chromosomes are stained with Hoescht to visualize DNA, and in B and D kinetochore-bound dynein is detected with an affinity-purified antibody that recognizes CHO cytoplasmic dynein. We also observe free vesicles and nonmembranous particles in this preparation by DIC microscopy and with electron microscopy (not shown), as well as vesicles bound to the chromosomes themselves (G,J). E–J show three representative chromosomes from this isolation procedure. E and H are DIC images, F and I detect DNA with Hoescht dye, and G and J are chromosomes stained with the lipophilic dye $DIOC_5$ (Molecular Probes). When used at 2.5 μg/ml this lipophilic dye incorporates into a variety of membranous organelles (Terasaki *et al.*, 1984). The observations that vesicles are associated with many chromosomes in this prep and that dynein is faintly detectable on the chromosome surface (B and D) may explain why MTs are occasionally seen to bind to the chromosome arms as well as to the primary constriction during our assay (Coue *et al.*, 1991). Bar = 1 μm.

9. Pellet cells (350g, 3 minutes, 4°C) and add 2–3 ml of lysis buffer again, resuspending gently.

10. Transfer cell suspension to a 15-ml glass Dounce homogenizer and gently shear cells first with a loose (A) glass pestle for 10 strokes, then with a tight (B) glass pestle for 20 strokes. Take care not to foam the solution while homogenizing.

11. Immediately make a slide preparation of the homogenate to examine the chromosomes with fluorescence optics (5 μl homogenate + 1 μl of 10 μg/ml Hoescht). If many of the chromosomes in the prep remain clustered together, continue to homogenize for 10–15 more passes until most are dispersed from

this cytoskeletal "web". *Note: Lysis and homogenization (steps 8–11) should not take longer than 10 minutes; otherwise chromosomes begin to unravel and lose structure.*

12. Gently layer the lysis mixture onto one or two 15-ml sucrose step gradients (depending on the original yield of mitotic cells) and centrifuge at 4°C in a swinging bucket rotor (Sorvall HB-4) at 2500 *g*, for 15 minutes. This spin will layer the smallest chromosomes (0.2–0.5 μm) at the 30–50% sucrose interface and the remainder will sediment on the 50–60% sucrose interface.

13. After the spin, aspirate almost to the 30–50% sucrose interface and collect a sample to examine later. Again, this sample is composed of small CHO chromosomes as well as a lot of cellular debris.

14. Carefully aspirate to the 50–60% interface and (with a Pasteur pipet) collect the white flocculent material that has sedimented there. *This layer is enriched with chromosomes.*The concentration of chromosomes in the final sample can be quite high (10^9/ml) if one is careful to reduce the amount of sucrose gradient buffer collected with the chromosomes at this step.

15. Examine the preservation of chromosomes collected from the 50–60% sucrose interface with fluorescence optics, again using Hoescht dye at a final concentration of 1–2 μg/ml.

16. Use chromosomes fresh or aliquot (into 15-μl samples), quick-freeze in liquid nitrogen, and store at -70°C. Chromosomes can be used for up to at least 4 months. Once an aliquot is thawed it should not be refrozen. A standard prep yields about 500 μl of concentrated chromosomes, enough for over 100 assays.

Buffers for Chromosome Isolation: All solutions should be fresh, unless stated otherwise.

Swelling buffer contains 5 m*M* Pipes, pH 7.2, 0.5 m*M* ethylenediaminetetraacetic acid (EDTA), 10 m*M* NaCl, and 5 m*M* MgCl$_2$.

Lysis buffer contains 20 m*M* Pipes, pH 7.2, 2 m*M* EDTA, 2 m*M* MgCl$_2$, 1 m*M* spermidine, 0.5 m*M* spermine, and 0.1% β-mercaptoethanol. Saturate the solution with digitonin (Sigma) by adding to 0.1%, stirring for several hours, and centrifuging to remove insoluble detergent. This buffer can be made batchwise and stored frozen at -20°C in 10-ml aliquots. Before the prep, thaw lysis buffer and add protease inhibitors.

Sucrose gradient buffer (SGB) contains 10 m*M* Pipes, pH 7.2, 1 m*M* EDTA, 2 m*M* MgCl$_2$, 0.25 m*M* spermidine, 0.1 m*M* spermine 0.1% β-mercaptoethanol, and 30, 50, or 60% sucrose. Just before beginning the preparation make up step gradients in two 15-ml glass centrifuge tubes (Corex). The volumes of the step gradient should be as follows: 3 ml 60% SGB, 3 ml 50% SGB, and 4 ml 30% SGB. Keep them at 4°C until needed.

The following concentrations of protease inhibitors should be used in the lysis and gradient buffers: 5 μg/ml α_2-macroglobulin, 0.2 mg/ml soybean trypsin inhibitor, 10 μg/ml leupeptin, 1 μg/ml aprotinin, 1 μg/ml pepstatin, 10 μg/ml *p*-tosyl-L-arginine methyl ester, 10 μg/ml benzamidine.

D. Further Characterization of Assay Components: Insignificant Levels of Contaminating ATP

We have characterized the assay components further by examining the average concentration of ATP in each component. We were curious to know this, because it was conceivable that even though we do not add ATP to the system in the MT depolymerization-powered assay, sufficient ATP-contaminated the preps to serve as a source of fuel for mechanoenzymes present within the chromosomes. Using a luciferin–luciferase ATP assay sensitive to the picomolar range (Sigma), we measured the ATP concentrations of the individual components in the assay (Table I). The maximal ATP concentration present in the entire system at the time of movement is ~5 nM. This concentration is significantly low, considering that the known K_m for cytoplasmic dynein is well over three orders of magnitude greater (Shpetner *et al.*, 1988). Regardless, these low levels of ATP do not appear to be important in our assay, as movements still occur in the presence of apyrase, which converts ATP to ADP and AMP (Coue *et al.*, 1991).

E. Preparation of Axonemes from *Chlamydomonas*

Axonemes from *Chlamydomonas reinhardtii* are used in the second motility assay, which tests the activity for motors to generate MT sliding. In the presence of a motor that has sliding activity, MTs grown off these axonemes will be translocated along the surface of the tethered (pellicle) MTs. We prepare axonemes from *Chlamydomonas* precisely as described by Huang *et al.* (1979). Briefly, cells are grown to density on plates and then transferred to liquid culture to generate flagella. Cells are harvested and then deciliated with a pH shock of 4.5 on ice. The pH is brought up to 7.4 and cell debris is separated from free flagella by pelleting it through a sucrose cushion (20%). The flagella remain in the supernatant and are then demembranated with 0.1% Nonidet-P40. The axonemes are washed several times, divided into 5-μl aliquots, and quick-frozen in nitrogen. These axonemes (stored at $-70°C$) are competent to seed MT growth in ~15 μM tubulin.

Table I
Concentrations of ATP in Assay Components

Assay component	[ATP] at attachment	[ATP] at movement[a]
Tetrahymena pellicles	Less than picomolar levels	$<<<$ pM
Tubulin + 1 mM GTP	500 nM	0.5–5.0 nM
chromosome preparation	$\leq$10 nM	0.01–0.1 nM

[a] After 2–3 chamber vol have been perfused. Each chamber volume dilutes solutes approximately 10-fold (Coue *et al.*, 1991).

In Vitro Motility Assay I: Inducing Chromosome and Vesicle Motions with Microtubule Depolymerization

With the components prepared above (chromosomes, tubulin, and pellicles), one is now able to perform this *in vitro* motility assay. Other supplies needed for the assay are:

Light microscope with DIC optics (optional fluorescence), stage prewarmed to ~32°C (using air curtain incubator or other regulated heating source).

Syringe (1 cc) with 18-gauge needle filled with high-vacuum grease to make streaks on coverslip prior to constructing the perfusion chamber.

Humid chamber made from a covered Petri dish with moist paper towel at the bottom and Parafilm on top of the towel.

Melted VALAP (a 1:1:2 mixture of vaseline, lanolin, and paraffin) and a camel hair paint brush for application.

Solutions: (1) PMED (80 mM Pipes, pH 6.9, 2 mM MgCl$_2$, 1 mM EGTA, 2 mM dithiothreitol, DTT) at 4°C and room temperature and (2) 10 mM GTP (to promote tubulin polymerization).

1. Thaw one aliquot of each assay component (pellicles, tubulin, chromosomes, 10 mM GTP).

2. Add 5 μl of pellicles to an 18-mm coverslip that has two streaks of vacuum grease about 4 mm apart.

3. Incubate coverslip in the humid petri chamber for 5 minutes on ice to allow pellicles to adsorb onto the coverslip surface.

4. Prepare a perfusion chamber by inverting the coverslip onto a microscope slide and sealing the edges with VALAP. Construct wells and a thin barrier at the edges of the coverslip to prevent the perfusate from interfering with the optics (as shown in Fig. 3A). Chamber volume should be approximately 6–10 μl.

5. Rinse the chamber with 2–3 vol of cold PME. Mount the slide onto the microscope stage that has been prewarmed to 32°C.

6. Perfuse 15 μl of 30 μM tubulin + 1 mM GTP in PME to initiate MT polymerization.

7. Allow roughly 15 minutes for the pellicle-initiated MTs to polymerize until they reach about 15–20 μm in length. Take care not to allow overgrowth of MTs at this stage; overly dense arrays of pellicle-initiated MTs bind excessively onto chromosomes and can interfere with movements. Also, if MTs are polymerized with tubulin concentrations over 30 μM, many MTs will form spontaneously in solution and inhibit chromosomes from binding to pellicle-initiated MTs.

8. Allow chromosomes (diluted 50-fold) to flow into the chamber. (First dilute chromosomes 1:4 with PME buffer; then add 1 μl of diluted chromosomes to 9 μl of 30 μM tubulin.)

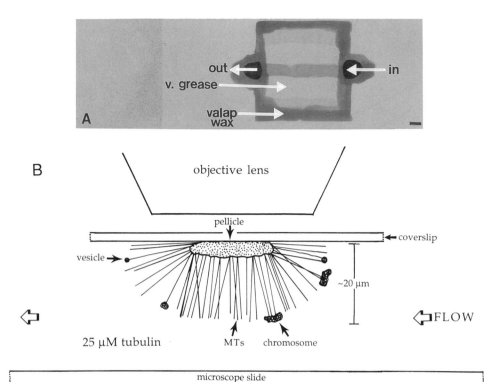

Fig. 3 (A) Light microscope perfusion chamber. This simple chamber is constructed with an 18-mm coverslip in which a channel is created with two streaks of high-vacuum grease. The coverslip (with pellicles adsorbed) is then inverted onto a microscope slide, gently tapped down, and sealed with VALAP in a manner that forms wells for perfusion. VALAP is also "painted" across the edge of the coverslip to prevent the perfusate from running onto the top of the coverslip. To modify this chamber for an inverted microscope, the microscope slide is simply substituted by a large coverslip (60 × 25 mm) and pellicles are adsorbed onto this coverslip surface. Bar = 2 mm. (B) Anatomy of the *in vitro* assay. This diagram illustrates the *in vitro* assay we use to study movements driven by MT depolymerization. Pellicles adsorbed to the top of the chamber (the coverslip surface) nucleate the assembly of MTs from 25 μM tubulin at 32 °C. These MTs radiate their plus ends into the chamber (Heidemann and McIntosh, 1980). The distal ends of the MTs bind objects from the chromosome preparation. Movement occurs when the MTs are induced to depolymerize.

9. Allow chromosomes to bind to pellicle-initiated MTs for about 3 minutes. Inverting the chamber during this incubation may increase the number of chromosomes bound to the tethered MTs (Fig. 3B).

10. Optional step: Perfuse 10 μl of the tubulin solution from step 6 that also contains 5 μg/ml Hoescht 33258 to be better able to identify bound chromosomes with fluorescence optics. This dye does not interfere with motility.

11. Select a bound chromosome or vesicle and begin to record the experiment on video while perfusing tubulin-free PME (to reduce the soluble tubulin concentration) or 25 μM tubulin containing up to millimolar concentrations of calcium salts (to destabilize the MT polymer). Diluting the tubulin by flowing 2–3 chamber vol of buffer at rates of 100–200 μm/s will typically yield movement within a few minutes (Fig. 4).

12. If the bound object does not move when all other MTs in the field of view are beginning to depolymerize, it is possible that the bound object is stabilizing the MTs. If this is the case and no movement is apparent within a few minutes of perfusion, scan the chamber for other objects that may be moving during this time. This step (knowing if an object will move and deciding whether to move on to another object) is empirical and may take practice. From a well-preserved chromosome prep, we find that nearly half of all bound objects will move with the depolymerizing MT end at least a few micrometers, if not all the way to the pellicle. Often, attachments that are completely stable to tubulin dilution may still move with calcium-induced MT disassembly.

IV. *In Vitro* Motility Assay II: Testing Microtubule Sliding Activity of Motor Enzymes

The assay described here was first reported in Nislow *et al.* (1992). Refer to that particular report for the molecular details of the motor enzyme we have assayed, MKLP-1. While we were testing the orientation of MTs in a bundle caused by the addition of MKLP-1, a kinesin-like protein, we detected this form of motility quite serendipitously. We were initially examining the polarity of bundling of MTs from two closely spaced pellicles. Antiparallel bundling would be observed between the MTs of neighboring pellicles, and parallel bundles would form on individual pellicles. MKLP-1 specifically bundled antiparallel MTs in an ATP-sensitive fashion, and after ATP addition we observed that some contaminating *Tetrahymena* axonemes moved outwardly along a "track" of tethered pellicle MTs. In subsequent experiments we supplemented the assays with *Chlamydomonas* axonemes to increase the frequency of observing this motility. *Chlamydomonas* axonemes also have the distinct advantage of possessing an internal polarity marker so that one is able to discern the directionality of movement. The assay involves first nucleating MT growth from pellicles that have been adsorbed onto the coverslip of the light microscope chamber shown in Fig. 3A. Axonemes are then added to the chamber with purified tubulin and are allowed to grow MTs for several minutes. When recombinant MKLP-1 is perfused into the chamber, free axonemes bind the tethered MTs of the pellicle. The tubulin concentration is reduced and pellicle-initiated MTs that are not crosslinked to axonemal MTs depolymerize. ATP is added after several minutes, causing the axonemes to move outwardly along the pellicle MTs. This

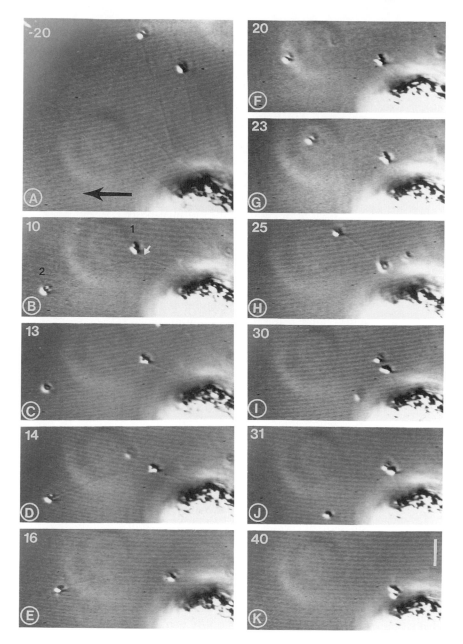

Fig. 4 Sequence of chromosome movement. These two small chromosomes (B, 1 and 2) bound Hoescht dye (not shown) and moved against the direction of buffer flow (arrow in A), each with an average speed of 69 μm/min, and then moved at 36 μm/min after they merged together (J). The rate of chromosome movement in our assay depends on both the temperature and the rate at which tubulin is diluted from the chamber. In this sequence, chromosome 1 moves toward the pellicle while bound to a distinct bundle of MTs (arrow in B) until it appears to attach laterally to the bundle of MTs to which chromosome 2 is attached (F). Chromosome 2 begins moving in D along an arc and reaches chromosome 1 (in J), and finally both move in toward the pellicle (K). Bar = 5 μm.

form of motility occurs with MKLP-1, a kinesin-like protein suspected to generate MT sliding at the interzone of mitotic spindles (Nislow *et al.*, 1992). In general, this assay may detect a unique activity of some motors that are able to crosslink two MTs asymmetrically, generating motility with one end and remaining tightly bound to MTs with the other. We have performed this assay only with bacterially expressed protein, but it should be useful for testing proteins isolated from any source. Below we outline more explicitly how to perform this assay.

Use the same microscope setup (DIC), accessories (VALAP, vacuum grease syringe, air curtain incubator, and humid Petri chamber), and buffer (PMED) described in Section III for motility assay I.

1. Thaw one aliquot of each assay component (pellicles, tubulin, chromosomes, 10 mM GTP, 100 mM ATP stock, your motor of interest).

2. Adsorb pellicles to the coverslip that will form the top of a perfusion chamber; 5 μl pellicles should be added to an 18-mm coverslip that has two streaks of vacuum grease about 4 mm apart. Incubate this coverslip in a humid petri chamber for about 5 minutes.

3. Prepare the perfusion chamber as described in step 4 of assay I (and Fig. 3A). Rinse unbound pellicles out of the chamber with 2–3 vol of cold PMED and mount the slide onto a prewarmed microscope stage (32°C).

4. Perfuse 15 μl of 30 μM tubulin + 1 mM GTP in PMED to grow MTs off the pellicles. Allow MTs to grow for about 15–20 minutes (or otherwise until they reach about 20 μm in length).

5. Dilute the *Chlamydomonas* axonemes 10-fold into PMED; then add 1 μl of these diluted axonemes into 9 μl of 20 μM tubulin and perfuse this solution into the chamber. Allow 3–5 minutes for axonemes to nucleate MT growth.

6. Perfuse 10 μl of the motor enzyme of interest into the chamber (in buffer with 20 μM tubulin). At this stage axonemes should bind to the tethered MTs if the enzyme is able to crosslink MTs; allow 5 minutes for this binding to occur.

7. Perfuse 15 μl of 5 mM MgATP in PMED (15 mM MgATP if a concentrated bacterial cell lysate is being tested) + 10 μM tubulin into the chamber. This concentration of tubulin will cause the depolymerization of pellicle-initiated MTs that are not stabilized by crosslinked MTs (either from axonemal MTs or MTs from other pellicles). These conditions reduce the population of MTs in the assay significantly, so motility is more easily observable.

8. Begin recording the assay (via a video camera onto videotape or laser disk) and slowly scan the chamber for bound axonemes that are moving along the tethered MTs. Bear in mind that movements may be sufficiently slow that they can be detected only on reviewing the sequence in time lapse (Fig. 5).

9. Control experiments should be performed either by not adding the motor fraction to the assay (for native protein preps) or by adding a bacterial lysate from cells *not* expressing the motor enzyme (for recombinant protein preps).

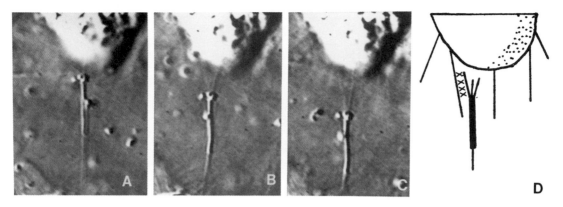

Fig. 5 A sequence of MT–MT sliding. This *Chlamydomonas* axoneme is moving out toward the plus ends of the tethered (pellicle) MTs after we have added the kinesin-like protein MKLP-1 and ATP. The splayed end of the axoneme designates the fast-growing (or plus-end) end, so this is an example of antiparallel MT sliding. The velocity of movement is always constant and this particular axoneme moved 7 μm in 67 seconds (A–C). D is an interpretation of the image in B, with MKLP-1 shown as an X.

V. Other Procedures

A. Fluorescence Observations

We have been able to nucleate rhodamine-labeled MTs from pellicles in an attempt to resolve individual MTs. We use labeled tubulin made with the method of Hyman *et al.* (1991). This is a useful adjunct, as imaging MTs several micrometers into the chamber can be a problem with DIC optics. Confocal microscopy, however, may be necessary to optimally eliminate the out-of-focus fluorescence generated by the radial array of derivatized MTs.

B. Optimal Fixation of Pellicle-Initiated Microtubules

If one wishes to preserve the interactions of chromosomes or vesicles with the MTs we find that adequate fixation of MTs can be achieved with very brief treatment to low concentrations of glutaraldehyde as described by Cross and Williams (1991). Treatments with higher than 0.1% glutaraldehyde for longer than 2 minutes cause the pellicle-initiated MTs to kink and/or break. After fixation, one can process a sample for conventional immunolabeling of MTs, but we find that the final preservation is greatly improved by fixing a sample previously nucleated with rhodamine-labeled MTs. The amount of processing through the perfusion chamber is thereby reduced and MT rearrangements are less likely to occur. Figure 6 shows a pellicle nucleated with rhodamine-labeled MTs.

Fig. 6 Image of fluorescent pellicle-initiated MTs. This pellicle was nucleated with rhodamine-labeled tubulin and then fixed wth 0.1% glutaraldehyde for 1 minute in the perfusion chamber. After the preparation was rinsed with 20 chamber vol of PME, 1 mg/ml *p*-phenylenediamine in PME was added to the chamber to slow photobleaching. This micrograph was taken from a conventional fluorescence microscope. Bar = 5 μm.

V. Additional Remarks

We have described two different assays that use MTs tethered to *Tetrahymena* pellicles. The advantage of this system in general is that these anchored MTs are of known polarity, with their plus ends radiating distally (Heidemann and McIntosh, 1980). Also, this radial array of MTs is sufficiently large (compared with axonemal or centrosomal MT arrays) that the efficiency of capturing added chromosomes or axonemes is greatly increased. Because these assays are performed in real time in a perfusion chamber, conditions can be easily varied at any time throughout the course of the experiment.

In the first assay, we examine the movement of objects bound to the shortening ends of depolymerizing MTs. We do yet understand the precise mechanism by which this movement occurs. To study how the "MT disassembly motor" may power chromosome and vesicle movements one would like to identify the molecules responsible for binding these objects to the ends of shortening MTs as the polymers disassemble. One approach we are taking to understand this form of motility is to simplify the assay even further and identify conditions in which latex microspheres coated with motor proteins and/or other microtubule-associated proteins can move. So far, we have been successful in reconstituting depolymerization-dependent motility of latex spheres coated with a kinesin (manuscript in preparation). Such assays with microspheres may shed light on the mechanism used for this novel form of motility.

The second assay described here allows one to test the ability of a motor enzyme to slide MTs past each other. So far, this assay works with a kinesin-like protein, MKLP-1, that is a likely candidate to be involved in spindle elongation (from its interzonal localization). This assay may therefore be of general use in testing the MT sliding activity of other motors.

In addition to using these types of assays to examine motility powered *exclusively* by either MT depolymerization or mechanoenzymes, one could conceivably study both forms of motility in concert. This unique feature of our system may prove useful in investigations aimed at assessing the relative contributions of MT dynamics and motor enzymes during various forms of MT-mediated motility.

Acknowledgments

We thank Eugeni Vaisberg for antibodies to HeLa cytoplasmic dynein. Many thanks are also due to Corey Nislow and Mike Koonce for carefully reading the manuscript. This work was supported by an NIGMS predoctoral fellowship (RFA-GM-9101) to Vivian Lombillo, a grant from La Foundation de la Recherche Medicale to Martine Coue, and a grant from the NIH (GM-33787) to J. R. McIntosh.

References

Berkowitz, S. A., Katagiri, J., Binder, H. K., and Williams, R. C., Jr. (1977). Separation and characterization of microtubule proteins from calf brain. *Biochemistry* **16,** 5610–5617.

Coue, M., Lombillo, V. A., and McIntosh, J. R. (1991). Microtubule depolymerization promotes particle and chromosome movement *in vitro*. *J. Cell Biol.* **112,** 1165–1175.

Cross, A. R., and Williams, R. C., Jr. (1991). Kinky microtubules: Bending and breaking induced by fixation *in vitro* with glutaraldehyde and formaldehyde. *Cell Motil. Cytoskel.* **20,** 272–278.

Heidemann, S. R., and McIntosh, J. R. (1980). Visualization of the structural polarity of microtubules. *Nature (London)* **286,** 517–519.

Huang, B., Piperno, G., and Luck, D. J. L. (1979). Paralyzed flagella mutants of *Chlamydomonas reinhaardtii*. *J. Biol. Chem.* **254,** 3091–3099.

Hyman, A., Drechsel, D., Kellogg, D., Salser, S., Sawin, K., Steffen, P., Wordeman, L., and Mitchison, T. (1991). Preparation of modified tubulins. *In* "Methods in Enzymology" (R. B. Vallee, ed.), Vol. 196, pp. 478–485. Academic Press, San Diego.

Inoue, S. (1981). Cell division and the mitotic spindle. *J. Cell Biol.* **91,** 131s–147s.

Koshland, D. E., Mitchison, T. J., and Kirschner, M. W. (1988). Polewards chromosome movement driven by microtubule depolymerization *in vitro*. *Nature (London)* **331,** 499–504.

Lewis, C. D., and Laemmli, U. K. (1982). Higher order metaphase chromosome structure: Evidence for metalloprotein interactions. *Cell* **29,** 171–181.

Mitchison, T. J., and Kirschner, M. W. (1985). Properties of the kinetochore *in vitro*. I. Microtubule nucleation and tubulin binding. *J. Cell Biol.* **101,** 755–765.

Nislow, C., Lombillo, V. A., Kuriyama, R., and McIntosh, J. R. (1992). A plus-end-directed motor enzyme that moves antiparallel microtubules *in vitro* localizes to the interzone of mitotic spindles. *Nature (London)* **359,** 543–547.

Pfarr, C. M., Coue, M., Grissom, P. M., Hays, T. S., Porter, M. E., and McIntosh, J. R. (1990). Cytoplasmic dyein is localized to the kinetochore during mitosis. *Nature (London)* **345,** 263–265.

Shpetner, H. S., Paschal, B. M., and Vallee, R. B. (1988). Characterization of the microtubule activated ATPase brain cytoplasmic dynein (MAP 1C). *J. Cell Biol.* **107,** 1001–1009.

Terasaki, M., Song, J., Wong, J. R., Weiss, M. J., and Chen, L. B. (1984). Localization of endoplasmic reticulum in living and glutaraldehyde-fixed cells with fluorescent dyes. *Cell* **38,** 101–108.

Weisenberg, R. C. (1972). Microtubule formation *in vitro* in solutions containing low calcium concentrations. *Science* **177,** 1104–1105.

Williams, R. C., Jr., and Detrich, H. W., III. (1979). Separation of tubulin from microtubule-associated proteins on phosphocellulose. Accompanying alterations in concentrations of buffer components. *Biochemistry* **18,** 2499–2503.

CHAPTER 12

Use of ATP Analogs in Motor Assays

Takashi Shimizu,★ Yoko Y. Toyoshima,†,[1] and Ronald D. Vale‡

★ Research Institute of Bioscience and Human-Technology
Higashi, Tsukuba, Ibaraki 305, Japan

† Department of Biology
Ochanomizu University
Ohtsuka, Bunkyo-ku, Tokyo 112, Japan

‡ Department of Pharmacology
University of California, San Francisco
San Francisco, California 94143

I. Introduction
II. Purity Check and Purification of ATP Analogs
III. Chemistry of ATP Analogs
    A. Adenine Analogs
    B. Modification of Ribose Moiety
    C. Triphosphate Modification
IV. Additional Precautions in Using ATP Analogs
V. Uses of ATP Analogs
    A. Probes of Conformational Changes
    B. Specificity of *in Vitro* Translocation by Molecular Motors
References

I. Introduction

Besides the fact that enzymes exert their functions under conditions in which most artificial catalysts do not work (mild temperatures, neutral aqueous media, etc.), one important characteristic of enzymes as catalysts is that they exert

---

[1] Present address: Department of Pure and Applied Sciences, College of Arts and Sciences, University of Tokyo, Meguro-ku, Tokyo 113, Japan.

their catalytic activities toward a narrow range of substances (substrates). Each substrate out of such a restricted spectrum is catalyzed by the enzyme at a specific rate under a certain condition. Quantitation of such rates is called the substrate specificity.

A molecular motor is an enzyme (ATPase) which, of course, has its own substrate specificity. The natural substrate for motors known so far is ATP, but motors can use some ATP analogs as well to some extent. The motile activity of a molecular motor can be quantitated by assaying *in vitro* translocation of filaments (microtubules for dynein and kinesin and F-actin for myosin); for example, translocation speed at a certain concentration of ATP analog (e.g., 1 mM) is determined by this assay for the comparison to that by ATP itself. Thus, we could define the substrate specificity of the motile activity of a molecular motor.

The primary idea in the use of ATP analogs for motor assays is use of a battery of ATP analogs to determine the substrate specificity of an unknown motor in a cellular process and then to identify it by comparing it with the substrate specificities of representative motors. Fortunately, the ATP analogs of the three motor families—skeletal heavy meromyosin, brain kinesin, and ciliary dynein—are quite different from one another in terms of substrate specificity (Shimizu *et al.*, 1991); this strategy appears to have significance and has already been applied to the assay of the organelle-transporting motors of *Reticulomyxa*, a giant amoeba; the results strongly suggest that the motors are dynein-like (Schliwa *et al.*, 1991). Another application has been made to kinetochore motors (Hyman and Mitchison, 1991).

The ATP analogs can be examined with respect to their ability to support *in vitro* motility for other purposes as well. An analog that supports motility is generally a good substrate for the enzymatic activity of the molecular motor. For example, if a photoactivatable ATP analog like 8-azido-ATP supports the *in vitro* motility of a certain motor, it can be used to locate the active site on its polypeptides (Pfister *et al.*, 1985). Similarly, if a chromophoric or fluorescent ATP analog supports motility, it may be used for a kinetic study of the motor, with its spectroscopic property being a conformational change marker (Eccleston and Trentham, 1977; Rosenfeld and Taylor, 1984).

In this chapter, we describe the preparation, precautions, and uses of ATP analogs with respect to the study of molecular motors.

II. Purity Check and Purification of ATP Analogs

The most important precaution to bear in mind before using ATP analogs is to eliminate contamination by ATP. Because the molecular motors known so far, including myosin, have much higher affinity to ATP than other nucleoside triphosphates (NTPs) (Hackney and Clark, 1985; Hackney, 1988; Shimizu *et al.*,

1991), a little ATP contamination could be disastrous; for example, 1% ATP in 1 mM NTP is 10 μM ATP, which is more than sufficient to support motility by any motor. It is therefore very important to purify ATP analogs and to check their purity. It should be noted that commercial ADP usually contains a few percent ATP as a contaminant.

A nucleotide can be purified by DEAE-Sephadex A-25 column chromatography with a 0.1 to 0.6 M triethylammonium–bicarbonate buffer (pH 7.5) gradient; ATP and ADP elute at about 0.4 and 0.2 M buffer, respectively. Peak fractions of the nucleotide to be purified (detected by UV absorption) are combined, and water and the volatile buffer are removed by rotary evaporation. The resultant triethylammonium salt of a nucleotide is a gum or a syrup, and must be washed with methanol (dispersion of syrup in methanol and evaporation) several times to remove residual triethylamine. The nucleotide finally dispersed in water is a triethylammonium salt. To convert it to sodium or another salt, it must be passed through a small Dowex 50W (H$^+$ form) column, and the flow-through taken and pH-adjusted with the corresponding cation. A MonoQ or Dowex 1 column can also be used for purification of certain nucleotides.

Purity can be checked by thin-layer chromatography (TLC) on a polyethyleneimine (PEI)–cellulose plate (available from Macherey-Nagel or Baker) with 0.75 M phosphate buffer (pH 3.4) as a developing medium. Such a plate can be cut into a size appropriate to the purpose with scissors. This TLC method is very convenient and is completed within an hour, but quantitation and detection of less than 1% contamination will be quite difficult. More commonly, high-performance liquid chromatography (HPLC) with a C$_{18}$ reverse-phase column (4.6 × 250 mm, 5-μm particles; available from many companies) is used for the purity check. A 0.1 M potassium phosphate buffer (pH 7) with 25 mM tetrabutylammonium sulfate and a mixture of this buffer solution (70%) and methanol (30%) are prepared. A linear gradient from 0 to 30% methanol is suitable to elute nucleotides to examine their purity. If HPLC is used without a gradient maker, isocratic solution of the buffer with 12 to 18% methanol will be adequate.

Use of luciferin–firefly luciferase assay for examination of ATP contamination may not be suitable except for nonhydrolyzable analogs such as AMPPNP. Each NTP can be a substrate of luciferase to a certain extent, even if luciferase exhibits extremely high specificity for ATP itself. Unless the NTP is pure, it is impossible to estimate the extent of ATP contamination in NTP, because the magnitude of light emission with an NTP alone is not known.

III. Chemistry of ATP Analogs

ATP consists of three moieties: adenine, ribose, and triphosphate (Fig. 1). Modification of each moiety results in different ATP analogs. Some analogs are available from commercial sources, and Table I lists those company names and some spectral data.

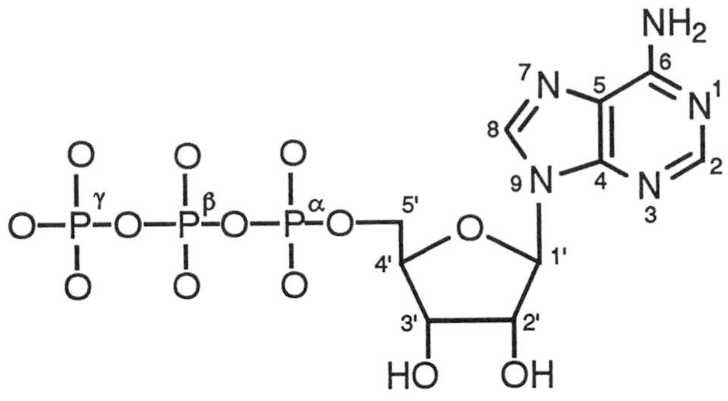

Fig. 1 Structure of ATP. Because of the tetrahedral nature of phosphorus atoms in the triphosphate moiety, negative charges and double bonds are omitted.

A. Adenine Analogs

Guanosine, cytidine, uridine, inosine, thymidine, and their analogs may not be regarded as adenosine analogs, but the triphosphate forms of those are also listed in Table I because of their commercial availability. Except for ITP and CTP, those NTPs seem to be practically free from ATP contamination. ITP usually contains less than 1% ATP and CTP may have a trace amount. The ATP contamination can be removed by Dowex 1 column chromatography (Shimizu, 1987). 8-Azido-ATP and 2-azido-ATP can be used to locate the adenine-binding sites of motors, although 2-azido-ATP is in equilibrium with other forms so that only about 45% is photoactive (Omoto and Nakamaye, 1989). Etheno-ATP and FTP are fluorescent (Rosenfeld and Taylor, 1984; Jackson and Bagshaw, 1988). 6-Thio-ITP is an ATP analog with a characteristic near-UV spectrum that changes on its binding to active sites of enzymes (Eccleston and Trentham, 1977). Chemical structures of modified adenines are shown in Fig. 2. Most adenine-modified ATP analogs commercially supplied are likely to have a little ATP contamination. One method that may greatly reduce ATP contamination in an ATP analog involves use of the high specificity of *Tetrahymena* ciliary 22 S dynein for ATP: an analog is subjected to hydrolysis by this enzyme until about half of the triphosphate is converted to the diphosphate form (when most of the contaminating ATP is already hydrolyzed), and then ethylenediaminetetraacetic acid (EDTA) is added in excess of Mg^{2+} to stop the dynein reaction, which is followed by chromatography as described above to obtain the purified ATP analog. By this method, about 1% ATP in 8-azido-ATP was reduced to less than 0.1% (Shimizu *et al.*, 1991).

B. Modification of Ribose Moiety

Naturally occurring deoxy derivatives of ATP can be included in this category of ATP analogs (3'-deoxy-ATP is also called cordycepin triphosphate). These

Table I
ATP Analogs

	λ_{max} (nm)	ε	Source
ATP	259 nm	15,400	Many companies
GTP	253	13,700	Many companies
ITP	249	12,200	Many companies
CTP	271	9,000	Many companies
UTP	262	10,000	Many companies
dTTP	267	9,600	Many companies
6-Thio-ITP	322	22,500 (pH 4.5)	PL Biochemicals
8-Bromo-ATP	263	15,000	Sigma
8-Azido-ATP	280	13,300	Sigma
FTP	295	9,500	Sigma[a]
Etheno-ATP	275	6,000	Sigma, Molecular Probes
Dialdehyde ATP	(Same as ATP)		Sigma
Dialcohol ATP	(Same as ATP)		Sigma
3′-dATP	(Same as ATP)		Sigma
Dideoxy ATP	(Same as ATP)		Pharmacia
AMPPNP	(Same as ATP)		Sigma, Boehringer-Mannheim
AMPPCP	(Same at ATP)		Sigma, Boehringer-Mannheim
ATPγS	(Same as ATP)		Sigma, Boehringer-Mannheim
(Sp)ATPαS	(Same as ATP)		Boehringer-Mannheim
Purine riboside triphosphate	262	5,900 (pH 1)	Not commercially available
N^6-Methyl-ATP	266	15,900	Not commercially available
N^6,N^6-Dimethyl-ATP	275	18,800	Not commercially available

[a] Quoted as inquiry.

deoxy derivatives, which are commercially available, are practically free from ATP contamination.

Another type of ribose-modified ATP analog is obtained by periodate treatment. This treatment breaks 2′C–3′C bond, forming ATP 2′,3′-dialdehyde (Fig. 3). This dialdehyde can be reduced with sodium borohydride to ATP 2′,3′-dialcohol. The dialdehyde analog of ATP is sometimes used to locate the ATP (ribose) binding domain of a protein, as aldehydes and carboxyl groups of proteins can form Schiff base covalent bonds. So far, this method has not been used for molecular motors. Both types of ATP analogs are commercially available, but it is rather easy to prepare them from ATP (King and Coleman, 1983). As usual, care must be taken to ensure purity and avoid breakdown.

Introduction of groups to the ribose moiety by ester linkage through 3′(2′)-O is readily possible and effective in making functional analogs of ATP. Some examples include benzoylbenzoyl-ATP (Bz$_2$ATP) (Mahmood *et al.*, 1987) and

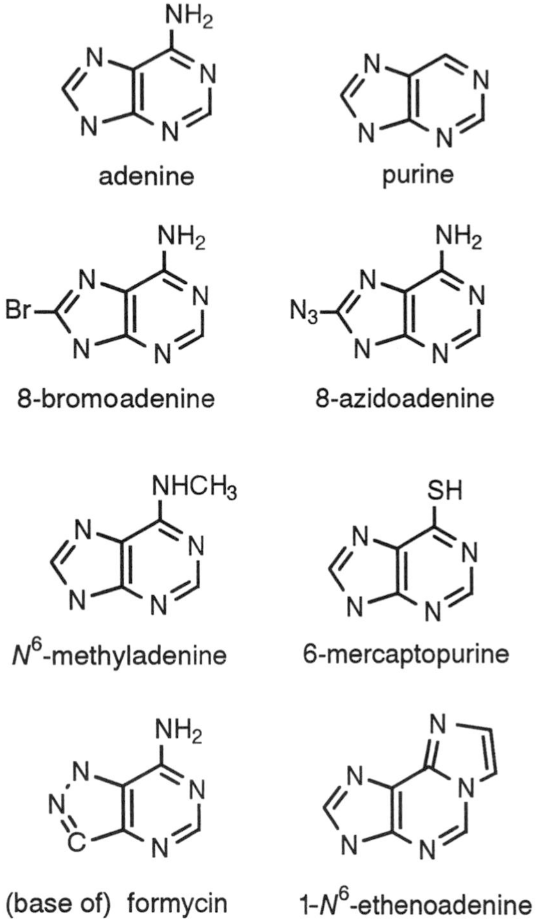

Fig. 2 Structures of modified adenines.

arylazido-ATP (Guillory and Jeng, 1977), which are photoactivatable analogs; bis-ATP (Munson *et al.*, 1986), which has two ATP molecules connected at two ribose moieties; and anthraniloyl-ATP (Ant-ATP) and methylanthraniloyl-ATP (Mant-ATP) (Hiratsuka, 1983; Inaba *et al.*, 1989), which are highly fluorescent. One drawback of this type of analog is that the ester linkages are susceptible to hydrolysis by both acid and, especially, alkali, and ATP is formed as a breakdown product. Though they are stable at a slightly acidic pH (e.g., 5–7), the usual 1% ATP contamination, even after purification, is difficult to eliminate so that one has to be careful in doing experiments and in interpreting results.

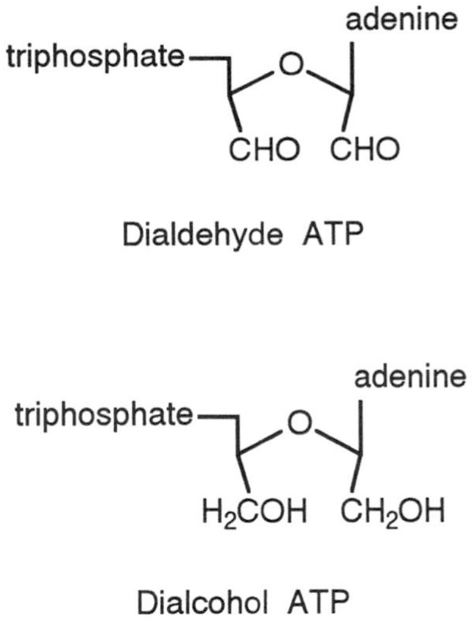

Dialdehyde ATP

Dialcohol ATP

Fig. 3 Structures of dialdehyde ATP and dialcohol ATP.

C. Triphosphate Modification

The triphosphate moiety has two types of oxygen atoms: bridging and nonbridging (see Fig. 1). Replacing the bridging oxygen between β- and γ-phosphorus with NH or CH_2 makes AMPPNP or AMPPCP, respectively. These are nonhydrolyzable and they do not support ATP-related enzyme activities, including motility. Commercial products may contain 1 to 2% ATP as a contaminant, which can be removed by DEAE-Sephadex chromatography. Addition of hexokinase and glucose for the elimination of ATP contamination in AMPPNP or AMPPCP may not be effective, because hexokinase is inhibited by either analog. The Firefly luciferase assay for ATP contamination is particularly effective for these nonhydrolyzable analogs.

Replacing one nonbridging oxygen with a sulfur atom leads to phosphorothioate analogs (Eckstein and Goody, 1976; Eckstein, 1983). Depending on the phosphorus atom to which the sulfur atom attaches, the analog is called ATPαS, ATPβS, or ATPγS. As phosphorous in phosphate is tetrahedral, two diastereomers, Sp and Rp, exist for ATPαS and for ATPβS. γ-Phosphorus of ATPγS has two identical oxygen atoms so that no stereoisomers of ATPγS exist. ATPγS and (Sp)ATPαA are commercially available, but again, contamination of ATP must be checked. To purify and check the purity of ATPγS, MonoQ with 0.3 M ammonium bicarbonate may be better than DEAE-Sephadex A-25. ATPγS is stable at a neutral or slightly alkaline pH but unstable at an acidic pH.

IV. Additional Precautions in Using ATP Analogs

As tubulin and actin are guanine and adenine nucleotide-binding proteins, respectively, microtubule and F-actin suspensions have some free nucleotides. It is therefore very important to remove those nucleotides from the suspension before use in *in vitro* motility assay with ATP analogs. Usually, microtubules or F-actin stabilized by taxol or phalloidin can be pelleted by centrifugation and resuspended in a nucleotide-free solution. This cycle can be repeated to ensure elimination of inherent free nucleotide. To assay the turnover (hydrolysis) of molecular motors to analogs, this method for removing inherent ATP or GTP is reasonable. Another method for removing GTP and ATP in motility assays is perfusion of microtubules or actin into a flow chamber to form a rigor complex with motors. Fresh buffer solution containing the analog can then be applied to wash the GTP or ATP out of the flow chamber.

Another point is to avoid contamination of ATP in other nucleotide solutions. Our own experience indicates contamination can occur in an unexpected manner, and ATP (and deoxy-ATPs) must be handled separately from other NTPs, preferably on other benches.

ATP is more stable than one might think; breakdown to ADP and phosphate is negligible even after 24 hours at 25°C and pH 7. Most other NTPs are also stable and can be used for motility assays just as ATP. Some analogs are not as stable as ATP at acidic or alkaline pH as described above. At any rate, it is still advisable to avoid exposing your ATP or analogs to nonneutral pH, high temperature, and frequent freeze–thawing. Photosensitive analogs such as 8-azido-ATP have to be protected from UV light including near-UV light.

V. Uses of ATP Analogs

A. Probes of Conformational Changes

Fluorescent or chromophoric ATP analogs can be used for structural or kinetic studies of molecular motors. Though conformational changes of proteins can be traced by many physicochemical means, fluorescence or absorption spectroscopy has an advantage over other means in terms of ease in measurements. More importantly, changes can be traced very quickly, which is extremely useful for rapid kinetic studies. Protein fluorescence as a conformation marker is often used for the same purpose. With some enzyme proteins like 22S dynein, however, protein fluorescence does not change on mixing with ATP, and in such a case those ATP analogs may be useful. Furthermore, even if a motor enzyme changes its own fluorescence on conformational changes (like myosin does), those analogs may provide additional information on the motor protein conformation.

One important point for such analogs is that the absorption, excitation or emission maximum should be distant from the protein absorption peak, that is, 280 nm. Etheno-ATP has an absorption maximum at 275 nm, but is still useful because it can be excited by light of 310–330 nm. These analogs change their spectral properties not only on binding to the enzyme active site but also on conformational change of the active site and its vicinity to the protein. With fluorescent analogs, the intensity change in fluorescence is most obvious. This change is due mainly to a change in polarity around the chromophore.

Another point to be mentioned is that for such an investigation to be possible, a considerable fraction of added nucleotide must bind to the enzyme protein. For example, if a fluorescent analog is used, free nucleotide molecules in solution will emit some fluorescence, which will reduce the magnitude of fluorescence intensity change from the nucleotide molecules bound to the protein. If only a small fraction of the analog added is expected to bind to the enzyme protein, such a spectroscopic change could be detected by a stopped-flow method.

At any rate, a variety of ATP analogs have already been used for the study of molecular motors, especially skeletal myosin, and some are cited herein again for reference: skeletal myosin (or its subfragments), 6-thio-ATP (Eccleston and Trentham, 1977), etheno-ATP (Rosenfeld and Taylor, 1984), FTP (Jackson and Bagshaw, 1988), and Ant-ATP and Mant-ATP (Hiratsuka, 1983). Ant-ATP and Mant-ATP are also used for studies of dynein (Inaba *et al.*, 1989).

B. Specificity of *in Vitro* Translocation by Molecular Motors

As reported elsewhere (Shimizu *et al.*, 1991), ciliary dyneins exhibit high specificity, whereas brain kinesin and skeletal heavy meromyosin have rather broad specificity, when *in vitro* filament translocation is assayed. Brain kinesin and heavy meromyosin are also different from each other in terms of specificity; at 1 mM nucleotide, the variation in speed of translocation is large with kinesin, whereas it is rather small with heavy meromyosin.

In more detail (Table II), heavy meromyosin can be distinguished by the fact that its motility is supported by (Rp)ATPαS but not by 8-bromo-ATP, 8-azido-ATP, or ATPγS. Kinesin motility is supported by most analogs, including GTP, ITP, UTP, and CTP, with a large variation of speed. Though dyneins are not supported by most analogs, it is conspicuous that 8-azido-ATP supports some dyneins including 14 S dynein from *Tetrahymena* cilia. Thus, these three representative motors can be differentiated by substrate specificity, which can be regarded as a sort of "nucleotide fingerprint."

The degree of specificity for a nucleotide fingerprint remains to be determined. For example, how similar or different are motors belonging to a certain family or superfamily (Vale and Goldstein, 1990) with respect to substrate specificity? As for actin filament motors, the specificity of myosin I (more specifically, intestinal brush border myosin I) is quite similar to that of skeletal

Table II
Substrate Specificities of Molecular Motors with Some Nucleoside
Triphosphates at 1 mM^a

	Filament Translocation Speed (μm/s)			
	Heavy Meromyosin	Kinesin	22 S Dynein	14 S Dynein
ATP	5.2	0.42	4.52	4.3
Dideoxy-ATP	5.4	0.29	0.94	2.9
Monomethyl ATP	3.8	0.20	0.28	0.73
Dimethyl-ATP	3.0	0.086	—[b]	—
8-Bromo-ATP	—	0.013	—	—
8-Azido-ATP	—	—	—	1.2
Etheno-ATP	1.2	0.054	—	—
FTP	1.3	0.037	—	0.084
(Sp)ATPαS	10.4	0.077	0.83	1.2
(Rp)ATPαS	0.52	—	—	—
ATPγS	—	0.0091	0.21	0.52

[a] The translocation speed was assayed at 25°C for heavy meromyosin and dyneins and at room temperature (22–24°C) for kinesin as described (Shimizu *et al.*, 1991).

[b] Dash indicates that no motility was observed.

heavy meromyosin (T. Shimizu, Y. Y. Toyoshima, and K. Collins, unpublished observation). Cytoplasmic dynein may be regarded as a member of the dynein superfamily, and we find that brain dynein (MAP1C) exhibits almost the same specificity as ciliary dynein (T. Shimizu and Y.Y. Toyoshima, unpublished observation). On the other hand, the ncd protein, a member of the kinesin superfamily, moves in a direction opposite that of kinesin (Walker *et al.*, 1990; McDonald *et al.*, 1990), and its substrate specificity is different from that of kinesin but similar to that of dynein (T. Shimizu, R. D. Vale, and J. M. Scholey, unpublished observation). Therefore, two interesting questions arise: Does each member of the kinesin superfamily have a characteristic substrate specificity? Is the specificity correlated with the direction of movement?

Acknowledgments

This work is supported by a grant-in-aid from the Agency of Industrial Science and Technology. We thank Dr. S. Ohashi and Dr. K. Furusawa for their advice and discussion and Ms. Iseko Akui for secretarial assistance.

References

Eccleston, J. F., and Trentham, D. R. (1977). The interaction of chromophoric nucleotides with subfragment 1 of myosin. *Biochem. J.* **163**, 15–29.

Eckstein, F. (1983). Phosphorothioate analogues of nucleotides—Tools for the investigation of biochemical processes. *Angew. Chem., Int. Ed. Engl.* **22**, 423–439.

Eckstein, F., and Goody, R. S. (1976). Synthesis and properties of diastereoisomers of adenosine 5'-(O-1-thiotriphosphate) and adenosine 5'-(O-2-thiotriphosphate). *Biochemistry* **15**, 1685–1691.

Guillory, R. J., and Jeng, S. J. (1977). Arylazido nucleotide analogs in a photoaffinity approach to receptor site labeling. *In* "Methods in Enzymology" (W. Jakoby and M. Wilchek, eds.), Vol. 46, pp. 259–288. Academic Press, New York.

Hackney, D. D. (1988). Kinesin ATPase: Rate-limiting ADP release. *Proc. Natl. Acad. Sci. U.S.A.* **85**, 6314–6318.

Hackney, D. D., and Clark, P. K. (1985). Steady state kinetics at high enzyme concentration: The myosin MgATPase. *J. Biol. Chem.* **260**, 5505–5510.

Hiratsuka, T. (1983). New ribose-modified fluorescent analogs of adenine and guanine nucleotides available as substrates for various enzymes. *Biochim. Biophys. Acta* **742**, 496–508.

Hyman, A. A., and Mitchison, T. J. (1991). Two different microtubule-based motor activities with opposite polarities in kinetochore. *Nature (London)* **351**, 206–211.

Inaba, K., Okuno, M., and Mohri, H. (1989). Anthranyloyl ATP, a fluorescent analog of ATP, as a substrate for dynein ATPase and flagellar motility. *Arch. Biochem. Biophys.* **274**, 209–215.

Jackson, A. P., and Bagshaw, C. R. (1988). Kinetic trapping of intermediates of the scallop heavy meromyosin adenosine triphosphatase reaction revealed by formycin nucleotides. *Biochem. J.* **251**, 527–540.

King, M. M., and Coleman, R. F. (1983). Affinity labeling of nicotinamide adenine dinucleotide dependent isocitrate dehydrogenase by the 2', 3'-dialdehyde derivative of adenosine 5'-diphosphate. Evidence for the formation of an unusual reaction product. *Biochemistry* **22**, 1656–1665.

Mahmood, R., Cremo, C., Nakamaye, K. L., and Yount, R. G. (1987). The interaction and photolabeling of myosin subfragment 1 with 3'(2')-O-(4-benzoyl)benzoyl-adenosine 5'-triphosphate. *J. Biol. Chem.* **262**, 14479–14486.

McDonald, H. B., Stewart, R. J., and Goldstein, L. S. B. (1990). The kinesin-like ncd protein of *Drosophila* is a minus end-directed microtubule motor. *Cell (Cambridge, Mass.)* **63**, 1159–1165.

Munson, K. B., Smerdon, M. J., and Yount, R. G. (1986). Cross-linking of myosin subfragment 1 and heavy meromyosin by use of vanadate and a bis(adenosine 5'-triphosphate) analogue. *Biochemistry* **25**, 7640–7650.

Omoto, C. K., and Nakamaye, K. (1989). ATP analogs substituted on the 2-position as substrates for dynein ATPase activity. *Biochim. Biophys. Acta.* **999**, 221–224.

Pfister, K. K., Haley, B. E., and Witman, G. B. (1985). Labeling of *Chlamydomonas* 18 S dynein polypeptides by 8-azidoadenosine 5'-triphosphate, a photoaffinity analog of ATP. *J. Biol. Chem.* **260**, 12844–12850.

Rosenfeld, S. S., and Taylor, E. W. (1984). Reactions of 1-N^6-ethenoadenosine nucleotides with myosin subfragment 1 and acto-subfragment 1 of skeletal and smooth muscle. *J. Biol. Chem.* **259**, 11920–11929.

Schliwa, M., Shimizu, T., Vale, R. D., and Euteneuer, U. (1991). Nucleotide specificities of anterograde and retrograde organelle transport in *Reticulomyxa* are indistinguishable. *J. Cell Biol.* **112**, 1199–1203.

Shimizu, T. (1987). The substrate specificity of dynein from *Tetrahymena* cilia. *J. Biochem. (Tokyo)* **102**, 1159–1165.

Shimizu, T., Furusawa, K., Ohashi, S., Toyoshima, Y.Y., Okuno, M., Malik, F., and Vale, R. D. (1991). Nucleotide specificity of the enzymatic and motile activities of dynein, kinesin, and heavy meromyosin. *J. Cell Biol.* **112**, 1189–1197.

Vale, R. D., and Goldstein, L. S. B. (1990). One motor, many tails: An expanding repertoire of force-generating enzymes. *Cell (Cambridge, Mass.)* **60**, 883–885.

Walker, R. A., Salmon, E. D., and Endow, S. A. (1990). The *Drosophila claret* segregation protein is a minus-end directed motor molecule. *Nature (London)* **347**, 780–782.

CHAPTER 13

Myosin–Mediated Vesicular Transport in the Extruded Cytoplasm of Characean Algae Cells

Bechara Kachar,† Raul Urrutia,†,* Marcelo N. Rivolta,† and Mark A. McNiven*

† Laboratory of Cellular Biology
National Institute for Deafness and
 Other Communication Disorders
National Institutes of Health
Bethesda, Maryland 20892

*Center for Basic Research in Digestive Diseases
Mayo Clinic
Rochester, Minnesota 55905

I. Introduction
II. Extruded Algal Cytoplasm Preparations
III. Methods for the Direct Visualization of Myosin-Mediated Organelle Movements
IV. Observation of Myosin-Mediated Vesicular Movement *in Vitro*
V. Identification of a Soluble Pool of Myosins in *Nitella* Cytoplasm
VI. Concluding Remarks
 References

I. Introduction

Myosin is a ubiquitous enzyme that supports important motile activities in the eukaryotic cell such as chemotaxis, phagocytosis, pseudopod and lamellopod extension, and cytokinesis (Korn and Hammer, 1990; Pollard *et al.*, 1991; Warrick and Spudich, 1987). Recently, one class of monomeric myosins, or ''minimyosins'' (Adams and Pollard, 1986, 1989; Miyata *et al.*, 1989; Garcia *et al.*, 1989), has been shown to bind to membrane lipids, implicating them in the movement of vesicles along actin filaments. There is, however, no direct evidence that this myosin is involved in the translocation of organelles *in vivo*.

The giant cells of the algae *Chara* and *Nitella* have been used as classical models to study cytoplasmic streaming (Kamiya, 1981). These cells exhibit vigorous cytoplasmic streaming produced by the movement of the endoplasmic reticulum along actin bundles present in the cortical cytoplasm (Kachar and Reese, 1988). Cytoplasmic organelles can also be visualized moving continuously in these algal cells at a velocity of approximately 60 μm s^{-1}. These algal cells are particularly useful systems for reconstituting the actin-based translocation of organelles *in vitro* because the cytoplasm can be extruded easily and the movement visualized directly using video-enhanced interference microscopy. We have developed a stable and reproducible, *in vitro*, experimental system that facilitates the study of myosin and actin-based organelle movement using video and electron microscopy of extruded cytoplasm of the giant cells of the characean algae (Kachar and Reese, 1988; Kachar, 1985b). With this assay we have determined that the characean algal cytoplasm displays at least two myosin-mediated organelle movements. One population of vesicles moves along actin bundles at an average speed of 11 μm s^{-1}, while the other vesicle class is translocated at velocities up to 60 μm s^{-1}. The transport of these organelles is exceptionally rapid and unprecedented in other systems. On the basis of three-dimensional reconstruction of the cytoplasm obtained by fast-freeze electron microscopy, we have predicted that the myosin-dependent movements of the endoplasmic reticulum in these cells are responsible for the streaming of the whole cytoplasm (Kachar and Reese, 1988). These motile activities can be successfully recovered in a soluble form on fractionation of the *Nitella* cytoplasm and provide a powerful tool to study the role of myosin in the translocation of membrane-bound organelles.

II. Extruded Algal Cytoplasm Preparations

Chara corallina and *Nitella flexilis* are two of the most commonly used characean algae and are easy to maintain in the laboratory. They can be collected from local ponds or obtained from suppliers such as Carolina Biological Company. In these species the individual internodal cells grow as cylinders with lengths that can reach up to 10–20 cm or more and diameters up to 1 mm. These giant internodal cells are isolated from neighboring cells, then rinsed and incubated for 20 minutes in a buffer containing 4 mM ethylene glycol bis(β-aminoethyl ether) N, N'-tetraacetic acid (EGTA), 25 mM KCl, 4 mM MgCl$_2$, and 5 mM imidazole at pH 7.0. An internodal cell is then blotted on tissue paper and transected at one end to drain the contents of the central vacuole. The cell is lifted at one end with a pair of forceps and, with the aid of forceps that have tips covered by Teflon tubing (Fig. 1), a cylinder of cytoplasm is gently extruded into a drop of isotonic buffer (4 mM EGTA, 25 mM KCl, 4 mM MgCl$_2$, 5 mM Tris–HCl, and 200 mM sucrose at pH 7.6) on a thin microscope slide. The cylinder of cytoplasm without the cell wall is a protoplast. This protoplast has an

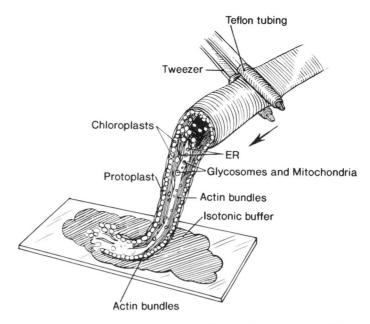

Fig. 1 Diagram showing physical manipulation of *Chara* cells to obtain a cell-free cytoplasmic preparation to study myosin-based vesicle movement.

intact plasma membrane that prevents dispersion of the cytoplasm contents. Subsequently, the buffer is replaced with fresh isotonic buffer solution containing 2 mM adenosine triphosphate (ATP) by slow and continuous addition while removing solution from the edges of a drop. Fine needles are used to break and disorganize the cytoplasm until its components are dispersed into a small drop of buffer, which is covered with a coverglass and sealed with Vaseline. Samples are then analyzed by video-enhanced interference contrast microscopy. This preparation provides all that is needed for reconstituting the movement of organelles along the native actin bundles (Fig. 2).

III. Methods for the Direct Visualization of Myosin-Mediated Organelle Movements

Organelles moving along actin bundles can be visualized using differential interference contrast (DIC) optics via conventional Koehler illumination, fiberoptic illumination, or critical illumination. For critical illumination (Kachar and Reese, 1988), a Zeiss Axiomat microscope is used in the DIC mode equipped with an internally corrected 100X, 1.3 numerical aperture (NA) plana-

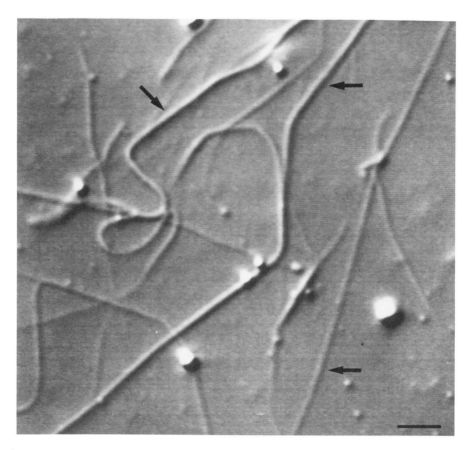

Fig. 2 Video micrograph of a dissociated *Chara* cytoplasmic preparation. In this preparation, the actin filaments remain packed as bundles while attaching to the glass surface (arrows). Cytoplasmic membranous organelles exhibit Brownian motion while floating in free solution or are fixed in place, attached either to bundles or to the glass surface. Bar = 1 μm.

pochromatic objective. The aperture of a 1.4NA condenser is fully illuminated with a 100-W mercury lamp. The light arc is focused in the specimen plane, achieving a fully illuminated aperture. The optical image is projected out of the camera port of the microscope onto a Newicon Dage-MTI 70 video camera (Dage-MTI, Michigan City, IN). In this manner, high-resolution images of organelles translocating along native actin bundles can be obtained without the need for digital image processing. Asymmetric illumination techniques become preferable to DIC when the field of view contains highly birefringent objects such as the chloroplasts (Kachar, 1985a). Video-enhanced interference analysis of algal organelle movements can be complemented with fast-freeze electron microscopy studies (Kachar and Reese, 1988).

IV. Observation of Myosin-Mediated Vesicular Movement in Vitro

The extruded cytoplasm contains areas where actin bundles and vesicular organelles are the only components present. Typically, the actin cables visualized are approximately 100 μm long and 0.2 and 0.3 μm in diameter and appear either free or attached to the glass surface while displaying sharp bends along their length (Figs. 2 and 3). In the absence of ATP, a large percentage of

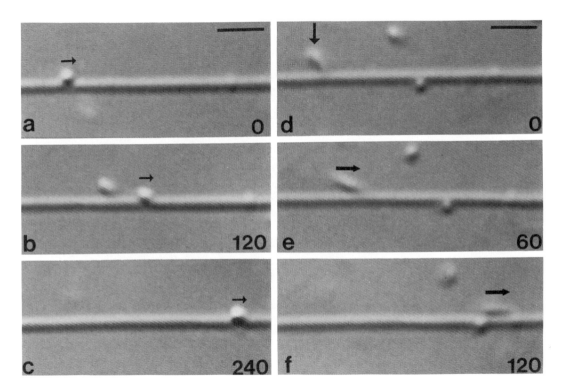

Fig. 3 Translocation of membranous organelles along actin bundles in extruded algal cytoplasm at different velocities. (a–c) Video micrographs showing a vesicular organelle moving at an average velocity of 11 μm s^{-1}. (d–f) Micrograph of a small elongated membranous organelle moving along an actin bundle at 60 μm s^{-1}. The latter type of organelles seem to have originated from the fragmentation of the endoplasmic reticulum during dissociation of the *Chara* cytoplasm. Bars = 1 μm. Time is in milliseconds.

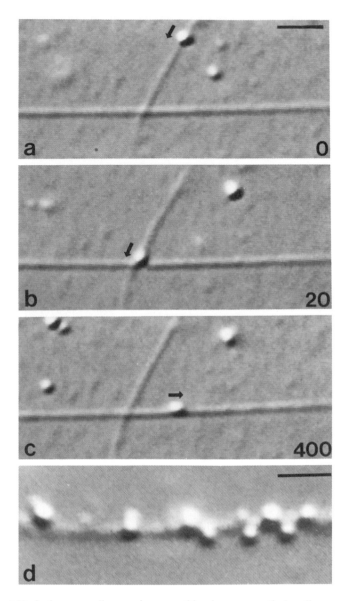

Fig. 4 (a–c) Vesicular organelles moving smoothly along one actin bundle can often be seen attaching to another bundle and moving away from the original track. (d) Immotile organelles are often visualized in a "rigor state" attached to an actin cable in the absence of ATP. Bars = 1 μm. Time is in milliseconds.

organelles are visualized bound to the actin cables (Fig. 4); however, when ATP is added almost all of the organelles present in the field move smoothly along the surface of the cables (Figs. 3 and 4). Usually it is difficult to determine from these images the precise organelle class that is moving, although most of the vesicles present in the field have the morphological features characteristic of glycosomes and translocate along the actin cables with an average speed of 11.2–0.2 μm s^{-1}. Electron microscopy shows the close associations of these vesicular organelles with the actin cables (Fig. 5). Unfortunately, in direct-

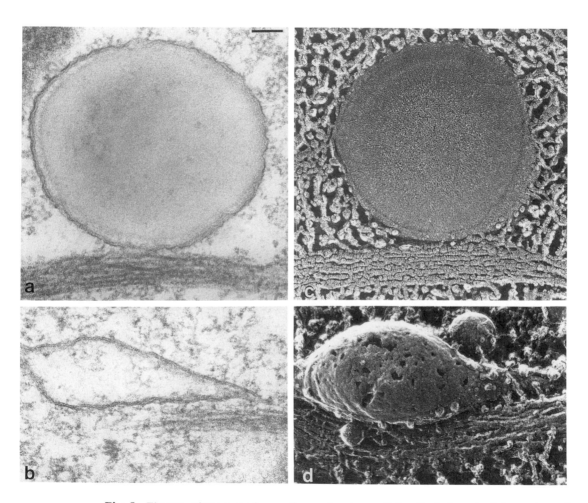

Fig. 5 Electron microscopy of organelles moving along actin bundles in the extruded cytoplasm of the *Nitella* cell. Nonfilamentous structures can be detected at the interface between the organelle and the actin bundle by either thin-section (a,b) or fast-freeze electron microscopy (c,d). The surface of the organelle appeared extensively deformed by the shearing force generated during the translocation (b,d). Bar = 0.1 μm.

freeze preparations (Figs. 5c and 5d), the granular material that forms the background of the freeze-etched cytoplasm does not permit detailed definition of the myosin molecules that mediate the interaction.

A second population of organelles which have a low contrast appearance and a tubular shape of variable size can also be seen moving along the same actin tracks at a rate of 62.3–1.1 μm s^{-1} (Figs. 3d–f). Freeze-fracture (Fig. 5c) and thin-section (Fig. 5b) electron microscopy after direct freezing of extruded cytoplasm show these elongated tubulovesicular organelles intimately associated with the actin bundles. The site of interaction of these tubular organelles shows extensive deformation produced by the sheer force generated during translocation (Figs. 5b and 5d). In the extruded cytoplasm, these tubular organelles often appear as large anastomosing networks that form multiple interactions with several adjacent actin cables. Each interacting site generates pulling forces along the surface of the actin cables oriented in different directions, which generates continuous deformations of the network (Fig. 6). Electron microscopy identifies these networks of tubules and cisternae as components of the endoplasmic reticulum by the presence of polyribosomes on their surface. On the basis of complementary video microscopy and fast-freeze electron microscopy of the intact algae, we have proposed a model in which the endoplasmic reticulum slides along the parallel stationary cortical actin cables. This sliding is mediated by myosin and moves other enmeshed organelles to support the streaming of the entire cytoplasm (Kachar and Reese, 1988).

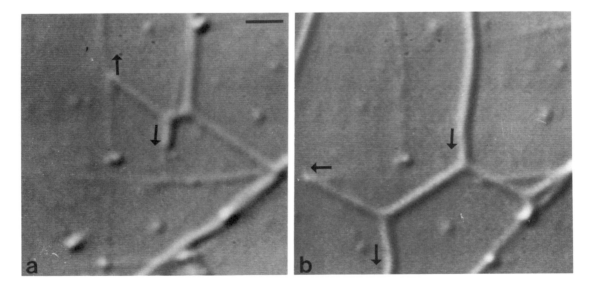

Fig. 6 Tubulovesicular membrane network translocating along the native actin bundles. Such networks, which may be part of the endoplasmic reticulum, move smoothly along the actin bundles rapidly at rates similar to the speed of the cytoplasmic streaming *in vivo*. Bar = 1.6 μm.

V. Identification of a Soluble Pool of Myosins in *Nitella* Cytoplasm

In addition to vesicular movements, naked actin bundles can be seen moving continuously along the surface of the glass coverslip for several hours (Fig. 7), suggesting that a soluble pool of myosin exists in the algal cytoplasm.

For preparation of a soluble cytoplasmic fraction, the cytoplasm of *Nitella* cells are extruded in a volume of a buffer containing 200 mM sucrose, 25 mM KCl, 25 mM NaF, 2.5 mM dithiothreitol, 2 mM ethylenediaminetetraacetic acid (EDTA), 0.5 mM phenylmethylsulfonyl fluoride (PMSF), and 10 μg/ml of pepstatin, leupeptin, and aprotinin, in a 20 mM imidazole buffer, pH 7.2. Usually the cytoplasm extruded from 50–100 cells is pooled and centrifuged at 15,000g for 20 seconds using an Eppendorf microfuge. The pellet containing nuclei and chloroplasts is discarded, and the supernatant is again centrifuged at 150,000g for 30 minutes at 4°C in a Beckman TL-100 ultracentrifuge to obtain a soluble cytoplasmic fraction. The absence of organelle and filamentous material can be assayed by video-enhanced interference microscopy. The presence of myosin in this fraction is clearly demonstrated by assaying the gliding of fluorescent actin filaments as described by Kron and Spudich (1986), as well as by determining the K^{2+} EDTA and actin-activated Mg^{2+} ATPase activities (Rivolta *et al.*, 1990). Visualization of the fluorescently labeled actin filaments moving on the glass surface is achieved using a Zeiss Axiomat microscope equipped with a 100X, 1.3NA objective. The image is projected onto a Newvicon or SIT video camera (Dage-MTI) and recorded in an optical memory disk recorder (Panasonic).

The soluble fraction from *Nitella* cytoplasm also supports the movement of fluorescently labeled actin filaments on the glass surface (Figs. 8a and b) with a bimodal speed distribution similar to that of the vesicular movements observed in the crude cytoplasm. Most of the filaments move smoothly on the glass surface at an average speed of approximately 11 μm s^{-1}, whereas the remaining filaments move at a rate of 60 μm s^{-1}. Another type of movement was produced when one third of a filament became stuck to the glass, while the remaining part either waived or underwent vigorous oscillatory motion (Figs.8c–k). Fragmentation of the filaments is also observed frequently and may occur as a result of a filament segment sticking to the glass surface while active myosin heads pull and break the filament.

VI. Concluding Remarks

This chapter has described a detailed method for the *in vitro* reconstitution and study of the native movement of organelles which are supported by myosins along the actin bundles of characean algae cells. In the extruded algal cell cytoplasm, different types of organelles can be visualized directly by video

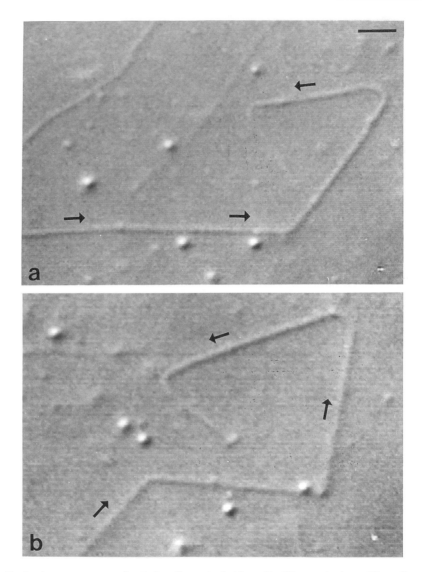

Fig. 7 *In vitro* movements of actin bundles extruded from the *Chara* cytoplasm. The actin cables can be visualized free or attached to the glass surface and displaying sharp bends along their length. In the presence of ATP, the actin bundles can be seen moving on the glass surface in the direction indicated by the arrows while displaying sharp bends along its length. Bar = 1.4 μm.

microscopy moving along the actin cables at different speeds. Video-enhanced interference light microscopy has also been complemented with fast-freeze electron microscopy to demonstrate that the endoplasmic reticulum translocates along actin cables in the ectoplasmic zone to move organelles present in

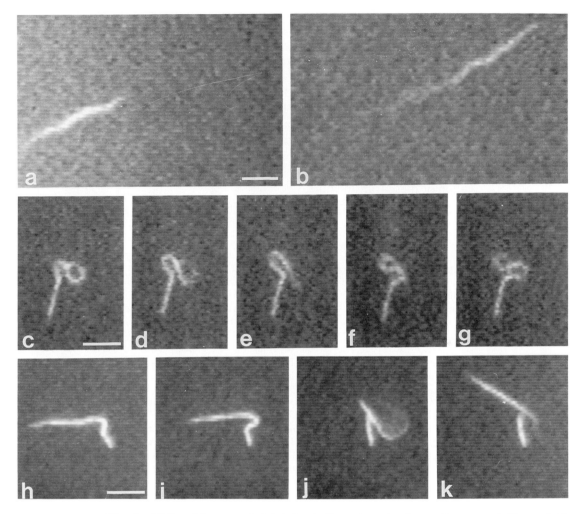

Fig. 8 Gliding of fluorescently labeled actin filaments on the glass surface supported by a soluble fraction of the *Nitella* cytoplasm. In this assay, fluorescently labeled rabbit actin filaments are incubated with a *Nitella*-soluble cytoplasmic fraction and visualized by video microscopy. (a,b) Micrographs showing the smooth movement of the fluorescently labeled filaments on the glass surface. Most of the filaments present in the field moved smoothly on the glass surface; other filaments displayed either beating or oscillatory movements (c,d). Bars = (a,b) 2.0 μm and (c–k) 3.0 μm.

the actin-free endoplasm. We believe that this phenomenon constitutes the basic mechanism of the cytoplasmic streaming in these cells.

Presently, we do not know what type of myosin is responsible for this activity. Kato and Tonomura (1977) isolated a filamentous form of myosin from *Nitella* with physical and enzymatic properties similar to those of myosin II; however, its mechanochemical properties were not studied. It seems unlikely that a myosin II is responsible for the fast movement of organelles along the actin bundles as no filamentous structures are resolved at the organelle–actin bundle interphase. Myosin II has been visualized in filamentous form in the cytoplasm of *Acanthamoeba castellanii* (Baines and Korn, 1990) and *Dictyostelium discoideum* (Yomura and Fukui, 1985) and should be detected in our electron micrographs of the moving organelles. Therefore, it is likely that nonconventional or monomeric myosins may be powering these movements. The mechanochemical activities of these myosins can be recovered easily in the soluble cytoplasmic fraction of *Nitella* and should be useful as an assay for the purification of these translocators.

References

Adams, R. J., and Pollard, R. D. (1986). Propulsion of organelles isolated from *Acanthamoeba* along actin filaments by myosin I. *Nature (London)* **322**, 754–756.

Adams, R. J., and Pollard, T. D. (1989). Binding of myosin I to membrane lipids. *Nature (London)* **340**, 565–568.

Baines, I. C., and Korn, E. D. (1990). Localization of myosin IC and myosin II in *Acanthamoeba castellanii* by indirect immunofluorescence and immunogold electron microscopy. *J. Cell Biol.* **11**, 1895–1904.

Garcia, A. E., Coudrier, E., Carboni, J., Anderson, J., Vanderkerckhove, M. S., Mooseker, M., Louvard, D., and Airpin, M. (1989). Partial deduced sequence of the 110-kD-calmodulin complex of the avian intestinal microvillous shows that this mechanoenzyme is a member of the myosin-I family. *J. Cell Biol.* **109**, 2895–2903.

Kachar, B. (1985a). Asymmetric illumination contrast: A method of image formation for video light microscopy. *Science* **227**, 766–768.

Kachar, B. (1985b). Direct visualization of organelle movement along actin filaments dissociated from characean algae. *Science* **227**, 1355–1357.

Kachar, B., and Reese, T. S. (1988). The mechanism of cytoplasmic streaming in characean algal cell sliding of the endoplasmic reticulum along actin filaments. *J. Cell Biol.* **106**, 1545–1552.

Kamiya, N. (1981). Physical and chemical basis of cytoplasmic streaming. *Annu. Rev. Plant Physiol.* **32**, 205–236.

Kato, T., and Tonomura, Y. (1977). Identification of myosin in *Nitella* flexilis. *Biochemistry* **82**, 777–782.

Korn, E. D., and Hammer, J. A., III (1990). Myosin I. *Curr. Top. Cell Biol.* **2**, 57–61.

Kron, S., and Spudich, J. A. (1986). Fluorescent actin filaments move on myosin fixed to a glass surface. *Proc. Natl. Acad. Sci. U.S.A.* **83**, 6272–6276.

Miyata, H., Bowers, B., and Korn, E. D. (1989). Plasma membrane association of *Acanthamoeba* myosin I. *J. Cell Biol.* **109**, 1519–1528.

Pollard, T. D., Doberstein, S. K., and Zot, H. G. (1991). Myosin I. *Annu. Rev. Phys.* **53**, 653–681.

Rivolta, M. N., Urrutia, R., Sellers, J., and Kachar, B. (1990). Preliminary characterization of an actin based organelle motor from *Nitella*. *Biophys J.* **57**, 535a.

Warrick, H. M., and Spudich, J. A. (1987). Myosin structure and function in cell motility. *Annu. Rev. Cell Biol.* **3**, 379–421.

Yomura, F., and Fukui, Y. (1985). Reversible cyclic AMP-dependent change in distribution of myosin thick filaments in *Dictyostelium*. *Nature (London)* **314**, 194–196.

CHAPTER 14

Assay of Vesicle Motility in Squid Axoplasm

Scott T. Brady, Bruce W. Richards, and Philip L. Leopold[1]

Department of Cell Biology and Neuroscience
University of Texas Southwestern Medical Center
Dallas, Texas 75235

Marine Biological Laboratory
Woods Hole, Massachusetts 02543

I. Extrusion of Isolated Axoplasm
II. Physiological Buffers
III. Video Microscopy
IV. Pitfalls
V. Summary
   References

Many recent advances in our understanding of cell biology can be traced to the development of video microscopy for direct observation of cellular processes. One example of these advances is our vastly improved understanding of intracellular motility at the molecular level. The prototype for such studies was video microscopic analysis of fast axonal transport in axoplasm isolated from the squid giant axon (Allen *et al.*, 1982; Brady *et al.*, 1982, 1985). Experiments using this preparation led to discovery of the kinesins, the newest class of molecular motors (Brady, 1985; Vale *et al.*, 1985), and provided a basis for identifying other motile activities (Lye *et al.*, 1987; Paschal *et al.*, 1987; Shpetner *et al.*, 1987). Although squid giant axons are not widely available, the unique characteristics of this preparation ensure that it will continue as an

[1] Present address: Department of Pathology, Columbia University College of Physicians and Surgeons, New York, New York 10032.

invaluable experimental model. Moreover, consideration of the strengths and limitations of the isolated axoplasm model for studying motility provides a basis for evaluating the utility of motility assays developed subsequently.

The giant axon of the Atlantic squid *Loligo pealeii* is used routinely for preparation of isolated axoplasm (''extruded'' axoplasm). Careful extrusion of *Loligo* axons yields a cylinder of axoplasm that maintains the structural integrity of the cytoskeletal elements and organelles (Bear *et al.*, 1937; Morris and Lasek, 1982). Giant axons from other squid species may be employed if their diameter is sufficiently large (see below), and giant axons from a marine fanworm, *Myxicola infundibulum*, may be suitable for some purposes (Breuer *et al.*, 1988). Although other giant axons exist, no other source of gaint axons suitable for extrusion has been identified. Most giant axons in nature are too small, too inaccessible, or too fragile to be extruded. For example, other mollusks such as *Aplysia* have giant neurons, but their axonal diameters are below the minimum size for extrusion and are stellate in shape. Structurally intact axoplasm cannot be extruded from such axons, although fractions enriched in axoplasmic proteins may be obtained. Comparable problems are encountered with teleost Mauthner axons and other invertebrate axons. Although fractions enriched in axoplasm can sometimes be obtained, the structural integrity of such fractions has been lost and their utility is limited to biochemical analyses.

The high content of neurofilaments in *Myxicola* axons makes it possible to pull the axoplasm from this axon intact. Microtubule domains in *Myxicola* axoplasm are sparse and segregated from one another by neurofilaments. These domains constitute channels for transport of membranous organelles, which may be useful for some analyses. Descriptions of motility in *Myxicola* axoplasm have appeared in the literature (Breuer *et al.*, 1988), but these preparations will not be discussed further except to note that variability between animals may affect the success rate for extraction of axoplasm. Anecdotal reports suggest that axoplasm from specimens collected in British waters may be more readily extracted than axoplasm from worms collected in Nova Scotia.

I. Extrusion of Isolated Axoplasm

Medium to large squid with translucent mantles (0.3–0.5 m long) are selected and decapitated (Brady *et al.*, 1985). The mantle is split along the dorsal ridge and pinned to a dissection table with running seawater. Internal organs are removed and the pen is carefully dissected out. The skin is removed to improve visibility and the mantle is placed on a light table with a low-power stereo dissection microscope. Indirect, transmitted light is adjusted to enhance visualization of the medial giant axon located superficially on either side of the midline. Fine dissection scissors (such as FST Trident scissors) and Dumont No. 5-45 forceps are used to free the nerve bundle from the muscle, starting distal to the stellate ganglion and extending to where the nerve moves deeper into the muscle mass (approximately 2–3 cm). Some connective tissue and associated axons may be removed at this stage, but most are removed subsequently. Cotton thread (~10 cm) is used to tie off the nerve adjacent to the stellate ganglion and near the most distal free segment. Contrasting colors (i.e., black for distal and white for proximal) are used to ensure that orientation of the axon is maintained throughout experimental manipulation and preparation. The ganglion is dissected free from the mantle and the distal nerve is transected. The nerve should be handled by the threads to avoid compression damage to the giant axon during handling.

The biomechanics of axoplasm requires that axons be greater than 100 μm in diameter for routine mechanical extrusion. This represents a minimum diameter from which limited amounts of material can be obtained, so larger diameters are preferred. Typical diameters for axons from *Loligos* with mantle lengths greater than 0.3 m is 400–500 μm, which produces approximately 2.5 μl per centimeter of giant axon. Axons from other species of squid may not be proportionately as large.

After dissection, the nerve is transferred to a standard 100-mm Petri dish containing ice-cold filtered seawater (0.45-μm-pore filters). The nerve should be kept at 0–4°C for this step. Some investigators use Ca^{2+} free artificial seawater, but this is harder on the neuronal plasma membrane and neuronal metabolism. In addition, the use of Ca^{2+} free artificial seawater tends to obscure damage to the axon, which may affect transport (see below). The nerve bundle is held in place by placing a small amount of dental wax (Surgident Periphery Wax, Columbus Dental, St. Louis, MO) at opposite ends of the Petri dish and inserting 18-gauge syringe needles for securing the threads at either end. Stainless-steel syringe needles are preferred, because they resist the corrosive effects of seawater. Vannas-style iris scissors and No. 5 Dumont forceps (titanium is preferred to resist corrosion from exposure to seawater) are used to gently tease small axons and connective tissue away from the giant axon using dark-field illumination in a stereo dissection microscope. Care must be taken to avoid damage to the giant axon, particularly at the site of small collateral branches, which should be cut 0.10 mm from the axon. Collaterals can often be detected just proximal to regions in which axonal diameter is slightly reduced.

When extraneous tissue has been cleared down to the glial sheath, the giant axon should be carefully examined for cloudy spots, which indicate the position of small holes in the plasma membrane. These lesions result from entry of Ca^{2+} and associated proteolysis of axoplasmic structural elements. Axoplasm will leak from these holes during extrusion and disrupt axoplasmic organization, so axons with white patches in their middle regions should be discarded. Axons to be extruded are removed from filtered seawater, rinsed in a suitable intracellular buffer, and then blotted to remove excess fluid. The axon should be held by the thread attached to the distal end and cut just distal to the proximal thread. With the extended axon in place on an appropriate surface, the axon is compressed near the distal end by exerting gentle pressure on a section of PE-190 tubing and extruded by pulling on the distal thread to draw the axon under the tubing. A cylinder of axoplasm 2–2.5 cm long and 0.35–0.45 mm in diameter can be obtained routinely. For video microscopy, the axon is extruded and sandwiched between two coverslips to optimize optical resolution.

For video microscopy of fast axonal transport in an intact axoplasmic cylinder, the axon is extruded on a 24 × 50-mm 0-thickness coverslip between spacers (made by cutting 5 × 24-mm glass silvers from 0-thickness coverslips with a diamond pencil), which are secured with Dow-Corning Compound 111 or an appropriate substitute. High-vacuum silicon grease is not suitable as it will be extracted slowly into the buffer and may compromise analyses. Position of the spacers can be adjusted to minimize the volume of the incubation chamber. Typical volumes are 25–30 μl. The top coverslip (22 × 22 mm, 0 thickness) is secured with VALAP (1 : 1 : 1; Vaseline : lanolin : paraffin) kept fluid at 50–60°C and applied with a wooden applicator. Both ends of the chamber are left open to allow diffusion of oxygen and perfusion of experimental buffers.

II. Physiological Buffers

Choice of buffers for introduction of experimental agents is crucial, particularly for maintaining structural integrity of axoplasm (Brady *et al.*, 1985; Morris and Lasek, 1982). The axoplasm is highly sensitive to ionic strength and ionic composition. The ability to obtain pure cytoplasmic fractions without dilution has permitted biochemists to measure the small-molecular-weight composition of axoplasm with precision. This information has in turn been used to devise artificial buffers that mimic the biochemical properties of axoplasm. Several such buffers have been used for studies of axoplasmic structure and metabolism (Adams, 1982; Brady *et al.*, 1982; Morris and Lasek, 1982; Rubinson and Baker, 1979). These buffers are distinguished by the use of organic anions (amino acids and related compounds), low levels of chloride and other halides, high osmotic strength, low levels of free Ca^{2+}, and the presence of suitable reducing agents.

For motility assays, simplified versions of these physiological buffers have gained popularity and variations are widely used. The composition of the original buffer used in analysis of motility in isolated axoplasm, buffer, X, is shown in Table I. Full-strength buffer will maintain the structural integrity of axoplasm and minimize extraction of cytoskeletal elements. Organelle motility will continue for several hours in buffer X provided that perfusion volumes remain small (four- to fivefold dilution of the axoplasm). In some cases, extraction of cytoskeletal elements into the surrounding buffer is desired. In such cases, buffer X can be diluted by half to increase the number of intact microtubules that splay into the perfusion buffer from the periphery of the axon during the first 10–20 minutes. Higher dilutions or lower-ionic-strength buffers may have a deleteri-

Table I
Composition of Buffer X

Potassium aspartate	350 mM
Taurine	130 mM
Betaine	70 mM
Glycine	50 mM
Hepes (N-2-hydroxyethylpiperazine-N'-2-ethanesulfonic acid, adjusted to pH 7.2 with KOH)	20 mM
MgCl$_2$ (for effective free Mg^{2+} = 10 mM)	12.9 mM
Potassium EGTA [ethylene glycol bisaminoethyl ether) N, N'-tetraacetic acid, potassium salt as a Ca^{2+} buffer]	10 mM
CaCl$_2$ (for effective free Ca^{2+} = 65 nM)	3 mM
Glucose	1 mM
ATP (potassium salt)	1 mM

ous effect on motility and organization (Brady *et al.*, 1985). Diluted buffer X is referred to as buffer X/2 or motility buffer, although the latter name is misleading as motility does not seem to last as long as in full-strength buffer X.

Other buffer compositions may be used for special experimental needs, but care should be taken to avoid chaotropic agents (i.e., Cl$^-$ ions), which will disrupt axoplasmic cytoskeletal elements (Baumgold *et al.*, 1981), and some buffers may be incompatible with kinesin ATPase activity (i.e., Tris–HCl; for other examples, see Wagner *et al.*, 1989). The integrity of the organelle transport system appears to be preserved to the greatest extent in buffers that are modeled on the physiological composition of axoplasm, including high protein concentration.

III. Video Microscopy

Although a variety of light microscopes and video equipment may be used for video microscopy, investigators must recognize that the video is not a simple extension of light microscopy. The video and image processing are far more than a convenient way of making a record of an experiment (Allen, 1985; Allen and Allen, 1983; Allen *et al.*, 1981; Inoué, 1981, 1986). The configuration of the microscope and the considerations of what constitutes a suitable specimen are profoundly different from those for conventional light microscopy and photographic film. A failure to recognize these differences will lead to unsatisfactory images. We use an inverted Zeiss Axiomat with high-numerical-aperture differential interference contrast (DIC) optics, including a 100× planapochromatic objective (1.3NA) and matched condenser (1.4NA) with zoom set at 2.5×. This means a screen magnification of approximately 10,000× for an NTSC standard video signal giving a screen view of 24 μm. Light microscopy texts often discuss the concept of "empty magnification," because further magnification dims the image without providing more information; however, high-resolution video microscopy requires that the image of a diffraction-limited object have a diameter on the screen greater than two raster lines or detection may be marginal. Magnification, illumination, and contrast must be adjusted to provide this level of magnification.

Microscopes and optics made by other manufacturers can be used equally well if suitable optics are available and an optical path can be constructed with sufficient final magnification at the video screen. Several issues need to be taken into consideration when designing a microscope system for study of organelle transport along individual microtubules. DIC objectives should be 63X to 100X power with the highest available numerical aperture (1.3–1.4) with a matching high-numerical-aperture condenser. Planapochromatic objectives are preferred because they are most highly corrected, which will minimize optical aberration and optimize resolution; however, before selecting lenses, the objective should be examined for strain by examination of the back focal plane with crosssed polars and polarized light to ensure that light scatter is minimal. Strain will appear as a series of light and dark lines in the back focal plane. Any intermediate lens or zoom should be of similar high quality.

A short optical path with adequate illumination should be available and should be brighter than needed for direct viewing. The short optical path is desirable to reduce the number of glass surfaces which can degrade a high-contrast image and to ensure that adequate illumination is provided. Traditional light microscopy limits light to produce sufficient contrast, and images are generated near extinction. In contrast, high-resolution DIC video microscopy is optimal with the analyzer well away from extinction (20–22°) and a $\lambda/4$ compensator which produces much brighter images. Many video cameras are sensitive at low light levels and will produce good images of large objects at low light levels, but higher levels of illumination are needed to detect objects below the resolution limit of the microscope. A bright, even illumination such as that provided by a fiber optic scrambler (Inoué, 1986) and a 100-W mercury arc lamp is ideal for imaging fast axonal transport in isolated axoplasm.

The video and image processor used is a Hamamatsu Photonics Microscopy Model C1966 frame memory, but a variety of other systems now exist that can be used as well. Minimum requirements include two frame buffers and real-time computation. Not all systems are capable of real-time processing, and this is critical for motility assays. For recording of images, either ¾-in. U-matic videotape or S-VHS will give high-quality routine images. A laser video disk recorder provides the best preservation of high-resolution images in a standard video format and is particularly well suited for quantitative analysis. Although direct digitization of video images is technically feasible, the storage requirements for digital video frames are still prohibitive. Given the large number of frames generated by motility assays and the relative ease with which standard video signals can now be digitized for analysis, there is little incentive to use this method extensively.

A system equipped as described above is well suited to examine several elements of subcellular motility in detail. The real-time observation of moving organelles offers the opportunity to measure organelle velocities in the axoplasm or using *in vitro* motility systems. In addition, video microscopy provides information regarding the orientation of motion (e.g., anterograde versus retrograde in the axoplasm, plus-end versus minus-end-directed vesicle movement *in vitro*). Examples of the video data obtained from isolated axoplasm and quantitation of that motion can be seen in video supplements 1 and 2 to the journal *Cell Motility and the Cytoskeleton* (Brady *et al.*, 1983, 1990).

IV. Pitfalls

As any methodology becomes widely available, particularly one based on a novel technological advance, naivete about its limitations can lead to misuse. In most cases, problems stem either from a fundamental misunderstanding about intrinsic limits of the technique or from inappropriate use of the processing algorithms. Examples of these two classes of problems are (1) attempts to determine numbers of moving particles and (2) inappropriate frame averaging.

A statement about the number of moving objects in an experimental preparation is often desirable, but may not be possible. Typical values for resolution with high-quality optics with matched numerical apertures, monochromatic green light, high-resolution video cameras, and suitable image processing to enhance contrast are on the order of 200 nm. The depth of focus achieved with these same optics in DIC microscopy will be comparable, and both values approach the theoretical maximum obtainable with visible wavelengths of light; however, many objects of interest in isolated axoplasm, including translocating organelles and microtubules, are only 25–50 nm in diameter. Video microscopy allows detection of these small objects by enhancing light detection using the background subtract, custom gray level expansion, and gain features available for manipulating the digitized video image. Light scattered from objects below the diffraction limit projects an image of uniform size for a set of optics [the Airy disk for a sphere; see Fig. 1 for a schematic representation and see Inoué (1986) for a more rigorous discussion of the Airy disk and light microcopy resolution]. The size of this diffraction-inflated image is related to the resolution limit, because when two small objects lie closer than the limit of resolution, their Airy disks overlap, causing the two objects to appear as a single object (Fig. 2).

These principles clarify the problems associated with quantifying the number of moving organelles on a video image. Two particles moving in the same direction at the same velocity may have overlapping Airy disks, causing the two particles to appear as a single moving particle (Fig. 3). Similarly, particles

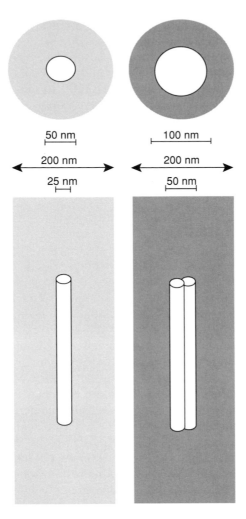

Fig. 1 Video microscopy is capable of detecting scattered light from small objects such as vesicles or microtubules. Objects below the limit of resolution for light microscopy may be visualized but the light scattered from the object is expanded to the diffraction limit. As a result, both 50- and 100-nm-diameter vesicles appear to be 200 nm in diameter, because light scattered by each object is spread over an area determined by diffraction of incident light, forming an image known as an Airy disk for a spherical object. The diffraction pattern produced by small objects extends beyond the borders of the 200-nm shaded areas indicated above, but shaded areas correspond to the apparent size of objects under these conditions. The size of the Airy disk limits resolution in the light microscope and varies according to wavelength of light and numerical aperture of the optics (Allen *et al.*, 1981; Allen and Allen, 1983; Inoué, 1981, 1986). The diffraction limit is typically about 200 nm for monochromatic green light and high-resolution optics. An organelle with larger size, but still below the resolution limit (100 nm versus 50 nm in this figure) will have the same-size Airy disk with a higher apparent contrast (shown above as darker shading). Contrast cannot be simply related to the size of a object, because an organelle with more birefringent contents (e.g., a chromaffin granule) or

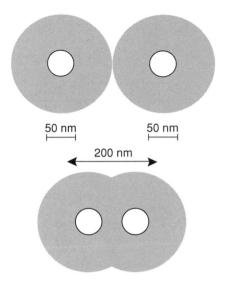

Fig. 2 The Airy disks of two 50-nm-diameter vesicles can be resolved only if the objects are more than 200 nm apart. Adjacent 50-nm-diameter vesicles will produce overlapping Airy disks which cannot be resolved as two individual objects.

moving along a microtubule appear to pass through one another even when they are several diameters apart. This limitation becomes more severe in complex preparations, such as isolated axoplasm, which contain many structures that diffract or scatter light. Given an optical section approximately 200 nm in depth, organelles may constantly enter and exit the plane of focus. The observer must remember that a population of organelles moving in the same direction (as in axonal transport) will tend to create a group of moving Airy disks, but the limits of resolution preclude a one-to-one correlation of Airy disks and organelles. Changes in the illumination, in the amount of scattered light, in the refractive index of the surrounding medium, or in the local thickness of the specimen can further influence the apparent number of moving particles, even when the number is unchanged. Unfortunately, when objects below the resolution limit are of interest, population properties such as velocity can be measured with high confidence, but quantifying the number of objects cannot be done reliably.

---

internal structure (e.g., a multilamellar body) may scatter more light, producing a higher-contrast image than a simple organelle with the same diameter. Similarly, light scattered from a single microtubule (25 nm diameter) will be detected as a 200-nm-wide rod. Two microtubules lying adjacent to each other will produce an image nearly identical in size but with somewhat greater contrast.

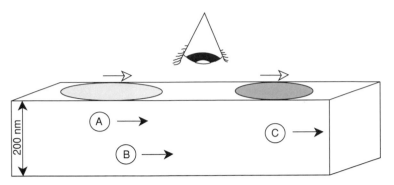

Fig. 3 The minimum thickness of an optical section in differential interference microscopy is approximately 200 nm, which increases the chance that two or more small objects will create overlapping Airy disks. Two 50-nm-diameter vesicles traveling in the same direction at the same velocity will be observed as a single Airy disk if the two vesicles lie within 200 nm of each other. Although vesicle C will be clearly resolved on the video screen, vesicles A and B will appear as a single, moving vesicle, thus complicating the determination of the number of moving vesicles. Further problems in quantitation arise from the probability that particles pass in and out of the optical section, whcih precludes tracking them for any distance. Although video microscopy is well suited to quantitate population characteristics such as velocity and direction, care should be taken in drawing conclusions about numbers of organelles in a complex biological sample.

The optical density of subcellular structures can be quantitated by comparing the optical density of the object with a standard curve for structures of known optical density and corresponding gray values (Inoué, 1986). Theoretically, this technique could be used to estimate the size of moving organelles that have diameters below the limits of optical resolution, but application of this approach is not trivial because information about the birefringence of objects or a uniform population of structures to be measured is needed.

For applications in which the limits of detection are being approached, the video system may introduce video noise, particularly random pixel noise familiar as "snow." The most common way of eliminating this noise is to perform a rolling average of 4–64 video frames. Frame averaging effectively eliminates this noise by showing only those signals that appear in most frames being averaged. Random pixel noise is removed, giving an image that is clearer and more stable. For fixed specimens, this can be a powerful technique for enhancing detail, but for moving specimens the results of such processing can be highly misleading.

In effect, frame averaging acts as a high-frequency filter and smoothes all movement in an image by removing the jitter caused by thermal motion, that is, Brownian motion. This can create the illusion of directed motion from random walks. The use of frame averaging for documentation of motility is particularly pernicious, because the amount of motility in a preparation may be overestimated and artifactual motile activities identified.

V. Summary

Axoplasm prepared as described above will maintain high levels of fast axonal transport for 1–2 hours, although moderate decrements in the average velocity may be noted over time. The organelles and structures that can be detected in isolated axoplasm are as small as the 50-nm synaptic vesicles or 25-nm microtubules, well below the limits of resolution for light microscopy; however, in the center of the axoplasm, where the density of structures is high, individual microtubules are not readily distinguished and individual vesicles can be followed only for short distances before they move out of the plane of focus or are lost in the multitude of neighboring organelles. On the periphery of perfused axoplasm, each of these structures may be readily detected and analyzed, but some information is lost about the role of specific axoplasmic organization in normal transport processes. Fortunately, the juxtaposition in one preparation of essentially structurally intact axoplasm with the extracted individual microtubules transporting organelles provides a unique preparation for molecular dissection of intracellular transport (Brady *et al.*, 1985).

Acknowledgments

The authors thank Dr. Raymond J. Lasek, Dr. Robert D. Allen, Dr. George S. Bloom, and our other collaborators in the study of fast axonal transport in squid axoplasm, in particular the work of Dr. Raymond J. Lasek and co-workers on the structure of isolated axoplasm and Dr. Robert D. Allen and his co-workers on the development of critical approaches to video light microscopy. The untimely death of Dr. Allen in 1986 remains a considerable loss to the field. Support for this work was provided by grants from the Council for Tobacco Research, the National Institutes of Health, NS-23320 and NS-23868, and the National Science Foundation.

References

Adams, R. J. (1982). Organelle movement in axons depends on ATP. *Nature (London)* **297**, 327–329.

Allen, R. D. (1985). New observations on cellular architecture and dynamics by video-enhanced contrast optical microscopy. *Annu. Rev. Biophys. Biophys. Chem.* **14**, 265–290.

Allen, R. D., and Allen, N. S. (1983). Video enhanced microscopy with a computer frame memory. *J. Microsc. (Oxford)* **129**, 3–17.

Allen, R. D., Allen, N. S., and Travis, J. L. (1981). Video-enhanced contrast, differential interference contrast (AVEC-DIC) microscopy: A new method capable of analyzing microtubule related movement in the reticulopodial netowork of *Allogromia laticollaris*. *Cell Motil.* **1**, 291–302.

Allen, R. D., Metuzals, J., Tasaki, I., Brady, S. T., and Gilbert, S. P. (1982). Fast axonal transport in squid giant axon. *Science* **218**, 1127–1129.

Baumgold, J., Terakawa, S., Iwasa, K., and Gainer, H. (1981). Membrane associated cytoskeletal proteins in squid giant axons. *J. Neurochem.* **36,** 759–764.

Bear, R., Schmitt, F. O., and Young, J. Z. (1937). Investigations on the protein constituents of nerve axoplasm. *Proc. R. Soc. London Ser. B* **123,** 520–529.

Brady, S. T. (1985). A novel brain ATPase with properties expected for the fast axonal transport motor. *Nature (London)* **317,** 73–75.

Brady, S. T., Lasek, R. J., and Allen, R. D. (1982). Fast axonal transport in extruded axoplasm from squid giant axon. *Science* **218,** 1129–1131.

Brady, S. T., Lasek, R. J., and Allen, R. D. (1983). Fast axonal transport in extruded axoplasm from squid giant axon. *Cell Motil.* **3,** Video Suppl. 1.

Brady, S. T., Lasek, R. J., and Allen, R. D. (1985). Video microscopy of fast axonal transport in isolated axoplasm: A new model for study of molecular mechanisms. *Cell Motil.* **5,** 81–101.

Brady, S. T., Pfister, K. K., Leopold, P. L., and Bloom, G. S. (1990). Fast axonal transport in isolated axoplasm. *Cell Motil. Cytoskel.* **17,** Video Suppl. 2, 22.

Breuer, A. C., Eagles, P. A. M., Lynn, M. P., Atkinson, M. P., Gilbert, S. P., Weber, L., and Leatherman, J. (1988). Long term analysis of organelle translocation in isolated axoplasm of *Myxicola infundibulum*. *Cell Motil. Cytoskel.* **10,** 391–399.

Inoué, S. (1981). Video image processing greatly enhances contrast, quality, and speed in polarization based microscopy. *J. Cell Biol.* **89,** 346–356.

Inoué, S. (1986). "Video Microscopy." Plenum, New York.

Lye, J., Porter, M. E., Scholey, J. M., and McIntosh, J. R. (1987). Identification of a microtubule-based cytoplasmic motor in the nematode. *Caenorhabditis elegans*. *Cell* **51,** 309–318.

Morris, J., and Lasek, R. J. (1982). Stable polymers of the axonal cytoskeleton: The axoplasmic ghost. *J. Cell Biol.* **92,** 192–198.

Paschal, B. M., Shpetner, H. S., and Vallee, R. B. (1987). MAP1C is a microtubule-activated ATPase which translocates microtubules in vitro and has dynein-like properties. *J. Cell Biol.* **105,** 1273–1282.

Rubinson, K., and Baker, P. (1979). The flow properties of axoplasm in a defined chemical environment: Influence of anions and calcium. *Proc. R. Soc. London, Ser. B* **205,** 323–345.

Shpetner, H. S., Paschal, B. P., and Vallee, R. B. (1987). Characterization of the microtubule-activated ATPase of brain cytoplasmic dynein (MAP1C). *J. Cell Biol.* **107,** 1001–1009.

Vale, R. D., Reese, T. S., and Sheetz, M. P. (1985). Identification of a novel force-generating protein, kinesin, involved in microtubule-based motility. *Cell (Cambridge, Mass.)* **42,** 39–50.

Wagner, M. C., Pfister, K. K., Bloom, G. S., and Brady, S. T. (1989). Copurification of kinesin polypeptides with microtubule-stimulated Mg-ATPase activity and kinetic analysis of enzymatic processes. *Cell Motil. Cytoskel.* **12,** 195–215.

CHAPTER 15

Assay of Membrane Motility in Interphase and Metaphase *Xenopus* Extracts

Viki J. Allan

Structural Studies Division
Medical Research Council Laboratory of Molecular Biology
Hills Road, Cambridge, CB2 2QH, United Kingdom

I. Introduction
II. Preparation of *Xenopus* Extracts and Membrane Fractions
    A. General Requirements
    B. Interphase Extracts
    C. Metaphase Extracts
    D. Assay of Cell Cycle Status
    E. Preparation of Membrane Fractions and High-Speed Supernatants
III. Motility Assay
    A. Microscope Perfusion Chamber
    B. Source of Microtubules
    C. Observation of Motility in Interphase and Metaphase
    D. Assaying Activity of Plus-End- and Minus-End-Directed Motors
    E. Troubleshooting
IV. Data Aquisition, Handling, and Analysis
    A. Microscopy
    B. Measurement
    C. Transferring Images from Video Signals to Film
V. Conclusions and Prospects
References

I. Introduction

Microtubule-based motors play a number of important roles in the process of membrane traffic in higher eukaryotes. Many organelles in the cell depend on microtubules both to achieve and to maintain their position and organization within the cell. The Golgi apparatus, for example, reaches its perinuclear position by moving toward the minus ends of microtubules (Ho *et al.*, 1989), where it forms an extensive interconnected system of membrane stacks. If microtubules are depolymerized, the Golgi apparatus not only scatters, but also fragments into individual stacks, indicating that microtubule-based motors play a significant part in maintaining the structure of this organelle. Indeed, the formation of connections between adjacent elements of the Golgi apparatus has been observed to involve movement of Golgi-derived tubules along microtubules (Cooper *et al.*, 1990). Microtubules also promote membrane traffic by facilitating the movement of transport vesicles between different parts of the cell. The most obvious example of this process is seen in axonal transport, where newly formed synaptic vesicles bud off from the Golgi apparatus in the nerve cell body, and are transported to the nerve terminal by movement along microtubules (Allan *et al.*, 1991). Even in smaller, nonneuronal cells, transport between the Golgi apparatus and the plasma membrane may involve microtubule-based transport (Arnheiter *et al.*, 1984).

During mitosis in higher eukaryotes, the Golgi apparatus becomes fragmented into many small vesicles (Lucocq *et al.*, 1989) and scatters throughout the cell. Other organelles that interact with microtubules during interphase, such as endosomes, lysosomes, and mitochondria, are excluded from the mitotic spindle. At the same time, membrane traffic is inhibited along both exocytic and endocytic routes (e.g., Featherstone *et al.*, 1985; Berlin and Oliver, 1978; Pypaert *et al.*, 1987). As the main role for the microtubules and motors during mitosis is the formation of the spindle to ensure chromosome partitioning, it is possible that most membrane movement along microtubules is switched off. Such an inhibition might in turn contribute to the changes seen in membrane traffic and organelle structure during cell division. We have used a motility system in which the cell cycle status can be manipulated to study this possibility *in vitro* (Allan and Vale, 1991). In *Xenopus* egg extracts we found that membrane movement along microtubules is indeed greatly reduced in metaphase as compared with interphase.

Xenopus egg extracts have proved invaluable in studying complex cell cycle-dependent processes such as nuclear envelope disassembly and re-formation (e.g., Newport, 1987; Newport and Spann, 1987; Newmeyer and Wilson, 1991), DNA replication (e.g., Blow and Laskey, 1986; Hutchison *et al.*, 1987), microtubule dynamics (Belmont *et al.*, 1990; Verde *et al.*, 1990), and spindle formation (Sawin and Mitchison, 1991), as well as in investigating the regulation of the cell cycle itself (e.g., Murray and Kirschner, 1989; Lohka *et al.*, 1988; Glotzer *et al.*, 1991). The *Xenopus* egg contains enough stockpiled components to enable the first rapid divisions to involve a simple oscillation from S phase to M phase, without any G_1 or G_2 phases. At the most basic level, the position in the embryonic cell cycle depends on the concentrations of cyclin A and B, proteins that activate the main mitotic kinase, p34^{cdc2} or maturation-promoting factor

(Murray and Kirschner, 1989; Murray *et al.*, 1989). Laid *Xenopus* eggs are naturally arrested in metaphase of meiosis II because of the presence of cytostatic factor (CSF), which maintains high levels of p34^{cdc2} activity (Sagata *et al.*, 1989). On fertilization, an influx of calcium results in the inactivation of both CSF and p34^{cdc2}, releasing the egg from metaphase arrest and allowing progression into interphase. This consequence of fertilization can be mimicked by raising calcium artificially either by electric shock treatment (which permeabilizes the plasma membrane transiently, allowing entry of extracellular calcium), or by treatment with calcium and the calcium ionophore A-23187 (Newport, 1987).

A number of groups have produced *Xenopus* egg extracts in which the cell cycle progresses *in vitro* through a single cycle (Blow and Laskey, 1986; Felix *et al.*, 1989) or several cycles (Murray and Kirschner, 1989; Hutchison *et al.*, 1987). To study the effects of the cell cycle on membrane motility *in vitro*, however, we found it more convenient to use extracts that were arrested at specific points in the cell cycle. Lohka and Maller (1985) first developed methods for making concentrated egg extracts that are arrested in metaphase of meiosis II by including high levels of ethylene glycol bis (β-aminoethyl ether) *N,N′*-tetraacetic acid (EGTA) to maintain CSF activity. By reducing the EGTA concentration, Murray and co-workers (1989) produced meiotically arrested extracts (CSF extracts), which can subsequently be released from metaphase arrest by adding calcium *in vitro* in the absence of protein synthesis (Fig. 1). Interphase-arrested extracts can be made from activated eggs in the presence of cycloheximide, and these extracts can then be converted *in vitro* to a stable metaphase state simply by adding a nondegradable analog of sea urchin cyclin B, cyclinΔ90 (Glotzer *et al.*, 1991). It is therefore possible to make extracts arrested in either interphase or metaphase of meiosis II that can be subsequently converted *in vitro* to metaphase or interphase, respectively, in the absence of protein synthesis (summarized in Fig. 1). The production and use of these extracts to study the cell cycle regulation of microtubule-based motility are described in this chapter.

II. Preparation of *Xenopus* Extracts and Membrane Fractions

A. General Requirements

1. Solutions

Unless otherwise indicated, all chemicals were purchased from either Sigma Chemical Company, St. Louis, Missouri, or B.D.H., Poole, United Kingdom.

MMR: 100 mM NaCl, 2 mM KCl, 1 mM MgSO$_4$, 2 mM CaCl$_2$, 5 mM 4-(2-hydroxyethyl)-1-piperazineethanesulfonic acid (Hepes), 0.1 mM ethylenediaminetetraacetic acid (EDTA). Adjusted to pH 7.8 with NaOH.

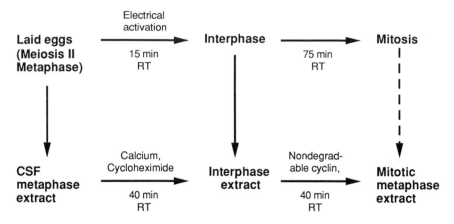

Fig. 1 Summary of the relationship between intact *Xenopus* eggs and egg extracts at different stages of the cell cycle. Laid *Xenopus* eggs are naturally arrested at metaphase of meiosis II, and will progress into interphase 15 minutes after fertilization or electrical activation (top line). After a further 75 minutes, the eggs enter metaphase of the first mitosis. Extracts made from meiotic (Section II,C,1) or interphase (Section II,B,1) eggs can be induced to enter the next cell cycle stage *in vitro* in the absence of protein synthesis (Sections II,B,3 and II,C,2).

XB (extract buffer): 100 mM KCl, 0.1 mM CaCl$_2$, 1 mM MgCl$_2$, 10 mM Hepes, 50 mM sucrose. Made as a 5× stock without adjusting pH, and kept at −20°C. The pH is titrated to 7.7 with KOH after dilution to 1×.

Cysteine: 2% (w/v) cysteine free base in H$_2$O. Made up just before use and adjusted to pH 7.8 with NaOH.

Acetate buffer: 100 mM potassium acetate, 3 mM magnesium acetate, 5 mM EGTA, 10 mM Hepes (biochemical grade, Fluka, Ronkonkoma, NY, or B.D.H., Poole, U.K.), 150 mM sucrose (Ultrapure, B.R.L., Gaithersburg, MD). Adjusted to pH 7.4 with KOH.

BRB80: 80 mM 1,4-piperazinediethanesulfonic acid (Pipes), 1 mM MgCl$_2$, 1 mM EGTA. Adjusted to pH 6.8 with KOH.

Histone kinase buffer: 80 mM β-glycerol phosphate, 15 mM MgCl$_2$, 20 mM EGTA, pH 7.3.

Histone H1 (Boehringer-Mannheim, 223549): 250 μg/ml in histone kinase buffer. Stored at −20°C.

ATP (vanadate-free, sodium salt; Sigma A5394): Stock of 100 mM is made to 100 mM MgCl$_2$ and titrated to approximately pH 7.0 with NaOH. Store at −20°C. [γ-^{32}P]ATP:4000 Ci/ml (35001X, ICN, Costa Mesa, CA).

20× energy mix: 150 mM creatine phosphate (Boehringer-Mannheim, 127574), 20 mM MgATP, 2 mM EGTA, pH 7.7. Stored in aliquots at −20°C.

GTP: 100 mM stock, stored at − 20°C.

Taxol: Taxol was a gift from Nancita R. Lomax at the National Cancer Institute, and was stored at −20°C as a 4 m*M* stock in dimethylsulfoxide (DMSO).

Protease inhibitors: 10 mg/ml leupeptin, pepstatin, and chymostatin in DMSO. Stored at −20°C.

Cytochalasin D: 1 mg/ml cytochalasin D in DMSO. Stored at −20°C. Cytochalasin B can be used instead at 10× the concentration throughout.

Cycloheximide: 10 mg/ml in water. Stored in aliquots at −20°C.

Versilube F-50 oil: This silicone oil is made by General Electric and can be purchased from Andpak-EMA, San Jose, California, or from Atochem Chimie U.K. Ltd., Newbury, United Kingdom.

Pregnant mare serum gonadotropin (PMSG): Purchased from either Calbiochem, San Diego, California, (367222), or from Intervet U.K., Cambridge, United Kingdom (Folligon). Made up at 200 U/ml in sterile water and stored at −20°C.

Human Chorionic gonadotropin (HCG): Purchased from either Sigma (CG-2) or from Intervet (Chorulon). Made up at 1200 U/ml in sterile water and stored at 4°C (keeps for at least 2 weeks).

2. Frogs

Extensive information on frog care and handling and a list of American suppliers are given in *Methods in Cell Biology,* Volume 36, which is devoted entirely to *Xenopus*. In the United Kingdom we purchase *Xenopus* from Blade Biological Company, Edenbridge.

1. Prime mature female frogs for ovulation by injecting with PMSG 3–12 days before you need the eggs. Inject 100 units into the dorsal lymph sac using a 27-gauge needle. Keep the frogs in a separate tank without feeding.

2. To stimulate ovulation, inject with human chorionic gonadotrophin. Inject 600 U into the dorsal lymph sac, about 14 hours before the eggs are needed.

3. Keep the frogs in separate containers after HCG injection, in 100 m*M* NaCl, overnight at 18°C. The frogs will start laying eggs after approximately 12 hours.

B. Interphase Extracts

1. Activated

We use extracts prepared using a slight modification of the method described by Murray and Kirschner (1989) for preparing cycling extracts, which is described in detail by Murray (1991). The main change is the inclusion of cycloheximide, which arrests the extracts in interphase.

1. Collect the laid eggs and transfer them to 1× MMR. Transfer the eggs to 10-cm glass Petri dishes. During these washes remove any shed frog skin and excrement. To handle the eggs, use a cut-off Pasteur pipet with an opening of about 3 mm. Smoothe the cut surface of the pipet using a Bunsen burner.

2. Wash the eggs twice with 0.1× MMR and let them sit for 5 minutes. This should ensure complete activation later.

3. Remove the jelly coat with fresh 2% cysteine (pH 7.8 with NaOH), which takes approximately 5 minutes. Swirl the eggs gently, replacing the cysteine several times. When the jelly has been removed, the eggs will pack tightly at the bottom of the dish. Remove any necrotic eggs with the cut-off Pasteur pipet.

4. Wash the eggs carefully two to four times with 0.2× MMR. The eggs are liable to burst very easily from now on.

5. This protocol requires a simple "activation chamber," which consists of an open-topped Perspex (plexiglass) box (10 (x) 10-cm base, 5 cm high) with two stainless-steel electrode plates that fit horizontally inside the chamber. Both electrodes have insulated stainless-steel strips attached at one side, which are vertical when the electrodes are in the chamber. These strips are connected to a 12-V ac power supply (e.g., as used for a microscope lamp) via electrode cables. Place the lower electrode in the chamber and cover with a 1-cm layer of 1% agarose in 0.2× MMR. After the agarose has set, fill the chamber with 0.2× MMR. Cover the agarose with a single layer of eggs; then carefully place the upper electrode in the chamber, avoiding trapping air bubbles. The upper electrode should have four small "feet" to prevent it coming into direct contact with the eggs. Activate with two 1- or 2-second pulses of 12-V (ac) power separated by a 5-sec interval. Start a timer. Watch for contraction of the animal pole pigment cap, which occurs after 4–5 minutes and is a sign of activation. Initially it is best to look for this contraction under a dissecting microscope, but with practice it can be spotted by eye, especially if the eggs are placed against a black background. If less than 80% of the eggs activate, the eggs should be shocked once more. The timing of the pulses required may vary depending on the particular power supply used, so it is worth performing a few trials. If the electric shocks are too long, the eggs will start to lose their sharp pigment boundary, and may eventually burst.

6. Transfer quickly but gently, using the cut-off Pasteur pipet, to fresh Petri dishes containing XB. Wash four times with XB and then twice with XB containing 100 μg/ml cycloheximide and 10 μg/ml protease inhibitors.

7. Add 1 ml XB/cycloheximide/protease inhibitors to an SW50 Ultraclear tube. Add cytochalasin D to 10 μg/ml to prevent actin polymerization. Transfer the eggs to the tube with the cut-off Pasteur pipet. To avoid diluting the cytochalasin D, allow the eggs to settle in the pipet before releasing them, so that the minimum of liquid is added to the tube. Remove as much of the buffer from the tube as possible after the eggs have settled. Add 1 ml of Versilube F-50 oil.

8. Spin at room temperature in a clinical centrifuge at 150g for 60 seconds and then 600g for 30 seconds to pack the eggs without lysing them. During this spin the Versilube displaces the XB from the spaces between the eggs, minimizing the dilution of the cytoplasm during the crushing spin.

9. Remove all of the buffer above the oil, and then as much of the oil itself as possible.

10. Continue to incubate at room temperature until 15 minutes postactivation, and then place the tube in an ice-water bath for a further 15 minutes.

11. Perform the crushing spin in a swinging bucket rotor, e.g., Sorval HB-4 rotor, 10 minutes, 10,000 rpm, 4°C. Use Falcon 10-ml white plastic culture tubes as adaptors. During this spin, the eggs will break and their contents will become stratified into lipid, cytoplasm, and a pellet composed of yolk and pigment granules.

12. Remove the cytoplasm from the tubes very slowly using a 1-ml syringe and 18-gauge needle inserted just above the pigment/yolk pellet. At times the Versilube density can be very similar to that of the cytoplasm. Add 10 μg/ml protease inhibitors, 1 μg/ml cytochalasin D, and 1/20th vol of the energy mix.

13. If there is significant contamination with pigment granules, lipid, or Versilube, clarify the extract by spinning in Eppendorf microfuge for 10 minutes at 4°C.

14. Add an extra 150 mM sucrose (from a 2.0 M stock) before freezing. Freeze in 50- to 100-μl aliquots in liquid nitrogen, and store at -80°C or under liquid nitrogen.

2. Extracts Prepared without EGTA

It is also possible to make extracts that appear to be in an interphase-like state without using electrical activation, simply by omitting EGTA from the buffers (Newmeyer and Wilson, 1991). For all experiments in which the cell cycle status needs to be clearly defined, it is preferable to use eggs that have been activated either electrically or by treatment with calcium ionophores. The EGTA-free extracts are simple to make, however, so for less critical applications they can be useful. I have used 100,000 g_{av} supernatants, made from CSF-arrested eggs homogenized in 2 vol of acetate buffer without EGTA, as a routine source of soluble components for motility assays using exogenous membranes (V. Allan, unpublished).

3. Conversion of Metaphase Extracts to Interphase *in Vitro*

Meiotic extracts prepared from metaphase-arrested eggs (see below) can be induced to enter interphase by adding 1/50th volume of 20 mM CaCl$_2$, 100 mM KCl, 1 mM MgCl$_2$ (to trigger destruction of CSF) and cycloheximide to

100 μg/ml (to inhibit cyclin synthesis), followed by incubation at 23°C for 30–45 minutes. After addition of EGTA to 0.4 mM, the sample is transferred to ice.

C. Metaphase Extracts

1. Cytostatic Factor Extracts

This procedure is similar in outline to that for making activated interphase eggs. Only the steps that differ are described in detail below. The most important difference is that the activation step is omitted, so that the eggs remain arrested in metaphase meiosis II, because of the presence of CSF.

1. Proceed as for making interphase extracts, following steps 1 and 3 but omitting step 2.

2. Remove any bad eggs with a cut-off Pasteur pipet. Wash the eggs carefully with 1× MMR. Eggs must be treated very gently during all washes or they will activate. If more than 20% of eggs activate, then the eggs should not be used.

3. Transfer to a fresh glass Petri dish containing XB. Wash four times with XB and then twice with XB containing 10 μg/ml protease inhibitors plus 5 mM EGTA and an extra 1 mM MgCl$_2$.

4. Add 1 ml XB/EGTA/protease inhibitors to an SW50 Ultraclear tube. Add cytochalasin D to 10 μg/ml. Proceed as for the activated extract, steps 7, 8, and 9, but not 10; i.e., keep the eggs at room temperature throughout.

5. Perform the crushing spin: Sorval HB-4 rotor, 10 minutes, 10,000 rpm, 16°C.

6. Carefully remove the cytoplasm from the tubes as described above, and transfer the extract to an ice bucket. Add 10 μg/ml protease inhibitors, 1 μg/ml cytochalasin D, and 1/20th vol of the energy mix.

7. CSF extracts are often clean enough to freeze at this point, but if there is contamination with pigment or lipid, clarify the extract as described in step 13 above. Freeze in 50- to 100-μl aliquots in liquid nitrogen, after the addition of 150 mM sucrose.

2. Conversion of Interphase Extracts to Metaphase *in Vitro*

Interphase extracts from activated eggs can be induced to enter mitosis by incubating for 45 minutes at 23°C with 1 mM cyclinΔ90, a nondegradable sea urchin cyclin B lacking the N-terminal 90 amino acids (Murray *et al.*, 1989; Glotzer *et al.*, 1991). CyclinΔ90 was kindly supplied by Michael Glotzer (Department of Biochemistry, University of California, San Francisco) as purified bacterially expressed protein. As the interphase extracts contain cycloheximide, the change in cell cycle status is driven solely by the presence of cyclin B, coupled with post-translational modifications.

D. Assay of Cell Cycle Status

1. Histone H1 Kinase Assay

Assaying the activity of histone H1 kinase is a convenient way of monitoring cell cycle status, as it has been shown that p34^{cdc2} is the kinase responsible for phosphorylating histone H1 (Langan *et al.*, 1989). Samples for assay (1 μl for undiluted extracts, 3 μl for high-speed supernatants) are taken in parallel with performing motility assays (see Section III,C,3), and are made up to 50 μl with histone kinase buffer plus protease inhibitors. The samples are then frozen in liquid nitrogen and stored at $-80°C$ until needed. The substrate mix for the assay contains 1 mM unlabeled ATP (from 100 mM ATP stock) and enough [γ-^{32}P]ATP to give a final specific activity of 0.25 μCi/μl, both added to 250 μg/ml histone H1 in histone kinase buffer. After thawing, 10 μl of the diluted sample mixture is mixed with 6 μl of the substrate mixture and incubated at room temperature for 15 minutes. The reaction is stopped by the addition of gel sample buffer and the samples are heated at 65°C for 20 minutes. After the samples are separated on 15% sodium dodecyl sulfate (SDS)–polyacrylamide gels, the front region of the gel is carefully removed, as it contains the unincorporated [^{32}P]ATP. The remaining gel is stained briefly with Coomassie blue and dried. Following autoradiography overnight, the histone bands are cut out of the gel and assessed by Cerenkov counting. Histone kinase activities are best expressed as picomoles of ATP incorporated per microliter of undiluted extract per minute, because standardization by original extract volume rather than by protein concentration allows direct comparison of histone kinase activities in extracts with those in high-speed supernatants.

2. Demembranated *Xenopus* Sperm Nuclei

The morphology of nuclei incubated in an extract provides a rapid assay for cell cycle status. In an unfractionated interphase extract, added demembranated sperm nuclei will decondense and gain a phase-dense nuclear envelope. In a metaphase extract, the chromosomes will condense fully, as visualized by labeling with Hoechst 33342 or DAPI labeling. A full protocol for preparing and using demembranated sperm nuclei is given by Murray (1991). Although sperm nuclei incubated with extract in a flow cell (see Section III) will often form nuclei or chromosomes, as appropriate, the nuclei do tend to stick to glass surfaces, so a more reliable method is to incubate the nuclei and extract in a microfuge tube, removing samples for fixation (Murray, 1991) at intervals.

E. Preparation of Membrane Fractions and High-Speed Supernatants

1. *Xenopus* Membranes and Supernatants

To study the activity of soluble motors alone, and also to test the effects of metaphase extracts on interphase organelles, for example, it is necessary to prepare high-speed supernatants from both interphase and metaphase extracts. The high metaphase $p34^{cdc2}$ activity can be lost on dilution of the extract in the absence of phosphatase inhibitors, such as β-glycerol phosphate. This inhibitor cannot be used for preparing supernatants, however, as it inhibits microtubule-based motility (V. Allan, unpublished results). The protocol shown below describes the preparation of high-speed supernatants by using minimal dilution of the extract in a physiological buffer containing ATP and creatine phosphate. In the process, the high-speed supernatants retain their metaphase nature.

Interphase and metaphase extracts are diluted on ice with 2 vol of acetate buffer plus energy mix. To prepare a soluble protein fraction, the diluted extracts are centrifuged either in an Airfuge (Beckman Instruments, Palo Alto, CA) at 28 psi for 30 minutes at 4°C or in a TLA 100 rotor at 55,000 rpm for 30 minutes at 4°C in a Beckman tabletop ultracentrifuge. Most of the supernatant is removed using a yellow tip, avoiding the region just above the pellet. To prepare organelles, the diluted extract is layered over 50 μl acetate buffer containing 300 mM sucrose plus energy mix and centrifuged as above. The opaque organelle layer on top of the clear amber pellet (which consists mainly of ribosomes and glycogen particles) is resuspended in acetate buffer plus energy mix to 40% of the original extract volume.

2. Rat Liver Golgi Fraction

We have observed cell cycle regulation of microtubule-based membrane movement when a rat liver Golgi fraction is used in place of *Xenopus* membranes (Allan and Vale, 1991). This indicates that the control mechanism may be conserved across vertebrate species. The rat liver fraction contains fragments of stacked Golgi cisternae, secretory vesicles, as well as other organelles such as endosomes. The purification method described below is a modification of that reported by Leelavathi and co-workers (1970).

1. Make up the following sucrose solutions in acetate buffer, using ultrapure sucrose (B.R.L., Gaithersburg, MD): 0.25 M, 1.1 M, 1.25 M, 1.4 M, and 2.0 M sucrose.

2. You will need a large, fed rat of either sex. Kill the rat and remove the liver as quickly as possible, and place in a beaker containing 0.25 M sucrose/ acetate buffer on ice.

3. Transfer the liver to a Petri dish and weigh it. Place the Petri dish on ice and mince the liver finely using two scalpels. All the following steps are performed at 4°C.

4. Put the minced liver in a 50-ml Falcon centrifuge tube, and add 1 vol of 0.25 M sucrose/acetate buffer (plus 10 μg/ml protease inhibitors) per gram wet weight.

5. Stir the liver to get it into suspension and then homogenize using a Polytron on the lowest setting. Turn on the motor and pass the probe slowly

down through the suspension, taking approximately 30 seconds to reach the bottom of the tube. Turn off the homogenizer before removing the probe from the homogenate. Repeat the procedure twice.

6. Spin the homogenate at 600g for 10 minutes at 4°C. Collect the supernatant, including the thick layer (which fills about half the tube) above the pellet proper.

7. Layer the homogenate over a cushion of at least 3.5 ml of 1.4 M sucrose/acetate in SW40/41 tubes.

8. Spin at 28,000 rpm (100,000g_{av}) for 60 minutes at 4°C.

9. Aspirate off most of the supernatant, and then collect the interface band using a plastic Pasteur pipet.

10. Dilute the interface fraction with ice-cold 1.25 M sucrose/acetate plus protease inhibitors (e.g., 2 ml fraction + 12 ml 1.25 M sucrose). Check the refractive index and adjust to 1.25 M with 2.0 M sucrose/acetate if necessary.

11. Make a gradient in SW25, 27, or 28 ultraclear tubes with the following steps: 10 ml 1.4 M sucrose/acetate; 15 ml sample fraction (1.25 M sucrose); 10 ml 1.1 M sucrose/acetate. Fill to the top of the tube with 0.25 M sucrose/acetate. The volume of each step can be altered as convenient, but the 1.1 M step is the most important as it will separate the crude Golgi fraction from the load fraction. To get a cleaner Golgi fraction at lower yield, include a 0.85 M step as well.

12. Spin at 65,000g_{av} for 90 minutes at 4°C

13. Collect the Golgi fraction from the 0.25/1.1 M sucrose interface in as small a volume as possible, and add more protease inhibitors. If there are aggregates, they can be broken up by pipetting through a yellow tip. Collect "smooth endoplasmic reticulum (ER)" from the 1.1 M/1.25 M interface.

14. Aliquot and freeze in liquid nitrogen. Store under liquid nitrogen or at −80°C

III. Motility Assay

Xenopus egg extracts are a convenient source of material for motility assays, as they contain not only motors and membranes, but also the soluble components which are needed to activate membrane movement along microtubules. The extracts also maintain their cell cycle status during the motility assays, which shows that such complex regulatory systems can remain active *in vitro*. In addition, these assays are performed at room temperature rather than at 37°C.

A. Microscope Perfusion Chamber

Motility can be observed by video-enhanced differential interference contrast (DIC) in very simple chambers made from microscope slides and glass coverslips (No.1,18 × 18 mm). The walls of the perfusion chamber are made from strips of vacuum grease (Apiezon M; Apiezon Products, London, U.K.) 1 cm apart which are extruded onto the slide through a squared-off 19-gauge needle. Spacers are provided either by strips of single-sided sellotape (Scotch tape) or fragments of No. 0 coverslips, placed between the vacuum grease and the edge of the slide. The former spacers give a flow cell volume of 7–10 μl, whereas the latter give a larger volume. It is also possible to omit the vacuum grease and use double-sided tape as both spacer and flow cell wall combined, to give a larger chamber volume. The choice of coverslip can have profound effects on the motility observed in the chamber, as some brands seem to permit much more active microtubule sliding on glass than others. In addition, coverslips that appear clean to the eye often have very uneven surfaces when viewed by DIC, presumably because of the treatments used during manufacturing. We are currently using coverslips from Chance Propper Ltd (Warley, U.K.) which, although apparently somewhat dirty by eye, give a clean field under the microscope. Gold-seal coverslips from Corning and Clay–Adams also give good results.

The obvious advantage of using an open-ended chamber is that solutions can be changed. This allows a sample to be fixed while under observation, for example, or an interphase supernatant to be replaced by a metaphase supernatant. Flow-through is achieved by gently wicking solution from one side of the flow cell using Whatman No. 1 filter paper while adding the exchange solution from the other side using a yellow tip. We generally use three to five times the flow cell volume per wash with buffer, or twice the volume if one extract is being replaced directly with another. Following washing and fixation (e.g., with 2% formaldehyde, 0.05% glutaraldehyde in BRB80, 15 minutes), it is possible to label components in the assay for fluorescence using antibodies or other fluorescent probes. We have used this method to visualize microtubules with antibodies at the same time as membrane tubules are labeled with the fluorescent dye $DiOC_6(3)$ (Allan and Vale, 1991).

B. Source of Microtubules

Interphase extracts will polymerize microtubules spontaneously when incubated in the flow cell. Metaphase extracts, in contrast, do not generally polymerize microtubules unless a small number of nucleating microtubules are supplied (see also Verde *et al.,* 1990). To provide similar numbers of microtubules for both interphase and metaphase samples, microtubule seeds can be supplied in two ways. First, microtubules can be stuck to the glass surface of the chamber before the extract is added. To do this, precoat the flow cell with 10 μl of 20-fold diluted extract by incubating for 5 minutes. Wash with 40 μl of acetate buffer, then flow through 20 μl of 100–250 μg/ml solution of taxol-stabilized, mammalian brain tubulin microtubules in acetate buffer. Incubate at 23°C for 2 minutes to bind the microtubules to the coated glass surface, wash with buffer, and flow through 15 μl of the sample to be tested. These exogenous microtubules are stable in interphase samples, whereas they are severed by metaphase extracts (Vale, 1991). The severed fragments then act as nucleating sites for polymerization of microtubules from endogenous *Xenopus* tubulin, which are more resistant to severing. In the second method, fragments of brain microtubules (seeds) are mixed directly with the diluted extract or high-speed supernatant to a final concentration of 1.25 μg/ml, where they act as nucleation centers for the polymerization of *Xenopus* tubulin. The seeds are made by shearing taxol-stabilized mammalian brain tubulin microtubules by passing them twice through a 27-gauge needle.

C. Observation of Motility in Interphase and Metaphase

1. Membrane Movement

The assay is started by perfusing in up to 10 μl of sample into the flow cell. Crude extracts are diluted with 2 vol of acetate buffer plus energy mix (plus microtubules as appropriate) and incubated in the flow cell for 5 minutes in a small humid chamber. During this time, microtubules polymerize and attach to the coverslip surface via motor proteins or microtubule-associated proteins which are bound to the glass. The movement of membranes along microtubules is assayed only on these attached microtubules. After immersion oil is applied both between the coverslip and the objective and between the slide and the condenser, the DIC microscope is focused by eye on the coverslip surface. If no membrane is present, it helps to focus on the vacuum grease first. The image can then be directed to the video camera (see Section IV).

In interphase extracts, three types of microtubule-based movement are observed (Allan and Vale, 1991): vesicle movement, membrane tubule extension, and microtubule gliding across the glass surface (see below). When two membrane tubules cross, they often appear to fuse (Dabora and Sheetz, 1988; Vale and Hotani, 1988; Allan and Vale, 1991), giving rise to polygonal membrane networks. When metaphase extracts are examined (Allan and Vale, 1991), very little vesicle movement or membrane tubule extension is seen, even though soluble motor proteins remain active (see below).

The large number of ribosomes and glycogen granules present in the unfractionated extracts degrade the DIC image significantly, making it difficult to see small vesicles moving. This problem can be resolved by combining high-speed supernatants and a crude egg membrane fraction (see Section II,D). By the use of high-speed supernatants, it is also possible to test other exogenous membranes in the assay, such as the rat liver Golgi fraction described above. In these experiments, high-speed supernatant (diluted 1/3 during preparation; Section II,E,1) is combined with 1/10th to 1/20th vol of crude *Xenopus* organelles or rat liver Golgi fraction. If the membrane concentration is too low, moving organelles will be seen only rarely; if the concentration if too high, the DIC image will suffer. After the mixture is introduced to the flow cell, the sample is incubated for 5–10 minutes in a humid chamber with the slide uppermost. This ensures that any large aggregates of membrane stick to the coverslip surface, and so will be visible during the motility assay. Although membrane motility will continue for at least 3 hours if the flow cell is kept in the humid chamber, we rarely use samples that are more than 1 hour old. If vesicle movement is being assayed, it is best to record sequences after 10 to 20 minutes of incubation, before the tubular membrane networks become too extensive. The membrane networks are generally recorded between 20 and 45 minutes of incubation, but the timing varies from extract to extract.

We have used these high-speed supernatants to show that the behavior of the added membrane fraction is determined by the cell cycle status of the soluble components of the assay (Allan and Vale, 1991). Metaphase supernatants inhibit tubular membrane network formation when either interphase or metaphase *Xenopus* membranes are added (Fig. 2), and this inhibition also extends to the rat liver Golgi membrane fraction. For all such mixing experiments, it is vital to assess the overall cell cycle status by measuring histone kinase activity, as described in Section III,C,3.

2. Soluble Motor Activity

Microtubule gliding occurs on the coverslip surface in crude extracts and in high speed supernatants, making it clear that there is active soluble motor protein present in both interphase and metaphase. Quantitating the relative amounts of gliding in different samples would be extremely difficult, so instead we have assayed soluble motor protein activity by preincubating carboxylated beads (0.11 μm, Polysciences No. 16688, diluted 1/100 from a 5% solid stock) in metaphase or interphase high-speed supernatant for 5 minutes at 23°C. The bead mixture can then be transferred to a flow cell and analyzed as usual by DIC. The relative activity of total soluble motors is measured by counting the number of bead movements per micrometer of microtubule per minute, as described in Section IV,B,2. If a polar source of microtubules is used (see Section III,C), it is clear that beads move predominantly toward the minus ends of microtubules in both interphase and metaphase (Allan and Vale, 1991).

3. Checking Cell Cycle Status

For all motility assays in which the cell cycle status of the extract or incubation mix needs to be known, parallel incubations should be set up in microcentrifuge tubes, and aliquots should be removed for histone kinase assays (Section II,D) at the same time as vesicle movement or tubule formation is recorded. Figure 2 gives an example of how histone kinase activity can be related to the amount of microtubule-based motility.

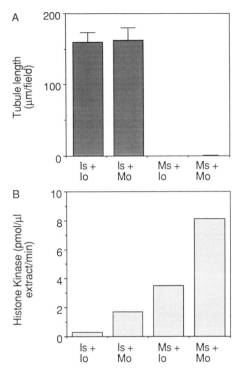

Fig. 2 The extent of membrane networks depends on the cell cycle status of the high-speed supernatant. Interphase soluble components (Is) promote active membrane tubule formation when combined with either interphase (Io) or metaphase (Mo) membranes (panel A). Metaphase soluble components (Ms), in contrast, fail to cause membrane tubule extension from either interphase or metaphase fractions (panel A). Each bar shows the average ($\pm$)S EM extent of membrane tubules in 25 random fields. The histone kinase activities of the same combinations show that the added membranes have some effect on the overall p34^{cdc2} activity (panel B). Even so, the combination mitotic supernatant + interphase membranes remains more mitotic, as determined by histone kinase activity, than does the combination interphase supernatant + metaphase membranes.

D. Assaying Activity of Plus-End- and Minus-End-Directed Motors

1. Directionality Assay

The most straightforward way of determining the direction of movement along microtubules is to use a source of microtubules with defined polarity. The three methods described below make use of recognizable nucleation centers (nuclei, axonemes, and centrosomes) from which microtubules extend plus ends outward. These assays establish whether the active motor is plus end or minus end directed, but cannot provide proof that the motors are kinesin or cytoplasmic dynein, respectively.

a. Demembranated Sperm Nuclei

The demembranated *Xenopus* sperm nuclei which can be used to show cell cycle status of the extracts (Section II,D,2) will also nucleate microtubules from their basal bodies *in vitro*. Dilute the sperm to about 5×10^5/ml in BRB80 and incubate for 15–20 minutes in a flow cell, coverslip downward, in a humid chamber. During this time, the nuclei will sink onto the coverslip surface and stick tightly to the glass. Wash with 3 vol of BRB80, and then flow in 10 μl of 2 mg/ml tubulin in BRB80 plus 1 mM GTP. Incubate the flow cells at 37°C for 20 minutes and then wash with 3 vol BRB80 plus 20 μM taxol (freshly diluted). (The microtubules are stable overnight if the flow cells are kept in a humid chamber at room temperature.) To avoid studying a diluted extract, 1.5–2 vol of sample should be perfused into the flow cell.

b. Axonemes

Salt-washed, dynein-depleted axonemes, prepared from sea urchin sperm as described (Gibbons and Fronk, 1979), are stored in 50% glycerol at −20°C. They can be used as nucleation centers for the polymerization of purified tubulin or mixtures of tubulin and *N*-ethylmaleimide (NEM)–tubulin (Hyman *et al.*, 1991). It is most convenient, however, to use the *Xenopus* tubulin in the extracts as follows. Dilute the axonemes in acetate buffer, and flow into the perfusion chamber. Incubate the chamber, coverslip down, for 15 minutes at room temperature in a humid chamber. Wash out the unbound axonemes with acetate buffer and then flow in the *Xenopus* extract or supernatant to be assayed. Microtubules then polymerize from one end of the axoneme within 5 minutes at room temperature, while the sample is being observed by DIC microscopy. Membrane or bead movement is best recorded within 10–20 minutes, before too many free microtubules polymerize. The microtubules polymerize only from the plus ends of the axoneme in these extracts, as can be shown by stabilizing the microtubules with taxol and then adding carboxylated beads coated with purified kinesin, (Allan and Vale, 1991), which always move away from the axoneme.

c. Centrosomes

Centrosomes can also be used to nucleate microtubules of defined polarity, but the methods involved are more complex than those given above and are not described here. Isolated centrosomes (prepared according to Mitchison and Kirschner, 1984) often do not nucleate well after being adsorbed to glass, so it is best to nucleate microtubules from the centrosomes in solution and then spin the asters through a sucrose cushion onto glass coverslips (Mitchison and Kirschner, 1984; Vale *et al.*, 1985).

2. Other

A number of other methods may be used to help define which motors are driving membrane movement along microtubules in the *in vitro* assay. The sensitivity of cytoplasmic dynein to vanadate treatment in the presence of ultraviolet light (Gibbons and Mocz, 1991) has been used to show that cytoplasmic dynein is involved in aster formation in metaphase *in vitro* (Verde *et al.*, 1991). It has recently been shown that individual purified motors can use a characteristic range of nucleotide analogs (Shimizu *et al.*, 1991), but this method is unlikely to be useful in crude extracts because of the large amounts of endogenous nucleotides present. The use of motor-specific antibodies either to inhibit motility directly (e.g., Ingold *et al.*, 1988) or to absorb out motors from a soluble fraction (Vale *et al.*, 1985) has not been tested so far in the *Xenopus* system.

E. Troubleshooting

A number of problems may be encountered using this assay, some of which are described below. The most common source of problems is the extract itself, as batches of eggs vary. At certain times of year (particularly in the summer), it can be difficult to obtain good eggs. The amount of motility seen in interphase extracts made from apparently good eggs can also vary, for unknown reasons.

One critical point to establish when a new extract is being tested, or when a different membrane fraction is added to a high-speed supernatant, is whether the vesicle or tubule movements observed are due to the membrane moving along the microtubule or whether the membrane is passively attached to a microtubule that is itself gliding across the glass surface. It is possible to distinguish these alternatives by close observation at low microtubule densities. It is also important to test the effects of any new buffers on the motility assay, as I have observed, for example, that the presence of more than 0.3 M sucrose in the assay decreases the amount of membrane movement along microtubules, as does the inclusion of β-glycerol phosphate or sodium pyrophosphate (V. Allan, unpublished results).

It is obviously important for comparisons of motility that the number of microtubules per field does not vary too much between samples. Although the protocols given in Section III,B should ensure this, some extracts will never give similar numbers in both interphase and metaphase states and so should not be used for quantitative comparison. It is also quite common for metaphase extracts to promote the formation of large asterlike groups of microtubules on the glass surface after 30–45 minutes (see also Verde *et al.*, 1991), which again makes comparisons with the same extract in interphase difficult. By comparing motility at earlier times, this problem can be avoided to a large extent.

Unadulterated extracts will maintain a particular cell cycle status in the motility assay *in vitro*, but it is vital to remember that the addition of motility inhibitors may affect the balance of kinases and phosphatases, and so alter histone kinase levels. Likewise, changing phosphorylation levels may have an indirect effect on motility, perhaps triggering the metaphase inhibition of membrane movement. It is therefore very important that histone kinase activities are also measured after any treatment that affects membrane motility.

====== **IV. Data Aquisition, Handling, and Analysis**

A. Microscopy

1. Differential Interference Contrast Microscopy

Many possible DIC microscope setups can be used to follow organelle movement along individual microtubules in real time (e.g., Inoué, 1986; Salmon *et al.*, 1989). I outline below two different arrangements I have used. In Dr. Ron Vale's laboratory (Department of Pharmacology, University of California, San Francisco), we have used (Allan and Vale, 1991) a Zeiss Axioplan equipped with DIC optics, a 100-W mercury arc lamp, glass heat filters, a 546-nm narrowband interference filter, a 1.4NA oil immersion condenser, and a 63X, 1.4NA Planapochromat objective. Images were projected onto the target of a Hamamatsu Newvicon camera using a 1.25x or 1.6x optivar and a 20x eyepiece. Contrast was enhanced with an Imaging Technology 151 processor, and images were recorded in real time using either a $\frac{3}{4}$-in. U-matic Sony VO-5800H or a $\frac{1}{2}$-in. Panasonic NV-8950 video recorder.

At the Laboratory of Molecular Biology (Cambridge, U.K.), we are using a system based on a Zeiss standard microscope (W. B. Amos, L. A. Amos, and V. J. Allan, unpublished results). The microscope is fitted with DIC optics, a 1.4NA achromatic aplanatic oil immersion condenser, and a plan 100X, 1.25NA lens. Light from a modified Oriel Scientific Ltd. (Leatherhead, Surrey, U.K.) 100-W lamp (fitted with Pyrex glass condenser lenses) is passed through both water and glass heat filters and a 546-nm narrowband interference filter, before being passed to the microscope via a 1-mm silica optical fiber. The optical fiber serves to "scramble" the light from the mercury arc, giving a more uniform filling of the back focal plane of the microscope, as is required for good Koehler illumination (e.g., Salmon *et al.*, 1989). The fibers are available from R. Knudsen (Marine Biology Laboratory, Woods Hole, MA). DIC images are passed through a 1.25x or 1.6x optivar and a 12x eyepiece, and are collected using a Hamamatsu Newvicon camera (C2400-07). Background subtraction, image contrast enhancement, and frame averaging are performed in real time using a Hamamatsu Argus 10 image processor. We routinely average two frames, and record onto S-VHS tape using a Panasonic S-VHS AG7330 recorder. Images are displayed on a Sony RGB color monitor (PVM 1442 QM).

2. Fluorescence

Fluorescently labeled samples in flow cells can be viewed using any microscope with epifluorescent illumination and appropriate filters. Images can be recorded directly onto 35-mm film. If higher magnification is required or if further image processing is desired, images can be collected using a low-light-level camera such as a Hamamatsu silicon-intensified-target camera (C-2400) or a Hamamatsu intensified CCD camera (C3505-17) coupled to an image processor. Single frames can be grabbed from the video signal as described below (Section IV,C). Fluorescent samples may also be viewed at high magnification on a confocal fluorescence microscope.

B. Measurement

1. Rate of Movement

Rates of movement are obtained from DIC sequences recorded onto videotape using an interactive AT-based computer program (Sheetz *et al.*, 1986). The velocity of bead, vesicle, and membrane tubule movement along microtubules, as well as rates of microtubule gliding over glass, can be measured easily in this way. The basic requirements for setting up this system are as follows:

1. A copy of the Measure program, which is available from Dr. S. M. Block, Rowland Institute for Science, Cambridge, Massachusetts

2. A Mark V video cursor board from Mike Walsh Electronics, San Dimas, California [FAX (818) 584 1654]

3. An IBM-PC, IBM-XT, IBM-AT, or IBM-compatible computer with at least 256 kb memory

4. A Panasonic AG-6300 or AG-7330 video recorder with the optional 34-pin interface fitted, which provides the Mark V board with video frame count information

This system was designed for NTSC signals, but the board and program can be successfully converted to PAL format (L. A. Amos and D. Cattermole, unpublished results: further details are available on request).

2. Amount of Vesicle Movement

To obtain some objective measure of vesicle movement in interphase compared with metaphase, we have quantitated the number of movements per micrometer of microtubule per minute. This can be performed on recorded sequences either by counting vesicle movements on individual microtubules, if the microtubule density is low, or by counting movements within a set region of the video field. In both cases, microtubules are traced from the monitor onto acetate sheets and their lengths are measured using a map measure. Exactly the same method can be used to quantitate the amount of bead movement along microtubules.

3. Tubular Membrane Network Extent

To compare the ability of interphase and metaphase extracts to form tubular membrane networks, some quantitation of network extent is needed. We have measured the lengths of membrane tubules (Allan and Vale, 1991) by recording about 25 random video fields per sample onto videotape, and then tracing the networks manually from the monitor onto acetate sheets. The length of tubule per sheet is then measured using a map measure, and converted to micrometers. The average total length of tubules is expressed per video field. It is much easier to measure tubule extent if the density of tubules is not too great. Using this method we have shown that the length of membrane tubules per field in interphase extracts is approximately 100-fold greater than in corresponding metaphase extracts (Allan and Vale, 1991).

4. Troubleshooting/Artifacts

There are two main causes of artifactual rates. Firstly, if the moving object displays discontinuous movement, then the overall rate will be too low. It is therefore important to check that the object to be tracked does not stop and start. Second, if a bead, vesicle, or tubule is moving along a microtubule that is itself gliding over the glass surface, then the rates of movement obtained will be inaccurate. By measuring rates from fields in which there are relatively few microtubules, it is possible to avoid those microtubules that are moving. Gliding microtubules also produce the most common artifact encountered when analyzing the number of moving beads or membranes, as such a structure, if it is simply stuck to a microtubule, will appear to be motile if the microtubule moves. With close observation, however, it is possible to distinguish truly moving particles from those being carried along by microtubules, particularly as the former are much more likely to switch from one microtubule to another. Again, it is best to analyze only those areas where microtubules are sparse. Alternatively, Schroer and Sheetz (1992) have described a method for preparing surfaces coated with stationary microtubules, which should avoid this problem.

The main difficulty encountered when measuring the extent of tubular networks is that part of the network also extends out of the plane of focus. Most of these tubules can be excluded from analysis if the focus is moved up and down while each field is being recorded.

C. Transferring Images from Video Signals to Film

It is sometimes necessary to transfer video images to 35-mm film to make slides or figures for publication. This can be achieved by taking photographs of the monitor screen, but this gives rise to prominent video lines on the film image. One way around this problem is to digitize video images and manipulate them as described below. It may be possible to use some image processors to grab frames and transfer them to a computer, but unless the image processor has two frame stores, it is not possible to use background subtraction at the same time as freezing images. We take the Argus 10 output signal (after background subtraction and contrast enhancement, if required) and grab frames using a Quick Capture board (Data Translation, Wokingham, Berks., U.K.) in an IBM-AT-compatible computer. It is also possible to use a Macintosh II computer with a frame grabber board for the same purpose. It is also sometimes very useful to be able to grab frames from videotape, although the quality of the images obtained will be lower than those captured live. Not all frame grabbers can lock onto the VCR synchronization signal successfully, so it is vital to test a particular board with your VCR. Some frame grabbers overcome this problem by containing a phase-locked loop.

The stored, digitized image is now ready for further image processing. A number of commercial programs are available, but it is also convenient to use IMAGE, which is a public domain program for the Macintosh II. Digitized images stored in a number of different formats on IBMs can be imported into IMAGE after conversion of the files to Macintosh format. If you are using a Macintosh with an appropriate frame grabber, the images can be directly imported into IMAGE. IMAGE can reduce or enlarge images and mount them side-by-side, allowing contruction of figure panels directly. Final images can then be transferred to film using a Macintosh slide maker or by photographing the Macintosh high-resolution screen directly. Alternatively, camera-ready copies can be printed directly from the file on a color laser writer (e.g., Canon color laser copier 300).

V. Conclusions and Prospects

Xenopus egg extracts have already been widely used to study cell cycle-regulated phenomena, including the control of microtubule dynamics (Belmont *et al.,* 1990; Verde *et al.,* 1990), spindle formation (Sawin and Mitchison, 1991), and microtubule-based membrane transport (Allan and Vale, 1991). The extracts provide a number of advantages for the study of such complex processes. First, all the necessary components for a wide range of functions are contained in the crude extracts; for the study of microtubule-related phenomena, these include tubulin, microtubule-associated proteins, motor proteins, and their activators, as well as membranes that are capable of binding active motor proteins. Second, membranes from different species can be used in place of *Xenopus* membranes to study motility (see Section III,C,1) and membrane fusion (Tuomikoski *et al.,* 1989). Third, the extracts work optimally at room temperature, and fourth, the extracts can be made in reasonable quantities, and for many uses can be frozen and thawed. *Xenopus* cell lines are also available (for a review, see Smith and Tata, 1991), which simplifies the screening of antibodies. In addition, more *Xenopus* proteins are being characterized using molecular biological methods, such as the kinesin-related protein Eg5 (Le Guellec *et al.,* 1991).

There are obviously some disadvantages in using *Xenopus* extracts, such as the effort and investment needed to set up a frog colony. In addition, frogs are not easily manipulated genetically. It is not trivial to obtain large amounts of material for biochemical purification, although it can be done (Lohka *et al.,*

1988), and the extracts themselves are always variable to some degree; however, by using an extract prepared from one batch of eggs and converting it between interphase and metaphase states *in vitro*, the latter problem is reduced, especially because the multiple aliquots can be kept frozen.

In conclusion, *Xenopus* extracts have already proved useful in studying many aspects of microtubule function and organization, and in the future may also provide information about complex processes such as the regulation of the direction of membrane movement.

Acknowledgments

I thank Dr. Ron Vale, with whom much of this work was carried out, for his support and enthusiasm. I am very grateful for help and advice from Dr. Andrew Murray, Dr. Michael Glotzer, and Dr. Mark Solomon on making egg extracts. I thank Michael Glotzer for providing expressed cyclin-Δ-90. I gratefully acknowledge the help of Dr. Brad Amos, Dr. Linda Amos, and Mr. David Cattermole in setting up the video microscope and velocity measuring system at the Laboratory of Molecular Biology. I thank Dr. Steven Block for providing the Measure program, and Dr. Linda Amos and Mr. Pete Brown for commenting on the manuscript.

References

Allan, V. J., and Vale, R. D. (1991). Cell cycle control of microtubule-based membrane transport and tubule formation *in vitro*. *J. Cell Biol.* **113,** 347–359.

Allan, V. J., Vale, R. D., and Navone, F. (1991). Microtubule-based organelle transport in neurons. *In* "The Neuronal Cytoskeleton" (R. D. Burgoyne, ed.), pp. 257–282. Wiley-Liss, New York.

Arnheiter, H., Dubois-Dalcq, M., and Lazzarini, R. A. (1984). Direct visualization of protein transport and processing in the living cell by microinjection of specific antibodies. *Cell (Cambridge, Mass.)* **39,** 99–109.

Belmont, L. D., Hyman, A. A., Sawin, K. E., and Mitchison, T. J. (1990). Real-time visualization of cell cycle dependent changes in microtubule dynamics in cytoplasmic extracts. *Cell (Cambridge, Mass.)* **62,** 579–589.

Berlin, R. D., and Oliver, J. M. (1978). Surface functions during mitosis. I. Phagocytosis, pinocytosis and mobility of surface-bound Con-A. *Cell (Cambridge, Mass.)* **15,** 327–341.

Blow, J. J., and Laskey, R. A. (1986). Initiation of DNA replication in nuclei and purified DNA by a cell-free extract of *Xenopus* eggs. *Cell (Cambridge, Mass.)* **47,** 577–587.

Cooper, M. S., Cornell-Bell, A. H., Chernjavsky, A., Dani, J. W., and Smith, S. J. (1990). Tubulovesicular processes emerge from trans-Golgi cisternae, extend along microtubules, and interlink adjacent trans-Golgi elements into a reticulum. *Cell (Cambridge, Mass.)* **61,** 135–145.

Dabora, S. L., and Sheetz, M. P. (1988). Microtubule dependent formation of a tubular vesicular network with characteristics of the endoplasmic reticulum from cultured cell extracts. *Cell (Cambridge, Mass.)* **54,** 27–35.

Featherstone, C., Griffiths, G., and Warren, G. (1985). Newly synthesized G protein of vesicular stomatitis virus is not transported to the Golgi complex in mitotic cells. *J. Cell Biol.* **101,** 2036–2046.

Felix, M.-A., Pines, J., Hunt, T., and Karsenti, E. (1989). A post-ribosomal supernatant from activated *Xenopus* eggs that displays post-translationally regulated oscillation of its cdc2 + mitotic kinase activity. *EMBO J.* **8,** 3059–3069.

Gibbons, I. R., and Fronk, E. (1979). A latent adenosine triphosphatase form of dynein 1 from sea urchin sperm flagella. *J. Biol. Chem.* **254,** 187–196.

Gibbons, I. R., and Mocz, G. (1991). Photocatalytic cleavage of proteins with vanadate and other transition metal complexes. *In* "Methods in Enzymology" (R. B. Vallee, ed.), Vol. 196, pp. 428–442. Academic Press, San Diego.

Glotzer, M., Murray, A. W., and Kirschner, M. W. (1991). Cyclin is degraded by the ubiquitin pathway. *Nature (London)* **349,** 132–138.

Ho, W. C., Allan, V. J., van Meer, G., Berger, E. G., and Kreis, T. E. (1989). Reclustering of scattered Golgi elements occurs along microtubules. *Eur. J. Cell Biol.* **48,** 250–263.

Hutchison, C. J., Cox, R., Drepaul, R. S., Gomperts, M., and Ford, C. C. (1987). Peridodic DNA synthesis in cell-free extracts of *Xenopus* eggs. *EMBO J.* **6,** 2003–2010.

Hyman, A., Drechsel, D., Kellogg, D., Salser, S., Sawin, K., Steffen, P., Wordeman, L., and Mitchison, T. (1991). Preparation of modified tubulins. *In* "Methods in Enzymology" (R. B. Vallee, ed.), Vol. 196, pp. 478–485. Academic Press, San Diego.

Ingold, A. L., Cohn, S. A., and Scholey, J. M. (1988). Inhibition of kinesin-driven microtubule motility by monoclonal antibodies to kinesin heavy chains. *J. Cell Biol.* **107,** 2657–2667.

Inoué, S. (1986). "Video Microscopy." Plenum, New York.

Langan, T. A., Gautier, J., Lohka, M., Hollingsworth, R., Moreno, S., Nurse, P., Maller, J., and Sclafani, R. A. (1989). Mammalian growth-associated H1 histone kinase: A homolog of cdc2 + /CDC28 protein kinase controlling mitotic entry in yeast and frog cells. *Mol. Cell Biol.* **9,** 3860–3868.

Leelavathi, D. E., Estes, L. W., Feingold, D. S., and Lombardi, B. (1970). Isolation of a Golgi-rich fraction from rat liver. *Biochim. Biophys. Acta* **211,** 124–138.

Le Guellec, R., Paris, J., Couturier, A., Roghi, C., and Philippe, M. (1991). Cloning by differential screening of a *Xenopus* cDNA that encodes a kinesin-related protein. *Mol. Cell Biol.* **11,** 3395–3398.

Lohka, M. J., and Maller, J. L. (1985). Induction of nuclear envelope breakdown, chromosome condensation, and spindle formation in cell-free extract. *J. Cell Biol.* **101,** 518–523.

Lohka, M. J., Hayes, M. K., and Maller, J. L. (1988). Purification of maturation-promoting factor, an intracellular regulator of early mitotic events. *Proc. Natl. Acad. U.S.A.* **85,** 3009–3013.

Lucocq, J. M., Berger, E. G., and Warren, G. (1989). Mitotic Golgi fragments in HeLa cells and their role in the reassembly pathway. *J. Cell Biol.* **109,** 463–474.

Mitchison, T., and Kirschner, M. (1984). Microtubule assembly nucleated by isolated centrosomes. *Nature (London)* **312,** 232–237.

Murray, A. W. (1991). Cell cycle extracts. *Methods Cell Biol.* **36,** 581–605.

Murray, A. W., and Kirschner, M. (1989). Cyclin synthesis drives the early embryonic cell cycle. *Nature (London)* **339,** 275–280.

Murray, A. W., Solomon, M., and Kirschner, M. W. (1989). The role of cyclin synthesis in the control of maturation promoting factor activity. *Nature (London)* **246,** 614–621.

Newmeyer, D. D., and Wilson, K. L. (1991). Egg extracts for nuclear import and nuclear assembly reactions. *Methods Cell Biol.* **36,** 607–634.

Newport, J. (1987). Nuclear reconstitution in vitro: Stages of assembly around protein-free DNA. *Cell (Cambridge, Mass.)* **48,** 205–217.

Newport, J., and Spann, T. (1987). Disassembly of the nucleus in mitotic extracts: membrane vesicularization, lamin disassembly, and chromosome condensation are independent processes. *Cell (Cambridge, Mass.)* **48,** 219–230.

Pypaert, M., Lucocq, J. M., and Warren, G. (1987). Coated pits in interphase and mitotic A431 cells. *Eur. J. Cell Biol.* **45,** 23–29.

Sagata, N., Watanabe, N., Vande Woude, G. F., and Ikawa, Y. (1989). The c-mos proto-oncogene product is a cytostatic factor (CSF) responsible for meiotic arrest in vertebrate eggs. *Nature (London)* **342,** 512–518.

Salmon, T., Walker, R. A., and Pryer, N. K. (1989). Video-enhanced differential interference contrast light microscopy. *BioTechniques* **7,** 624–633.

Sawin, K. E., and Mitchison, T. J. (1991). Mitotic spindle assembly by two different pathways *in vitro*. *J. Cell Biol.* **112,** 925–940.

Schroer, T. A., and Sheetz, M. P. (1992). Two activators of microtubule-based vesicle transport. *J. Cell Biol.* **115,** 1309–1318.

Sheetz, M. P., Block, S. M., and Spudich, J. A. (1986). Myosin movement *in vitro:* a quantitative assay using oriented actin cables from *Nitella. Meth. Enzymol.* **134,** 531–544.

Shimizu, T., Furusawa, K., Ohashi, S., Toyoshima, Y. Y., Okuno, M., Malik, F., and Vale, R. D. (1991). Nucleotide specificity of the enzymatic and motile activities of dynein, kinesin, and heavy meromyosin. *J. Cell Biol.* **112,** 1189–1197.

Smith, J. C., and Tata, J. R. (1991). *Xenopus* cell lines. *Methods Cell Biol.* **36,** 635–654.

Tuomikoski, T., Felix, M.-A., Dorée, M., and Gruenberg, J. (1989). Inhibition of endocytic vesicle fusion *in vitro* by the cell-cycle control protein kinase cdc2. *Nature (London)* **342,** 942–945.

Vale, R. D. (1991). Severing of stable microtubules by a mitotically activated protein in *Xenopus* egg extracts. *Cell (Cambridge, Mass.)* **64,** 827–839.

Vale, R. D., and Hotani, H. (1988). Formation of membrane networks *in vitro* by kinesin-driven microtubule movement. *J. Cell Biol.* **107,** 2233–2242.

Vale, R. D., Schnapp, B. J., Mitchison, T., Steuer, E., Reese, T. S., and Sheetz, M. P. (1985). Different axoplasmic proteins generate movement in opposite directions along microtubules *in vitro. Cell (Cambridge, Mass.)* **43,** 623–632.

Verde, F., Labbé, J., Dorée, M., and Karsenti, E. (1990). Regulation of microtubule dynamics by cdc2 protein kinase in cell-free extracts of *Xenopus* eggs. *Nature (London)* **343,** 233–238.

Verde, F., Berrez, J.-M., Antony, C., and Karsenti, E. (1991). Taxol-induced microtubule asters in mitotic extracts of *Xenopus* eggs: Requirement for phosphorylated factors and cytoplasmic dynein. *J. Cell Biol.* **112,** 1177–1187.

CHAPTER 16

Microtubule Motor-Dependent Formation of Tubulovesicular Networks from Endoplasmic Reticulum and Golgi Membranes

James M. McIlvain, Jr., Carilee Lamb, Sandra Dabora,★ and Michael P. Sheetz

Department of Cell Biology
Duke University Medical Center
Durham, North Carolina 27710

★ Department of Cell Biology
Washington University
St. Louis, Missouri 63130

I.   Introduction
II. Materials and Methods
     A. Stock Solutions
     B. Network Component Preparation
     C. Harvesting of Cells and Membrane Vesicles
     D. Motor Preparation
     E. Phosphocellulose-Purified Bovine Microtubule Preparation
     F. Network Formation
     G. Peripheral Membrane Extraction
     H. Microtubule Motor Depletion
     I. Microtubule Motor Purification
     J. Video-Enhanced Differential Interference Contrast Microscopy
III. Discussion
     A. Difficulties in Network Formation
     B. Gradients
     C. Extraction of Peripheral Membrane Proteins
     D. Motor Dependence of Network Formation
     References

I. Introduction

Many questions remain about specific functions within the endoplasmic reticulum, Golgi, and endosomes that are difficult to approach within cells or in lysed cell preparations. The dynamic nature of processes such as ion transport, protein compartmentalization, lateral protein movement, and sorting makes them difficult to study in cells. The *in vitro* networks provide sufficient spatial separation of these processes, allowing direct visual observation by immunofluoroesence (McIlvain and Sheetz, 1992) or single-particle tracking techniques (Gelles *et al.*, 1988).

A number of functions have been reconstituted in the *in vitro* networks, including protein translation and translocation, lateral protein movement, and calcium release. The major problem is defining the specific membrane fraction involved for each process. To solve this problem, gradient separation of the primary membrane fraction is necessary. The major difficulty encountered with membrane fractionation, however, is the loss of network-forming activity. Network formation is dramatically reduced when chick embryo fibroblast membranes are centrifuged through sucrose cushions. In general, we have found that with each experimental treatment, network-forming activity decreases.

We describe here methods for forming tubulovesicular networks, separating the membranes into at least two fractions with little loss of network-forming activity, and extracting peripheral membrane proteins. Finally, we present various problems encountered in working with the tubulovesicular membrane networks.

II. Materials and Methods

A. Stock Solutions

Freezing media
  Iscove's medium (Gibco)
  30% fetal calf serum (Gibco)
  10% Dimethylsulfoxide (DMSO) tissue culture grade (Sigma)
PMEE′ Plus buffer
  35 mM 1,4-piperazinediethane sulfonic acid (PIPES)
  5 mM MgSO$_4$·7H$_2$0
  1 mM ethylene glycol bis (β-aminoethyl ether) N, N' tetraacetic acid (EGTA),
    0.5 mM ethylenediamine tetraacetic acid (EDTA)
  pH 7.4 with KOH
  Add dithiothreitol (DTT, final concentration 1 mM) and protease inhibitors before use

Protease inhibitor cocktail
  Stock A (200×)
    To 100% ethanol add
      Pepstatin 0.2 mg/ml
      $N\alpha$-p-tosyl-L-arginine methyl ester (TAME) 2.0 mg/ml
      N-tosyl-L-phenylalanine chloromethyl ketone (TPCK) 2.0 mg/ml
    Store at −80°C
  Stock B (200×)
    To deionized distilled H_2O add
      Leupeptin 0.2 mg/ml
      Soybean trypsin inhibitor 2.0 mg/ml
    Store at −20°C
Phenylmethylsulfonyl fluoride (PMSF) stock solution (100×)
To 100% isopropyl alcohol add PMSF to a final concentration of 100 mM
ATP release buffer
  10 mM ATP, 10 mM $MgSO_4$, 1 mM GTP, and 20 μM taxol in PMEE' Plus
    Additional stock solutions required
  100 mM ATP (10×)
  200 mM GTP (200×)
  4 mM taxol in 100% DMSO (200×)
  Phosphocellulose-purified bovine tubulin (5 mg/ml)
  10 mM sodium vanadate
  SUK-4 (antikinesin)–Sepharose (Ingold *et al.,* 1988) and protein A–
Sepharose

B. Network Component Preparation

Several cell lines have been used to make networks by this procedure, such as chick embryo fibroblasts (CEF), African green monkey kidney cells (CV-1), Madin Darby canine kidney cells (MDCK) (McIlvain and Sheetz, 1992; Dabora and Sheetz, 1988), rat liver, and *Xenopus* egg extract (Allen *et al.,* 1981). We routinely use chick embryo fibroblasts obtained from 12-day-old embryos (Kelly and Schlessinger, 1978). Briefly, primary CEF cells isolated from embryos are grown to confluency (2–3 days) in minimum essential medium (MEM) containing Earle's salts (GIBCO), 5% fetal bovine serum, 40 units/ml penicillin, and 40 μg/ml streptomycin. The primary cells are then trypsinized and resuspended in 5 ml of freezing medium per roller bottle. Aliquots of 1 ml are frozen overnight at −80°C and then transferred to a liquid nitrogen freezer. Secondary cells are grown up using 2 ml of primary cells per roller bottle.

C. Harvesting of Cells and Membrane Vesicles

1. Harvest 2–3-day-old confluent secondary CEF cells from four 850-cm^2 roller bottles by trypsinizing with 12.5 ml of 0.05% trypsin–EDTA (GIBCO) per roller bottle for 15 minutes at 37°C.

2. Pipet cells into two ice-cold 50-ml conical tubes. Rinse each roller bottle with 12.5 ml of culture medium containing serum. Add to cells in the 50-ml conical tubes.

3. Pellet cells by centrifugation at 1000*g* for 10 minutes at 4°C. Resuspend the cell pellets with 30 ml of PMEE′ Plus total and combine in one 50-ml conical tube.

4. Centrifuge a second time (1000*g* for 10 minutes at 4°C). Discard the supernatant.

5. Determine the packed cell volume, generally 700–800 μl.

6. Resuspend the cells in an equal volume of PMEE′ Plus.

7. Homogenize cells in an ice-cold cell buster using a 8.004-mm-diameter ball (Balch and Rothman, 1985) and four or five full strokes (one stroke is equal to one down/up cycle of a syringe) per aliquot of cells.

8. Centrifuge at 1000*g* for 15 minutes at 4°C to remove nuclei and unbroken cells.

9. Collect the postnuclear supernatant (S1), avoid the lipid layer if present, and discard the nuclear pellet (P1).

10. Centrifuge the S1 at 100,000*g* for 30 minutes at 4°C.

11. Decant the supernatant (S2) and save. This is the motor fraction.

12. Resuspend the membrane pellet (P2) in 60 μl of PMEE′ Plus.

Fractionation of S1. We have performed preliminary fractionations of S1 on Nycodenz gradients and generated two visible membrane fractions. Both fractions are capable of forming membrane networks; however, further characterization is required to identify the cellular membrane components contained within these fractions.

D. Motor Preparation

1. Add 5 μl of taxol and GTP to each milliliter of S2 supernatant from step 11 above (final concentrations 20 μM and 1 mM, respectively). Incubate at 37°C for 15 minutes to polymerize endogenous microtubules.

2. Centrifuge the S2 for 5 minutes at 100,000*g* in a Beckman airfuge.

3. Collect the supernatant (S3); use neat in the network assay. Discard the pellet of endogenous microtubules.

Figure 1 provides an overview of network preparation.

E. Phosphocellulose–Purified Bovine Microtubule Preparation

1. Polymerize phosphocellulose-purified bovine tubulin (Williams and Lee, 1982) (5 mg/ml) by the addition of taxol and GTP (final concentrations 20 μM and 1 mM, respectively).

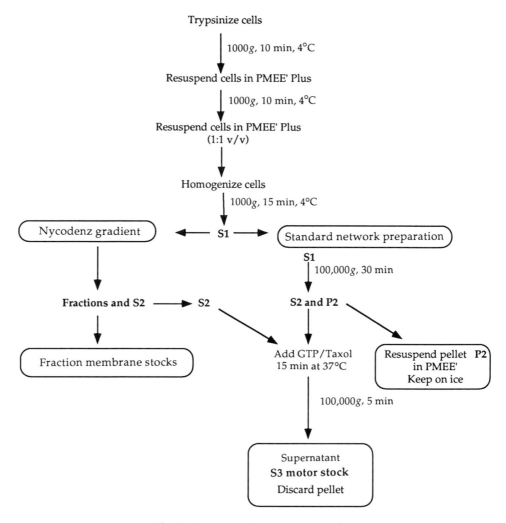

Fig. 1 Overview of network preparation.

2. Incubate at 37°C for 15 minutes.

3. Dilute 1 part freshly polymerized microtubules with 4 parts PMEE' Plus to form a working stock solution.

Microtubule density is critical to network formation, as too few microtubules result in little to no network formation, and too many may obscure observation of the tubulovesicular networks.

F. Network Formation

The standard network assay mixture consists of 1 μl of phosphocellulose-purified microtubule stock, 3 μl of S3, 1.2 μl of 10 mM ATP, and 1 μl of P2 (stock diluted 1 : 20 in PMEE' Plus), placed in order directly on a coverslip. We routinely use slotted aluminum slides that allow high-numerical-aperture microscopy. The samples are mounted between two coverslips, thereby minimizing the glass sample thickness. The standard network mixture is placed between parallel lines of high-vacuum silicon grease that form a sample chamber. The mixture is incubated in a humidified chamber for 1–2 hours at room temperature or 30–60 minutes at 37°C.

G. Peripheral Membrane Extraction

1. Add (final concentrations) 0.1 M sodium carbonate (pH 11.3) or 0.6 M potassium iodide to an aliquot of P2 membranes (Schroer *et al.*, 1989; Toyoshima *et al.*, 1992).

2. Incubate on ice for 5–10 minutes.

3. Centrifuge through a three-step sucrose gradient of 10, 15, and 20% sucrose in PMEE' Plus (w/v).

H. Microtubule Motor Depletion

The microtubule motors kinesin and cytoplasmic dynein can be selectively removed from the high-speed supernatant (S3) (Schroer *et al.*, 1989; Burkhardt *et al.*, 1993). These procedures can be performed sequentially, to deplete each motor.

1. Kinesin Depletion

1. Preequilibrate SUK-4 (antikinesin)–Sepharose and protein A–Sepharose (control) with PMEE' Plus.

2. Add 300 μl of S3 to 50 μl of each packed resin, and incubate for 1 hour at 4°C.

3. Remove the supernatant, and deplete a second time with fresh affinity resin as in step 2.

4. Remove the supernatant. Use in network assay as S3–kinesin or deplete the cytoplasmic dynein by UV sodium vanadate treatment.

2. Cytoplasmic Dynein Depletion

1. Incubate the S3 for 15 minutes with 2 mM ATP and 100 μM sodium vanadate.

2. Expose S3 to 365-nm UV for 50 minutes on ice, placing the UV lamp 7.6 cm above the sample.

I. Microtubule Motor Purification

1. Add 0.6 mg of phosphocellulose-purified tubulin per 1 ml of S2.

2. Add taxol and GTP (final concentrations 20 μM and 1 mM, respectively).

3. Incubate 20 minutes at 37°C.

4. Add AMPPNP and MgSO$_4$ (final concentration of 7 mM for both) to S2, and incubate at room temperature for 20 minutes.

5. Layer S2 onto a step gradient of 10 and 20% sucrose in PMEE' Plus with 1mM GTP and 20 μM taxol.

6. Centrifuge 60,000g for 30 minutes at 20°C.

7. Resuspend the microtubule pellets in 200 μl of ATP release buffer. Incubate at room temperature for 20 minutes.

8. Centrifuge for 10 minutes at 100,000g at room temperature.

9. Save the supernatant (the ATP release). Use it in the network assay to replace the S3. Discard the microtubule pellets.

J. Video–Enhanced Differential Interference Contrast Microscopy

Standard differential interference contrast (DIC) microscopy can be used to a limited extent to visualize the tubulovesicular networks. Because the tubulovesicular membranes are about 0.1–0.2 μm in diameter and network formation is highly dependent on microtubule motility, we recommend the use of video-enhanced DIC microscopy (Allen *et al.*, 1981; Schnapp, 1986).

III. Discussion

A. Difficulties in Network Formation

Although the CEF and CV1 P2 fractions have routinely formed networks, there are a number of relatively minor changes that can result in reduced network formation or complete inactivity. It is important to stress the need for fresh protease inhibitor cocktail; reduced protease inhibitor activity dramatically affects the active lifetime of the preparation. The best preparations last 2 to 3 days. Problems may also arise from differences in the extent of cell lysis. Incomplete lysis results in low concentrations of relevant membrane and supernatant fractions, and networks will not form. In addition, motility is readily lost on dilution of the S3.

Quantitative analysis of network formation and activity is difficult. Two key parameters are fusion and motility. Network formation is determined by the amount of membranes capable of fusing out of the total membrane volume. As the percentage of membranes capable of fusing within the cells may vary with time, we consistently use 2-day-old secondary cell cultures. We dilute the P2 membrane stock 1 in 20, which, when incubated 1–2 hours at room temperature, produces optimal networks 1–5 μm apart. Lower dilutions of the membrane stock (less than 1 in 10 in PMEE' Plus) produce a polygonal network too dense to work with.

Determining network activity by quantifying motility presents a similar dilemma. As with fusion, we expect only a small fraction of the total membrane vesicles in any cell to be capable of binding and moving on microtubules at any given time. One possible means of addressing this issue would be the use of laser optical tweezers (Kuo and Sheetz, 1992). The ability of random vesicles to bind and move along a microtubule could then be tested (see Burkhardt *et al.,* 1993); however, the costs associated with the optical tweezers limit their accessibility. Thus, relative vesicle motor activity is the easiest method to quantify network formation at this time (Schroer *et al.,* 1988; Schroer and Sheetz, 1991b).

B. Gradients

Classical gradient purification has been used to isolate the rat liver Golgi which retained the ability to form *in vitro* networks (Allan and Vale, 1991); however, centrifugation of CEF membranes through sucrose cushions often radically reduced network-forming activity. We found that the Nycodenz medium was generally suitable. This leads to further purification and identification of the CEF membrane fractions involved with network formation.

For each new membrane fraction it is important to characterize the membranes with standard membrane markers. We characterized membrane vesicles by extractions and protease treatments to develop an understanding of the nature of the surface components involved in motility.

C. Extraction of Peripheral Membrane Proteins

The membrane fractions prepared in these protocols have high concentrations of surface cytoskeletal and peripheral membrane proteins, including motor proteins. We have been able to remove many of these components with high-salt (potassium iodide) or alkaline (sodium carbonate) extractions. In our experience, the extraction with 0.1 M sodium carbonate (pH 11.3) (Toyoshima *et al.,* 1992) removed more peripheral proteins and gave more active membranes than 0.6 M potassium iodide extraction (Schroer *et al.,* 1988). These extractions are brief and followed by rapid washing of the membranes through a sucrose step gradient. It should be noted that a dramatic decrease in the buoyant density of the membranes occurs after extraction.

Important integral membrane proteins are required for organelle motility in our systems, as determined by the loss of motility with extraction and protease treatments. In addition to the extractions, we have tested for membrane protein dependence by treating membrane fractions with trypsin at 1 : 100 dilution (trypsin : membrane protein, for 30 minutes at room temperature).

Table I
Soluble Microtubule Motors and Other Soluble Factors Required in Network Formation

Soluble fraction	Network formation, unwashed P2 sonicated
S3 supernatant	+
ATP release	+
S3–kinesin	+
S3 UV vanadate	+
S3–kinesin, UV vanadate	−
S3–kinesin, UV vanadate, ATP release	+

D. Motor Dependence of Network Formation

Networks can be formed from either S2, S3, or an ATP release motor preparation (microtubule affinity purification procedure). The presence of motors is critical. When we tested the motor dependence of CEF network formation, no network formation was observed on removal of the motors, whereas either motor alone could support network formation (Table I). In other systems such as squid, the axoplasmic vesicles have been found to move without added motors, even after sodium carbonate extraction (Schnapp *et al.*, 1991).

It is important to point out the significance of carrier proteins to block the glass surface when dealing with low concentrations of soluble proteins in the assay. Significant fractions of the vesicles bind directly to the glass and are not available for motility or network formation. Schnapp and co-workers (1991) found that the addition of casein (0.1 mg/ml) dramatically reduced the adsorption of vesicles to the glass and preserved their motility.

It is possible to select for specific motors with different nucleotide and salt conditions (Schroer and Sheetz, 1991a,b). We generally use the ATP release outlined, for motility studies. Note, however, that if the motors are too highly purified, no motility is observed and no networks are formed. Thus, not only are motors necessary, but additional accessory proteins are required for motility (Schroer and Sheetz, 1991b) and subsequently network formation.

In summary, technical improvements continue to make it easier to prepare tubulovesicular networks *in vitro* for studies of subcellular membrane function. The advantage of a low membrane density and a relatively native environment makes it possible to study a wide variety of functions (McIlvain and Sheetz, 1992).

References

Allan, V. J., and Vale, R. D. (1991). *J. Cell Biol.* **113**(2), 347–359.
Allen, R. D., Allen, N. S., and Travis, J. L. (1981). *Cell Motil.* **1**, 291–302.

Balch, W., and Rothman, J. E. (1985). *Arch. Biochem. Biophys.* **240,** 413–425.

Burkhardt, J., McIlvain, J. M., Jr., Sheetz, M. P., and Argon, Y. (1993). *J. Cell Science* **104,** 151–162.

Dabora, S. L., and Sheetz, M. P. (1988). *Cell (Cambridge, Mass.)* **54,** 27–35.

Gelles, J., Schnapp, B. J., and Sheetz, M. P. (1988). *Nature (London)* **331,** 450–453.

Ingold, A. L., Cohn, S. A., and Scholey, J. M. (1988). *J. Cell Biol.* **107,** 2657–2667.

Kelly, B., and Schlessinger, M. J. (1978). *Cell (Cambridge, Mass.)* **15,** 1277–1286.

Kuo, S., and Sheetz, M. P. (1992). *Trends Cell Biol.* **2,** 116–118.

McIlvain, J. M., Jr., and Sheetz, M. P. (1992). *In* "Methods in Enzymology" (J. E. Rothman, ed.), Vol. 219, pp 72–80. Academic Press, San Diego.

Schnapp, B. J. (1986). *In* "Methods in Enzymology" (R. B. Vallee, ed.), vol. 134, pp 561–573. Academic Press, Orlando, FL.

Schnapp, B. J., Reese, T. S., and Bechtold, R. B. (1991). *J. Cell Biol.* **115,** 40a.

Schroer, T. A., and Sheetz, M. P. (1991a). *Annu. Rev. Physiol.* **53,** 629–652.

Schroer, T. A., and Sheetz, M. P. (1991b). *J. Cell Biol.* **115,** 1309–1318.

Schroer, T. A., Schnapp, B. J., Reese, T. S., and Sheetz, M. P. (1988). *J. Cell Biol.* **107,** 1785–1792.

Schroer, T. A., Steuer, E. R., and Sheetz, M. P. (1989). *Cell (Cambridge, Mass.)* **56,** 937–946.

Toyoshima, I., Yu, H., Stever, E. R., and Sheetz, M. P. (1992). *J. Cell Biol.* **118,** 1121–1131.

Williams, R. C., and Lee, J. C. (1982). *Nature (London)* **331,** 450–453.

CHAPTER 17

Cytoplasmic Extracts from the Eggs of Sea Urchins and Clams for the Study of Microtubule-Associated Motility and Bundling

Neal R. Gliksman,[1] **Stephen F. Parsons, and E. D. Salmon**

Department of Biology
University of North Carolina
Chapel Hill, North Carolina 27599

I. Introduction
II. Sea Urchin Egg Cytoplasmic Extracts
    A. Equipment
    B. Reagents for Extract Preparations
    C. Protocol
    D. Comments
III. Clam Egg Cytoplasmic Extracts
    A. Additional Equipment
    B. Reagents
    C. Protocol
IV. Visualizing the Microtubule-Associated Motility and Crosslinking
    A. Materials
    B. Protocol for a Slide Preparation
V. Example Results for Cytoplasmic Extracts
    References

[1] Present address: Department of Cell Biology, Duke University Medical Center, Durham, North Carolina 27710.

METHODS IN CELL BIOLOGY, VOL. 39

I. Introduction

Microtubules play active roles in cell locomotion, intracellular organelle transport, and organelle localization. Microtubule-based functions in the cell include pronuclear migration, intracellular organelle transport, the establishment and maintenance of membranous networks (endoplasmic reticulum and Golgi apparatus), and chromosome segregation by the mitotic or meiotic spindle. Many of these functions appear to be carried out by cytoplasmic microtubule-associated motor enzymes translocating along dynamic microtubules. Some of the microtubule-associated motor enzymes responsible for these functions have been identified as members of the kinesin and cytoplasmic dynein protein families, each of which has been purified from sea urchin eggs (Scholey *et al.*, 1984, 1985; Porter *et al.*, 1988). Many researchers are actively pursuing the roles of these enzymes using purified proteins. In this chapter we describe clarified cytoplasmic extracts made from the eggs of sea urchins or clams that are used to study microtubule-associated translocation and bundling in the presence of soluble cytoplasmic proteins. In the extracts, microtubules made from endogenous tubulin both translocate across coverslips and bundle to each other. In addition, the microtubules serve as a substrate for the bidirectional movement of organelles.

II. Sea Urchin Egg Cytoplasmic Extracts

Sea urchin oocytes and eggs are a rich source of the cytoskeletal components necessary for the rapid cell divisions following fertilization (Schroeder, 1986). Tubulin, microtubule-associated proteins, and microtubule motor proteins are three of the materials stockpiled for postfertilization events (Pfeffer *et al.*, 1976; Rose *et al.*, 1989). In the process of identifying *in vitro* conditions that would support dynamic microtubule assembly in interphase cytoplasm (Gliksman *et al.*, 1992; Simon *et al.*, 1992), we derived conditions that supported vigorous cytoplasmic dyneinlike motility and antiparallel microtubule–microtubule bundling in extracts of sea urchin eggs (Gliksman and Salmon, 1993). These cytoplasmic extracts are made from fertilized or unfertilized eggs as follows (Fig. 1):

A. Equipment

Disposable syringes (both 1 and 10 ml)

Two disposable needles (between 22 and 27 gauge)

Dissecting microscope or low-power (<20×) microscope

150-μm nylon mesh (105 μm for *Strongylocentrotus purpuratus*) (Tetko Inc., Elmsford, NY)

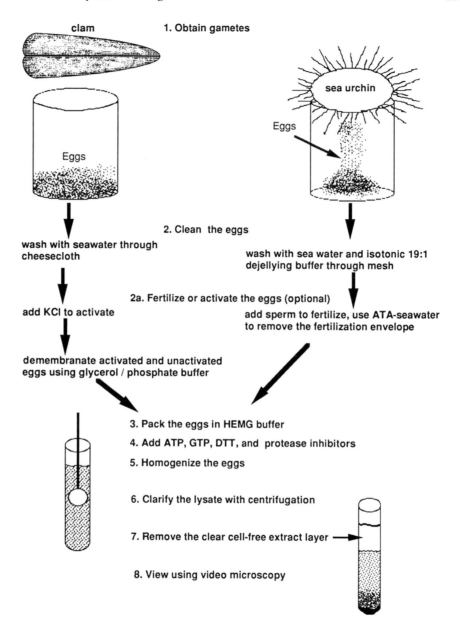

Fig. 1 Flowchart of the steps used to make a cytoplasmic extract from the eggs of sea urchins or clams.

Dounce-type hand tissue grinder/homogenizer (7 ml) (Wheaton Scientific, Millville, NJ)

Hand centrifuge or low-speed desktop clinical centrifuge for 15- and 50-ml conical centrifuge tubes

4–10 conical centrifuge tubes (15 and 50 ml)

Refrigerated ultracentrifuge (Beckman TL100 with TLA 100.3 and TL 55 rotors, or equivalent)

Aspirator

B. Reagents for Extract Preparations

Spawning buffer (5–20 ml required): 0.56 M KCl. Store at room temperature.

Isotonic 19:1 dejellying buffer (about 500 ml required): 0.53 M NaCl, 28 mM KCl, 5 mM Tris base, pH 8.0. Store at 4°C. Use at the appropriate temperature for the sea urchin species.[2]

Isotonic HEMG buffer (about 400 ml required): 180 mM potassium 4-(2-hydroxyethyl)-1-piperazineethanesulfonic acid (K-Hepes), 10 mM ethylene glycol bis(β-aminoethyl ether) N',N'-tetraacetic acid (EGTA), 10 mM MgCl$_2$, 0.5 M glycine, pH 7.35. Filter through 0.45-μm filter. Store and use at 4°C.

ATA seawater (if the eggs are to be fertilized; make up 500 ml at a time): 1 mM 3-amino-1,2,4-triazole in 0.45-μm filtered seawater. Make fresh at the appropriate temperature for the sea urchin species.[2]

Filtered seawater (0.45-μm Millipore filter or equivalent): Make up 1-liter batches. Store at 4°C and use at the appropriate temperature for the sea urchin species.[2]

Stocks of ATP (100 mM), GTP (50 mM), and DTT (100 mM) frozen in water or 50 mM Pipes, pH 7.2: Store frozen in 0.02- and 1-ml aliquots.

Stocks of protease inhibitors frozen in water: *p*-toluenesulfonyl-L-arginine methyl ester HCl (10 mg/ml), soybean trypsin inhibitor (10 mg/ml), benzamidine–HCl (10 mg/ml), aprotinin (10 mg/ml), and leupeptin (5 mg/ml). Store frozen in 20-μl aliquots.

100× stocks of protease inhibitors in ethanol: phenylmethylsulfonyl fluoride (saturated), pepstatin (1 mg/ml). Store at −20°C in less than 1-ml aliquots.

---

[2] Sea urchin eggs should be cultured and washed with seawater and isotonic 19:1 dejellying buffer which is close to physiological temperature. For *L. variegatus* eggs should be cultured and washed at 19–23°C; *L. pictus* eggs should be cultured and washed with much cooler (10–18°C) seawater and isotonic 19:1 dejellying buffer. If the eggs are to be fertilized, the eggs should be held at the same cool temperatures; otherwise the isotonic 19:1 dejellying buffer can be used on ice. Both seawater and isotonic 19:1 dejellying buffer should be kept on ice for *S. purpuratus* eggs.

C. Protocol

1. Obtain gametes. Shed sea urchin gametes by intracoelomic injection of 0.5–3 ml of spawning buffer through the peristomal membrane surrounding the mouth. In all three species the sperm will be white and the eggs will be various shades of yellow or orange. Concentrated sperm can be collected by pipetting directly from the gonadopores and should be stored immediately on ice with as little dilution as possible. Collect eggs by inverting each sea urchin over a 50- or 100-ml beaker (large enough for the urchin to sit on but not in the beaker) filled to the edge with seawater at an appropriate temperature for the sea urchin species.[1] Stop collecting the eggs when the shedding slows, as the later eggs tend to be immature. For consistency, exclude batches of eggs containing more than 5% immature oocytes, which have smaller egg diameters or large germinal vesicles. Generally, 30 ml of settled jellied eggs will yield 3–7 ml of dejellied packed eggs, the minimum volume necessary to make an extract.

2. Clean and demembranate the eggs. Wash the eggs three times in filtered seawater and twice in isotonic 19:1 dejellying buffer to remove contaminants and the egg jelly layer. The eggs are washed in seawater or other buffers by resuspending them in at least 10 times their volume and then filtering them through a 150-μm nylon mesh (105-μm for *S. purpuratus*). Finally, pellet the eggs softly using a hand centrifuge or a desktop clinical centrifuge at a low speed, and remove the supernatant by aspiration.

FOR FERTILIZED EGGS:

3. Sea urchin egg fertilization should be carried out in ATA seawater to allow for removal of the fertilization membrane (Showman and Foerder, 1979). Wash the sea urchin eggs and suspend in 50 ml of ATA seawater. Mix a few drops of concentrated sperm into 5 ml of ATA seawater and then add about 1 ml of the diluted sperm to the eggs; slowly stir or rock the solution to disperse the sperm among the eggs. Within 2–5 minutes of sperm addition or when nearly all of the eggs display a fertilization envelope, remove both the excess sperm and the fertilization envelopes by washing the eggs two or three times in ATA seawater using the 150-μm nylon mesh to strip the fertilization envelopes off of the embryos. 3-Amino-1,2,4-triazole can disrupt progression of the cell cycle; so wash the embryos in filtered seawater without 3-amino-1,2,4-triazole within 15 minutes of fertilization.

All subsequent steps should be performed on ice with ice-cold ingredients.

4. Wash unfertilized or fertilized eggs twice in isotonic HEMG buffer. Aspirate most of the supernatant of the second wash and pack the eggs using a hand-cranked centrifuge or desktop clinical centrifuge for 5–7 minutes at low speed. Pack the eggs tightly but do not crush the eggs. Aspirate the supernatant and uppermost 2 mm of eggs.

5. Transfer 3–7 ml of the packed, dejellied egg pellet into a precooled hand tissue grinder/homogenizer with a loose-fitting pestle. Add protease inhibitors and other ingredients to the following final concentrations: 1 mM ATP, 1 mM GTP, 1 mM dithiothreitol (DTT), 0.3 mM phenylmethylsulfonyl fluoride, 10 μg/ml p-toluenesulfonyl-L-arginine methyl ester HCl, 40 μg/ml soybean trypsin inhibitor, 1 μg/ml pepstatin, 1 μg/ml benzamidine–HCl, 10 μg/ml aprotinin, and 1 μg/ml leupeptin.

6. Homogenize the eggs until the majority of the eggs are lysed. Typically this takes between two and seven strokes, depending on the mortar and pestle being used, the sea urchin species, and the hardiness of the eggs. We use two or three strokes with the B pestle of a 7-ml Wheaton Dounce tissue grinder for eggs of *Lytechinus pictus* and *Lytechinus variegatus*. The A pestle can be used for *S. purpuratus*. Avoid creating air bubbles in the preparation, overhomogenizing the eggs, or grinding the eggs as these actions tend to increase contamination of the extract by yolk and DNA.

7. Clarify the homogenate by centrifugation. Spin the homogenate for 45 minutes at 50,000g (r_{av}) (Beckman TLA 100.3 rotor) at 3 °C. Carefully remove the cytoplasmic extract from the clear layer with a minimum of contamination from the top white lipid layer and the yellow pellet. We use a disposable syringe (1–5 ml) with a small-bore needle (22–27 gauge) and come in from the top of the centrifuge tube to remove the extract. If the initial extract is cloudy, as typically occurs with *S. purpuratus* eggs, reclarify the extract by spinning the extract for 60 minutes at 140,000g (r_{av}) (Beckman TL55 rotor) at 3°C. Five milliliters of packed, unbroken eggs typically produces 1 ml of clear cytoplasmic extract.

8. The cytoplasmic extract should be used immediately or aliquoted, shell-frozen in liquid nitrogen, and stored at −80°C. The motility properties of frozen extracts tend to be stable for at least 1 month.

D. Comments

Two types of seawater are commonly used in our laboratory for these procedures: MBL formula (Cavanaugh, 1956) and Instant Ocean Aquarium Systems (Mentor, OH).

The sea urchin sperm will retain its virility for several hours if it is kept cold and concentrated. The sperm can also be used for several days to make axoneme fragment microtubule nucleation sites (Pryer *et al.*, 1986).

The amount of time the eggs remain in 19:1 dejellying buffer must be limited, especially if the eggs are to be fertilized. Prolonged incubation in 19:1 dejellying buffer can inhibit post-fertilization development and makes egg membranes sticky.

The concentration of sperm used to fertilize the eggs should be checked each time with a small sample of eggs. The concentration should be low enough to prevent polyspermy, yet high enough to allow for efficient fertilization within 5 minutes of sperm addition.

A common mistake is to try to recover all of the clear cytoplasmic extract layer from the centrifuge tube. It is very easy to contaminate extracts with membranes or yolk material, decreasing the ability to visualize microtubules and microtubule-associated motility. Do not sacrifice the quality of your extract for quantity of extract.

III. Clam Egg Cytoplasmic Extracts

As with sea urchin eggs, the oocytes of the surf clam *Spisula solidissima* can be collected in quantities suitable for biochemical studies and are excellent cytological specimens. The mature *Spisula* oocyte has an intact germinal vesicle with a tetraploid complement of partially condensed chromosomes. On fertilization, the oocytes go through two meiotic divisions at 25 and 45 minutes, followed by zygotic, mitotic divisions. The proximity in time between fertilization (or artificial activation) and the first meiotic division allows the harvesting of cells in high cell cycle synchrony.

A. Additional Equipment

Cheesecloth

Flat knife

Scissors

Low-speed electric motor attached to a paddle (if the eggs are to be activated) (60 rpm or lower)

Desktop clinical centrifuge for 50-ml conical centrifuge tubes (1600 rpm max speed)

Light microscopic with 20× or 40× phase contrast

Refrigerated centrifuge

B. Reagents

Glycerol/phosphate wash buffer (200–500 ml required): 1 M glycerol, 10 mM Na_3PO_4, pH 8.1. Store at room temperature.

Filtered Ca^{2+}-free seawater (about 1 liter required for unactivated eggs): 436 mM NaCl, 9 mM KCl, 16 mM $MgSO_4 \cdot 7H_2O$, 34 mM $MgCl_2 \cdot 6H_2O$, 1 mM EGTA, 5 mM Tris, pH 8.1. Filter to 0.45 μm. Store at room temperature.

4 M KCl (5–10 ml required for activated eggs): Store at room temperature.

C. Protocol

1. Obtain gametes. Males and females are indistinguishable externally. Obtain gametes by first inserting a flat knife into the shell and cutting the two

adductor muscles which lie on either side of the hinge; open up the foot and snip open the gonads which surround the gut and hepatopancreas. Care should be taken not to perforate the gut when snipping through the gonads. Male and female gametes can be identified macroscopically by color and consistency: sperm appears white and milky; oocytes appear light pink and granular. Oocytes should be collected in 100 ml of filtered seawater; sperm, if needed, should be collected "dry" and kept on ice until use.

2. Clean and demembrane the oocytes. To separate oocytes from ovarian tissue, filter the oocyte suspension through four layers of cheesecloth wetted with seawater and wash the oocytes three times by settling, decanting, and resuspending in filtered seawater (50× volume of the oocytes).

FOR UNACTIVATED OOCYTES:

3. To prevent activation of the oocytes, wash them once in filtered Ca^{2+}-free seawater (Cavanaugh, 1956). The oocytes have a tough vitelline membrane which must first be softened using glycerol (Rebhun and Sharpless, 1964). After the wash in filtered Ca^{2+}-free seawater, sediment the oocytes using the highest setting in a clinical centrifuge for 1 minute, resuspend in glycerol/phosphate wash buffer, resediment, and resuspend in glycerol/phosphate wash buffer again. After the second resuspension in glycerol/phosphate wash buffer, check the oocytes using 40× phase-contrast microscopy. Once the vitelline membrane starts to "bleb," quickly sediment the oocytes, aspirate the glycerol/phosphate wash buffer, and resuspend in ice-cold isotonic HEMG buffer.

FOR ACTIVATED OOCYTES:

3. Extracts made from first meiotic metaphase oocytes must be activated in dilute suspension (10 μl of oocytes/ml of filtered seawater, suspended by slowly stirring the solution with an electrically powered paddle). Oocytes are activated by adding 1:100 4 M KCl to the seawater–egg solution. Confirm activation by the disappearance of the germinal vesicle at 8–10 minutes. The glycerol/phosphate wash steps (described above) must be carried out rapidly so that the activated eggs are in ice-cold isotonic HEMG buffer by 16 minutes after activation. This will ensure an extract with high histone H1 kinase activity (five to seven times greater than that of unactivated eggs).

All subsequent steps should be performed on ice with ice-cold ingredients.

4. To remove all of the glycerol, quickly wash activated or unactivated eggs in isotonic HEMG buffer, spin down the oocytes, resuspend again in isotonic HEMG buffer, and recentrifuge.

5. Transfer 3–7 ml of the packed, demembranated oocyte pellet into a precooled hand tissue grinder/homogenizer with a loose-fitting pestle. Add protease inhibitors, nucleotides, and DTT to the same final concentrations as were used for sea urchin egg cytoplasmic extracts (see above).

6. Homogenize the eggs until more than 80% of the oocytes are broken open. From 7 to 20 passes will be necessary depending on how well the vitelline membranes were softened during the glycerol/phosphate wash steps.

7. Clarify the extract. Extracts made from *Spisula* oocytes are more difficult to clarify than sea urchin egg extracts. We have found that two spins are necessary to yield extracts that can be used to view microtubule growth and particle motility with video-enhanced differential interference microscopy. Centrifuge the lysate at 10,000g (r_{max}) for 15 minutes (Beckman JS13.1 rotor or equivalent) at 4°C, and remove the middle cytoplasmic layer with a syringe and a 22-gauge needle by puncturing the side of the tube (Beckman 331372 centrifuge tube used with an insert for the Beckman JS13.1 rotor). Spin the cytoplasmic homogenate again at 100,000g (r_{max}) for 45 minutes (Beckman TL55 rotor) at 4°C. Carefully remove the cytoplasmic extract from the middle clear layer using a syringe inserted from the top of the tube, taking care not to collect any of the top lipid layer. Seven milliliters of packed eggs usually yields 300 to 500 μl of clarified extract.

8. Use the cytoplasmic extract immediately or aliquot, shell-freeze in liquid nitrogen, and store at −80°C.

IV. Visualizing the Microtubule–Associated Motility and Crosslinking

Recent advances in video microscopy techniques have made possible the visualization of single microtubules and associated organelles in extremely thin cells (Cassimeris *et al.,* 1988; Sammak and Borisy, 1988; Schulze and Kirschner, 1988), as purified components (Vale *et al.,* 1985; Horio and Hotani, 1986; Walker *et al.,* 1988), or in cytoplasmic extracts (Schnapp *et al.,* 1988; Belmont *et al.,* 1990; Gliksman *et al.,* 1992). The cytoplasmic extracts described here offer several unique advantages for observing microtubule-associated motility and microtubule crosslinking. First, the clarity of the extracts allows individual microtubules and small organelles to be observed in real time using video-enhanced differential interference contrast (VE-DIC) microscopy or video-enhanced dark-field microscopy. Second, no exogenous tubulin or taxol is needed in the preparations, as the endogenous tubulin in the extracts readily assembles microtubules from nucleation sites. Third, microtubule orientations are easily determined because microtubules assemble only from the plus ends of nucleation sites, and only the plus ends of the microtubules undergo dynamic instability (Gliksman *et al.,* 1992; Simon *et al.,* 1992; Suprenant, 1991). Fourth, the endogenous microtubule-associated proteins attach the majority of the microtubules to the glass coverslip surface. This simplifies the tracking of organelle movements along microtubules. A typical video assay using the sea urchin cytoplasmic extract is carried out as follows:

The only additional equipment needed is a VALAP melter (see Walker *et al.*, 1990).

A. Materials

Cleaned microscope slides and biologically clean coverslips (18 × 18 mm or 22 × 22 mm, thickness of 0 or 1) (Lutz and Inoué, 1986; Walker *et al.*, 1988): Dirty glass surfaces can inhibit microtubule gliding and can pose problems during microscopy by scattering light. We typically clean our coverslips using several washes with detergent in an ultrasonic bath or concentrated sulfuric acid as discussed by Walker *et al.* (1990). After sonication or acid washes, the coverslips are rinsed several times in distilled water and several times in 100% ethanol, resonicating between the rinses. Cleaned coverslips are stored in 100% ethanol in a closed container. Slides are usually washed with ethanol at the time of use. Both coverslips and slides are air-dried before use (Walker *et al.*, 1990).

Concentrated video assay buffer (5–50 μl required): Isotonic HEMG buffer containing 5 mM ATP, 2.5 mM GTP (from frozen 20× stocks), 250 μM cytochalasin B (from a 80× stock in ethanol at −20°C), and a low density of microtubule nucleation sites. Make fresh and store on ice.

Possible microtubule nucleation sites (frozen stocks):

1. Salt-extracted axoneme fragments from sea urchin sperm tails, *Chlamydomonas* flagella, or *Tetrahymena* flagella (Bergen and Borisy, 1980; Pryer *et al.*, 1986; Simon *et al.*, 1992)
2. Asymmetrical microtubule constructs (Hyman *et al.*, 1991)
3. Sea urchin sperm heads (Gliksman *et al.*, 1993)
4. Sea urchin sperm centriole complexes (G. Sluder, personal communication)

VALAP (<1 g required for several experiments): A 1:1:1 mixture of Vaseline, lanolin, and paraffin. Melt all of the ingredients together, mix, and store at room temperature.

B. Protocol for a Slide Preparation

1. On ice, mix the sea urchin egg cytoplasmic extract with video assay buffer containing nucleotides, cytochalasin B, and your choice of nucleation sites. We typically dilute the extract with video assay buffer and other reagents by only 4–10%.

2. Add the sea urchin extract sample to a clean, dry slide and quickly place a clean, dry coverslip on top, avoiding air bubbles as much as possible. Too little sample can increase air bubbles in the preparation, which scatter light and degrade the image through the microscope. Too much sample also increases light scatter and degrades the image. Typically, a 2-μl sample is used for an

18 × 18-mm coverslip and a 4- to 5-μl sample is used for a 22 × 22-mm coverslip. Avoid moving the coverslip once it is flat; movement of the coverslip almost always produces air bubbles in the preparation and builds up material on the glass surfaces.

3. Seal the edges of the coverslip to prevent evaporation and flow of the sample. Wax mixtures with low melting temperatures work well as sealants. We tend to use VALAP as a sealant.

4. Observe microtubules and organelles using video microscopy (differential interference contrast, dark-field, or fluorescence techniques). For the technique of high-resolution, VE-DIC microscopy, see Walker *et al.* (1990) and Schnapp *et al.* (1988). For fluorescence microscopy methods, see Belmont *et al.* (1990) and Hyman *et al.* (1991).

V. Example Results for Cytoplasmic Extracts

Several types of microtubule-associated motility may be observed in cytoplasmic extracts of sea urchin or clam eggs (Gliksman and Salmon, 1993). Microtubules and axoneme fragments glide across the coverslip and slide surfaces toward their plus ends (Fig. 2). Particulate organelles move bidirectionally on the microtubules and axoneme fragments (Fig. 3). Microtubules crosslink to other microtubules and to themselves when their microtubule lattices are oriented in opposite directions (Fig. 3a). Membrane fractions are pulled out along microtubules, forming membrane networks (Fig. 4), as was seen by Vale and

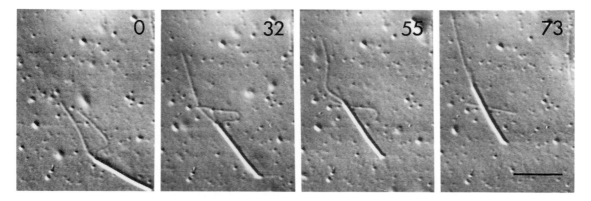

Fig. 2 Microtubules and an axoneme fragment glide upward across the coverslip surface. Microtubule motors from the sea urchin cytoplasmic extract attach to the glass surface and move both the sperm tail axoneme fragment and its microtubules toward their plus ends. Between 32 and 55 seconds, the microtubule on the right is pulled off of the axoneme. Bar = 5 μm. Time is in seconds.

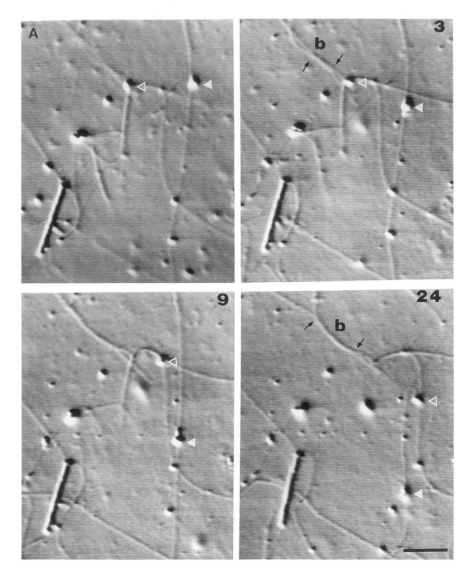

Fig. 3 Movement of endogenous particles along microtubules. (A) Endogenous particles (triangles) from a sea urchin cytoplasmic extract move along a microtubule network caused by microtubule–microtubule bundling. The extent of microtubule bundling (b) increases over time (arrows pointing to the edges of the overlap at 3 and 24 seconds). (b) Endogenous particles (triangles) from a clam cytoplasmic extract move in both directions along a single microtubule. Two particles (open triangles) move to the right toward the microtubule minus end. Another particle (closed triangle) moves slowly in the opposite direction. Bar = 5 μm. Time is in seconds.

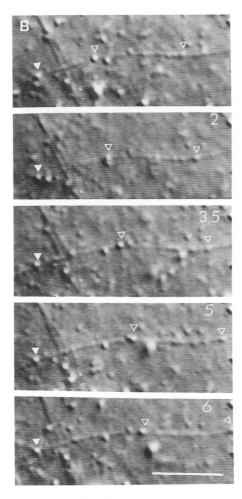

Fig. 3 *(continued)*

Hotani (1988). Thus the extracts serve as a useful model for studying aspects of microtubule-associated translocation in a cytoplasmic environment and, in addition, are very useful for the analysis of mechanisms regulating microtubule assembly (Simon *et al.*, 1992; Gliksman *et al.*, 1992).

Acknowledgments

We thank J. R. Simon and V. Petrie-Skeen for assistance during the development of the extract systems. We also thank B. A. Shirk and R. V. Skibbens for critical reading of this manuscript. This work was supported by National Institutes of Health Grant GM-24364.

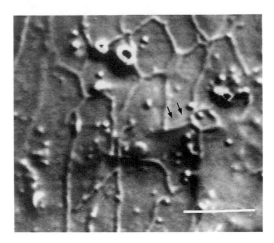

Fig. 4 Membrane networks formed by membrane movement along microtubules. Membrane contaminants of a sea urchin cytoplasmic extract form membrane networks as they are pulled out along the microtubules. The microtubules are not easily visible (arrows) because of the larger size and contrast of the membrane strands above them. Bar = 5 μm.

References

Belmont, L. D., Hyman, A. A., Sawin, K. E., and Mitchison, T. J. (1990). Real-time visualization of cell cycle dependent changes in microtubule dynamics in cytoplasmic extracts. *Cell (Cambridge, Mass.)* **62,** 579–589.

Bergen, L. G., and Borisy, G. G. (1980). Head-to-tail polymerization of microtubules in vitro. *J. Cell Biol.* **84,** 141–150.

Cassimeris, L. U., Pryer, N. K., and Salmon, E. D. (1988). Real-time observations of microtubule dynamic instability in living cells. *J. Cell Biol.* **107,** 2223–2231.

Cavanaugh, G. M., ed. (1956). ''Formulae and Methods IV of the Marine Biological Laboratory Chemical Room.'' Marine Biological Laboratory, Woods Hole, MA.

Gliksman, N. R., and Salmon, E. D. (1993). Microtubule-associated motility in cytoplasmic extracts of sea urchin eggs. *Cell Motil.* **24,** 167–178.

Gliksman, N. R., Parsons, S. F., and Salmon, E. D. (1992). Okadaic acid induces interphase to mitotic-like microtubule dynamic instability by inactivating rescue. *J. Cell Biol.* **119,** 1271–1276.

Gliksman, N. R. *et al.* (1993). In preparation.

Horio, T., and Hotani, H. (1986). Visualization of the dynamic instability of individual microtubules by darkfield microscopy. *Nature (London)* **321,** 605–607.

Hyman, A., Drechsel, D., Kellogg, D., Salser, S., Sawin, K., Steffen, P., Wordeman, L., and Mitchison, T. (1991). Preparation of modified tubulins. *In* ''Methods in Enzymology'' (R. B. Vallee, ed.), Vol. 196, pp. 478–485. Academic Press, San Diego.

Lutz, D. A., and Inoué, S. (1986). Techniques for observing living gametes and embryos. *Methods in Cell Biol.* **27,** 89–110.

Pfeffer, T. A., Asnes, C. F., and Wilson, L. (1976). Properties of tubulin in unfertilized sea urchin eggs: Quantitation and characterization by the colchicine-binding reaction. *J. Cell Biol.* **69,** 599–607.

Porter, M. E., Grissom, P. M., Scholey, J. M., Salmon, E. D., and McIntosh, J. R. (1988). Dynein isoforms in sea urchin eggs. *J. Biol. Chem.* **263,** 6759–6771.

Pryer, N. K., Wadsworth, P., and Salmon, E. D. (1986). Polarized microtubule gliding and particle saltations produced by soluble factors from sea urchin eggs and embryos. *Cell Motil.* **6**, 537–548.

Rose, P. M., Rothhacker, D. Q., and Penningroth, S. M. (1989). Quantitation of the dynein pool in unfertilized sea urchin eggs. *Biochim. Biophys. Acta* **990**, 31–39.

Rebhun, L. I., and Sharpless, T. K. (1964). Isolation of spindles from the surf clam *Spisula solidissima*. *J. Cell Biol.* **22**, 488–491.

Sammak, P. J., and Borisy, G. G. (1988). Direct observations of microtubule dynamics in living cells. *Nature (London)* **223**, 724–726.

Schnapp, B. J., Gelles, J., and Sheetz, M. P. (1988). Nanometer-scale measurements using video light microscopy. *Cell Motil.* **10**, 47–53.

Scholey, J. M., Neighbors, B., McIntosh, J. R., and Salmon, E. D. (1984). Isolation of microtubules and a dynein-like Mg ATPase from unfertilized sea urchin eggs. *J. Biol. Chem.* **259**, 6516–6525.

Scholey, J. M., Porter, M. E., Grissom, P. M., and McIntosh, J. R. (1985). Identification of kinesin in sea urchin eggs and evidence for its localization in the mitotic spindle. *Nature (London)* **315**, 483–486.

Schroeder, T. E., ed. (1986). ''Methods in Cell Biology,'' Vol. 27, Academic Press, Orlando, FL.

Schulze E., and Kirschner, M. (1988). New features of microtubule dynamics in living cells. *Nature (London)* **334**, 356–359.

Showman, R. M., and Foerder, C. A. (1979). Removal of the fertilization membrane of sea urchin embryos employing aminotriazole. *Exp. Cell Res.* **120**, 253–255.

Simon, J. R., Parsons, S. F., and Salmon, E. D. (1992). Buffer composition modulates the dynamic instability of microtubules assembled from sea urchin egg tubulin in vitro. *Cell Motil.* **21**, 1–14.

Suprenant, K. (1991). Unidirectional microtubule assembly in cell-free extracts of *Spisula solidissima* oocytes is regulated by subtle changes in pH. *Cell Motil.* **19**, 207–220.

Vale, R. D., and Hotani, H. (1988). Formation of membrane networks in vitro by kinesin-driven microtubule movement. *J. Cell Biol.* **107**, 2233–2241.

Vale, R. D., Reese, T. S., and Sheetz, M. P. (1985). Identification of a novel force-generating protein, kinesin, involved in microtubule-based motility. *Cell (Cambridge, Mass.)* **41**, 39–50.

Walker, R. A., O'Brien, E. T., Pryer, N. K., Soboeiro, M., Voter, W. A., Erickson, H. P., and Salmon, E. D. (1988). Dynamic instability of individual microtubules analyzed by video light microscopy: Rate constants and transition frequencies. *J. Cell Biol.* **107**, 1437–1448.

Walker, R. A., Gliksman, N. R., and Salmon, E. D. (1990). Using video-enhanced differential interference contrast microscopy to analyze the assembly dynamics of individual microtubules in real time. *In* ''Optical Microscopy for Biology'' (B. Herman and K. Jacobson, eds.), pp. 395–407. Wiley-Liss, New York.

CHAPTER 18

Kinesin–Mediated Vesicular Transport in a Biochemically Defined Assay

Raul Urrutia,*,‡ Douglas B. Murphy,† Bechara Kachar,* and Mark A. McNiven‡

*Laboratory of Cellular Biology
National Institute for Deafness and
  Other Communication Disorders
National Institutes of Health
Bethesda, Maryland 20892

†Department of Cell Biology and Anatomy
The Johns Hopkins Medical School
Baltimore, Maryland 21205

‡Center for Basic Research in Digestive Diseases
Mayo Clinic
Rochester, Minnesota 55905

I. Introduction
II. Assay Reagents and Components
   A. Purification of Kinesin and Tubulin
   B. Chromaffin Granule Isolation
   C. Motility Buffers
III. Motility Assay Procedures
   A. Vesicular Transport Assay
   B. Microtubule Aster Assay
IV. Kinesin-Mediated Dynamics *in Vitro*
V. Biological Implications of Kinesin-Supported Motility Assays
   A. Does Kinesin Work Alone?
   B. Effects of Salt Concentration on Kinesin Motility
   C. Kinesin-Mediated Microtubule Sliding during Aster Formation
VI. Summary
   References

I. Introduction

Kinesin is a mechanochemical enzyme that supports the unidirectional movements of latex beads on microtubules (MTs) *in vitro* and is believed to translocate cytoplasmic organelles toward the plus ends of MTs in cells (Vale *et al.*, 1985; Vale, 1987).

Despite these important observations all of the *in vitro* assays designed to examine kinesin-mediated organelle movements have been conducted using undefined axoplasmic or cellular extracts (Vale, 1987). Because these extracts are biochemically diverse it is not clear whether kinesin alone can support MT-based organelle movements or if additional cytoplasmic components are required. Schroer and colleagues (1988) have reported that kinesin is unable to initiate movements of purified vesicles unless axoplasmic supernatants are added subsequently to the assay, which suggests a requirement for undefined "soluble factors." Not surprisingly, the complexity of assays using cellular extracts makes it extremely difficult to study kinesin function, regulation, and its interaction with membranous organelles. Our goal was to establish a biochemically defined organelle motility assay reduced to the most basic components of vesicles, kinesin, and a MT substrate suspended in a simple MgATP motility buffer. We have used the chromaffin cells of the bovine adrenal medulla because they are one of the best characterized examples of vesicular secretion while containing large amounts of secretory vesicles and kinesin (Murofushi *et al.*, 1988). Here we describe a method in which exceptionally pure bovine adrenal and/or brain kinesin supports the ATP-dependent movements of purified chromaffin granule "ghosts" along MTs preattached to a substrate in the presence of a low-salt buffer, a condition that had been demonstrated previously to support high levels of MT-activated ATPase activity (Kachar *et al.*, 1987; Kuznetsov and Gelfand, 1986; Wagner *et al.*, 1989; Hackney, 1988). Under these conditions we find that kinesin supports the movements of granule ghosts. These movements are frequent in number and exhibit a velocity equal to that of kinesin-coated beads translocating along MTs. This suggests that kinesin alone in the presence of low salt can initiate and sustain the anterograde translocation of membranous organelles without the requirement of other soluble factors or proteins (Urrutia *et al.*, 1991). In addition, we found that purified kinesin promotes physical interactions between adjacent, free-floating MTs, leading to spontaneous formation of large astral arrays (Urrutia *et al.*, 1991). These asters supported the inward movements of kinesin-coated beads and chromaffin vesicles, indicating that they are uniformly polarized with the microtubule plus ends gathered at their centers. The methodology and biological implications of this system are discussed.

II. Assay Reagents and Components

A. Purification of Kinesin and Tubulin

All vesicle motility assays were performed using kinesin prepared by three different methods from two different bovine tissues. Kinesin was isolated from bovine adrenal medulla (Fig. 1a) by combining tripolyphosphate-induced MT binding (Kuznetsov and Gelfand, 1986) and chromatography as described by

Urrutia and Kachar (1992). Alternatively, bovine brain kinesin was isolated in 1,4-piperazinediethanesulfonic acid (Pipes) buffer with the ATP analog AMPPNP as described by Cohn *et al.* (1987) or was isolated in imidazole buffer with tripolyphosphate as described by Kuznetsov and Gelfand (1986). As discussed below, large numbers of vesicle movements are supported by all three of these kinesin preparations.

Very clean MTs are prepared by two cycles of assembly–disassembly (Mandelkow *et al.,* 1985). Contaminating MT-associated proteins (MAPs) were removed by a high-salt wash with 0.5 *M* Pipes followed by phosphocellulose (PC) chromatography (Mandelkow *et al.,* 1985) (Fig. 1b). PC–tubulin (2–4 mg/ml) is assembled into functional MTs by the addition of 2 m*M* $MgCl_2$ and 1 m*M* GTP (final concentrations), followed by a 10-minute incubation at 37°C. Subsequently, taxol is added to a concentration of 20 μM, followed by a second 37°C incubation for 10 minutes. The duration of the Mg^{2+} and GTP incubation prior to taxol addition determines the length of the MT polymer, with 1–3 minutes producing short MTs and 10-minute incubations resulting in longer MTs (Fig. 1c).

Perhaps one of the most important and challenging aspects of any MT-based vesicle assay is finding a dependable, reproducible, and benign method to fasten the MT substrate to the glass slide. This is critical for two reasons: first, unattached MTs drift away from the field of view during rinsing of the preparation; second, kinesin will move unfastened MTs along the glass slide, making it difficult to define whether a vesicle is moving independently along a fixed MT substrate or by association with a motile MT. To circumvent this problem, we have experimented with different reagents to coat coverslips and provide a sticky or charged substrate to which MTs can adhere. We have found the best substrate to date has been poly-L-ornithine (Sigma No. P-4538), the synthetic precursor to arginine. A 10 mg/ml solution of poly-L-ornithine in distilled H_2O, allowed to dry on a coverslip, provides a suitable substrate for MT attachment (Fig. 1c).

B. Chromaffin Granule Isolation

Adrenal chromaffin granules are prepared using sucrose density centrifugation of bovine adrenal medulla homogenate by the method of Smith and Winkler (1967). This chromaffin granule preparation has been determined to be greater than 95% pure by both light microscopy and thin-section electron microscopy (Fig. 1d). Immunoperoxidase blotting detected no associated kinesin, while these granules are nonmotile in the absence of added kinesin. Despite the fact that little or no kinesin is associated with these purified granules, it is best to use them immediately after isolation as motility is reduced after only 12 hours of storage.

Several technical problems are encountered when using large secretory organelles such as chromaffin granules in motility assays. First, the intact granules

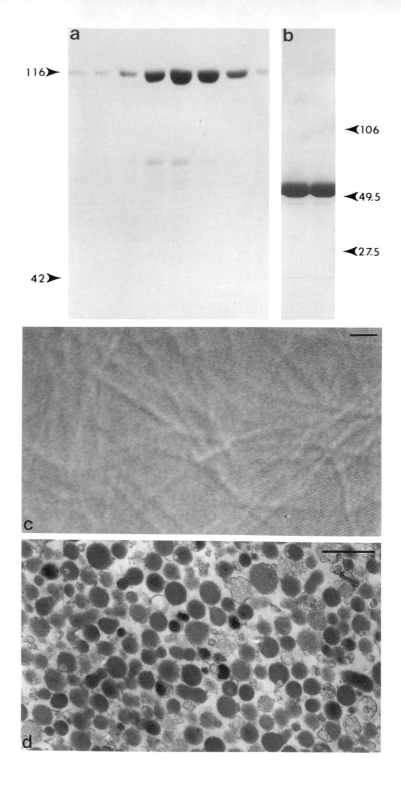

are dense and very refractile, thereby deteriorating the differential interference contrast (DIC) image of the supporting MTs; second, the excessive size of the chromaffin granules interferes with MT attachment because of Brownian motion and/or sticking to the poly-L-ornithine-coated substrate. Indeed, most, if not all, granules will adhere to the coverslip if additional modifications are not taken. To circumvent these problems we subject the granules to an additional washing and lysis step to produce granule membrane "ghosts" via incubation and centrifugation through 20 vol of 15 mM Tris–HCl, pH 7.00. This final procedure increases the purity of the granules while substantially reducing their size, refractivity, and capacity for sticking to the coverslip. The sticking of vesicle "ghosts" is reduced further by the inclusion of a protein carrier to the motility buffer. We use bovine serum albumin at 2 mg/ml, although others (Schnapp *et al.*, 1991) have found that 4–5 mg/ml casein in motility buffer is superior.

C. Motility Buffers

The motility buffer (MB) we use is a simple one and consists of 15 mM imidazole, pH 7.5, 2 mM ATP, 2 mM $MgCl_2$, 1 mM ethylene glycol bis (β-aminoethyl ether) N, N'-tetraacetic acid (EGTA), 30 μM taxol, and 2 mg/ml bovine serum albumin (BSA). This buffer is of low ionic strength, because, as discussed below, higher salt concentrations appear to limit the ATPase activity and vesicular motility properties of kinesin.

III. Motility Assay Procedures

A. Vesicular Transport Assay

1. Ten microliters of 10 mg/ml poly-L-ornithine is applied and spread to the central region of a 60 × 24-mm No. 0 coverslip and allowed to dry over a 4- to 5-minute period.

2. The area of dried poly-L-ornithine is rinsed with distilled water and blotted dry.

3. Five microliters of phosphocellulose-purified, taxol-stabilized MTs (2 mg/ml) is mixed with 5 μl of motility buffer (MB) minus BSA and added to the coverslip. Two parallel strips of vacuum grease which act as spacers are placed

Fig. 1 Components used for kinesin-mediated vesicle motility assay. (a) Sodium dodecyl sulfate–polyacrylamide gel of purified bovine adrenal medulla kinesin isolated by the method of Urrutia and Kachar (1992). (b) Sodium dodecyl sulfate–polyacrylamide gel of phosphocellulose-purified bovine brain tubulin isolated by the method of Mandelkow *et al.* (1985). (c) Differential interference contrast video micrograph of isolated taxol-stabilized microtubules attached to a poly-L-ornithine-coated glass coverslip. Bar = 1 μm. (d) Electron micrograph of purified bovine adrenal medulla chromaffin granules isolated by the methods of Smith and Winkler (1967). Bar = 5 μm.

on either side of the MT spot. Next, an additional 22-mm^2 No. 1 coverslip is placed over the MTs and pressed down onto the vacuum grease to produce a chamber with two open sides from which solutions can be perfused. To facilitate sticking, the MTs are allowed to incubate on the coverslip for approximately 5 minutes and then rinsed with 40 μl of MB (including 2 mg/ml BSA).

4. Two microliters of chromaffin granule ghosts is added to 10 μl of MB (plus BSA) and combined with 5 μl of kinesin (100–200 μg/ml) in a small Eppendorf vial and mixed gently.

5. The kinesin–vesicle solution is then added to the side of the MT-coated coverslip by capillary action with a pipet and filter paper.

6. Microscopic observations are made in the D/C mode using a Zeiss Axiomat microscope equipped with a Newvicon Dage MTI-70 video camera. Inclusion of two heat filters in the path of the mercury arc is important to promote longevity of the vesicle movements in the preparation which generally lasted up to 20 minutes. All manipulations are carried out at room temperature.

B. Microtubule Aster Assay

For the formation of MT asters *in vitro*, we use the same biological components, buffer, and light microscopy as for vesicle motility. The major difference between the two procedures is that MTs are not fixed to the glass coverslip with poly-L-ornithine while kinesin (5 μl of 100 μg/ml) is mixed with the taxol-polymerized MTs (5 μl of 2 mg/ml) prior to addition to the slide.

IV. Kinesin-Mediated Dynamics *in Vitro*

Analysis of video microscopic images show that vesicles in Brownian motion bind to MTs and translocate at rates of 0.4–0.6 μm/s in the presence of any of the three different kinesin preparations (Fig. 2). At any given moment 30–35% of the vesicles clearly show directed movements along the surface of MTs (Table I). The movements are unidirectional and continuous unless inhibited by interactions with other MTs or the glass surface. No directed movements occur in the absence of kinesin or in the absence of ATP, indicating that both components are required for translocation.

In addition to vesicle movements these preparations support an unexpected form of MT dynamics represented as the formation of MT asters. We originally

---

Fig. 2 Chromaffin granule "ghosts" translocate along taxol-stabilized microtubules in the presence of purified kinesin and ATP. In these video DIC images, numerous vesicular "ghosts" (numbers) can be seen moving along microtubules at a velocity of 0.4 to 0.6 μm/s. Bars = (a + b) 1 μm and (c–e) 0.6 μm. Time is in seconds.

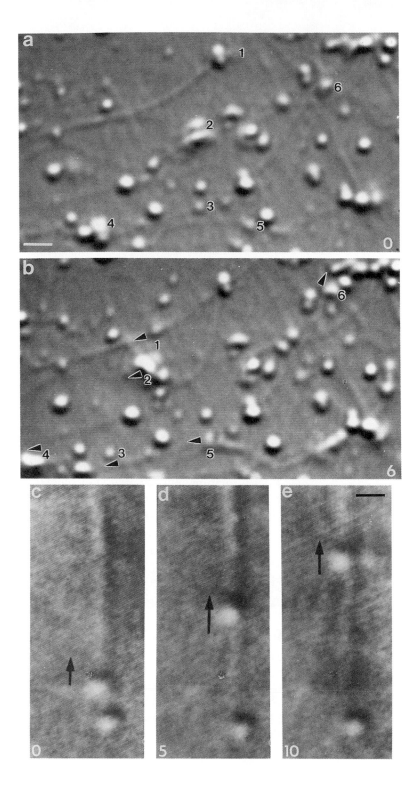

Table I
Membrane Vesicle Motility Supported
by Adrenal and Brain Kinesins[a]

Kinesin	Percentage	Velocity (μm/s)
Adrenal medulla	31 ($n = 240$)	0.58 ± 0.1
Brain		
(Tripolyphosphate)	34 ($n = 300$)	0.41 ± 0.1
(AMPPNP)	35 ($n = 220$)	0.40 ± 0.1

[a] Brain kinesin was prepared by binding to microtubules in the presence of tripolyphosphate (Kuznetsov and Gelfand, 1986; Urrutia *et al.*, 1991) or AMPPNP (Urrutia *et al.*, 1991). n = total number of membrane vesicles counted in eight different video fields viewed under 2 minutes per field.

observed the formation of asters in the presence of vesicle ghosts which produced large aggregates of MTs and vesicles (Fig. 3), although subsequent studies demonstrated that asters form with kinesin and MTs only (Figs. 4a–c). As for vesicle movements, aggregation of MTs required ATP and is inhibited by tripolyphosphate (Kuznetsov and Gelfand, 1986), indicating that it is due to kinesin. Polystyrene beads (Fig. 4d–f) added to the preparation travel unidirectionally to the centers of the asters, indicating that MTs are oriented with "minus ends out," the reverse orientation normally exhibited by astral MTs in cells. Thus, aster formation may be initiated through the dynamic interaction of two adjacent MTs that slide together via kinesin until both positive ends meet. This process is repeated and amplified by many adjacent MTs to form a unipolar aster.

V. Biological Implications of Kinesin–Supported Motility Assays

A. Does Kinesin Work Alone?

With the biochemically defined assay described above, we have demonstrated that kinesin has the capacity to move secretory vesicles along MTs while promoting dynamic interactions between adjacent MTs. An important question that needs to be addressed is whether these observations relate to conditions and motile phenomena in the intact cell. It has been postulated that kinesin requires additional membrane-bound or soluble factors (Schroer *et al.*, 1988) to support vesicle transport and that movement in the absence of such factors may

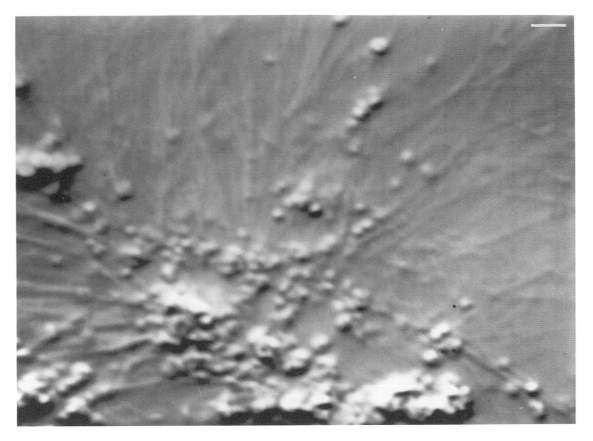

Fig. 3 Vesicles and free-floating microtubules interact dynamically to form large aggregates. A DIC video micrograph showing dynamic interactions between vesicle "ghosts" and microtubules to form clusters or aggregates. Vesicles are actively translocating inward to the center of the aggregate. Bar = 1.3 μm.

represent an artifical event similar to the kinesin-mediated movement of plastic beads. Our observations suggest that kinesin, MTs, and ATP are all that are required for vesicle translocation under low-ionic-strength conditions. To preserve the native surface properties of the vesicles we have not stripped their membranes with strong halide salts such as KI (Schroer *et al.*, 1988), which could denature surface proteins (von Hippel and Schleich, 1969). Furthermore, vesicle movement did not occur in the absence of both kinesin and ATP. Although our vesicles were not stripped beforehand with chaotropic salts, it is clear that under the conditions of low salt employed in our assays, soluble cytoplasmic proteins as such are not required for vesicle translocation in this defined *in vitro* system. These studies do not, however, preclude the possibility

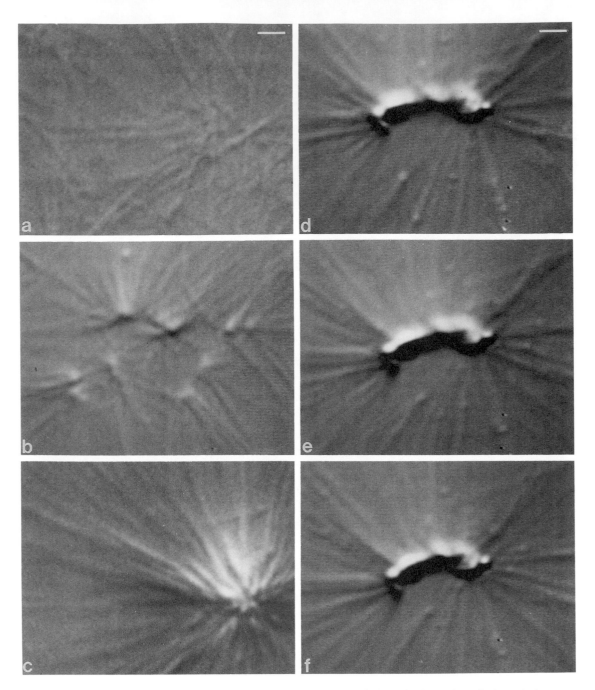

Fig. 4 Kinesin-mediated microtubule interactions induce the formation of microtubule asters. (a–c) DIC video micrographs showing a field of evenly distributed taxol-stabilized microtubules that have not been attached to the glass slide via poly-L-ornithine. Within 2–5 minutes, single microtubules interact dynamically with each other and become joined at their tips, forming small foci (b).

that a factor tightly associated with the granule membrane may act as a receptor to facilitate kinesin binding or that supplemental factors may be required at higher ionic strength. Indeed, our observations that chromaffin granule motility in the assay decreases substantially as the granules age, or are protease-treated, are consistent with a motor receptor model.

B. Effects of Salt Concentration on Kinesin Motility

We have found that both kinesin-mediated vesicle movements and aster formation are very sensitive to increases in the salt concentration of the motility buffer. To pursue this, we have examined the MT-activated kinesin ATPase and kinesin-dependent movements in the presence of 15 mM imidazole buffer containing different concentrations of KCl (Fig. 5). In this buffer, kinesin purified by all three different methods acted as an anterograde translocator to support bead movement and axonemal gliding. Both ATPase activity and bead movement are observed to significantly decrease with increasing salt concentration. At concentrations of KCl greater than 50–70 mM, microtubule activation of the ATPase activity is much reduced and bead movement is arrested. Inhibition is not specific for KCl, as potassium aspartate was similarly effective in inhibiting ATPase activity. Thus, the ATPase and motile activities of kinesin increase in a proportional manner with decreasing concentrations of salt, a property exhibited by the other mechanochemical enzymes: myosin (Moos, 1972; Stein *et al.,* 1981; Brenner *et al.,* 1982; Chalovich and Eisenberg, 1982; Williams *et al.,* 1984), myosin I (Albanesi *et al.,* 1983), flagellar dynein (Gibbons, 1966), and cytoplasmic dynein (Shpetner *et al.,* 1988).

Recent biochemical studies on the effects of salt on kinesin ATPase (Murphy and Wallis, 1992) indicate that the affinity of kinesin for MTs (K_{app}) is decreased 8.6-fold at 50 mM KCl, whereas the K_m for ATP (70 μM) is essentially unaffected by the presence or absence of 50 mM KCl. This indicates that the fundamental properties of ATP binding and hydrolysis are not significantly altered by moderate changes in the ionic strength. Because 50–70 mM salt inhibits kinesin motility and activity, it is not clear what cytoplasmic conditions and factors allow it to function within the cell.

C. Kinesin–Mediated Microtubule Sliding during Aster Formation

Although kinesin-mediated aster formation appears to involve sliding and recruitment of the plus ends of MTs into an organized focus, the mechanism of

---

Over the next 5 minutes, these foci coalesce into large asters (c). (d–f) Carboxylated latex beads added to the asters move into the asters' center. This demonstrates that the microtubules are polarized uniformly with the plus ends in the center and the minus ends extending outward. Bars = 1 μm.

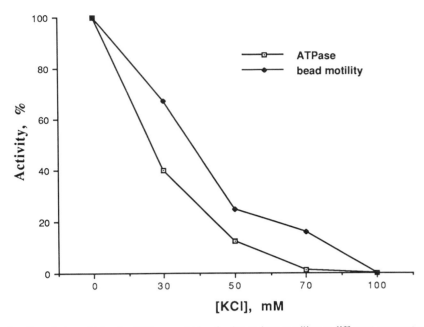

Fig. 5 Correlation of kinesin ATPase and kinesin-dependent motility at different concentrations of salt. Effect of KCl concentration on microtubule-activated MgATPase activity (□) and the movement of latex beads coated with kinesin (♦). Direction and maximum velocity of movement were not affected by the concentrations of salt. Relative values of motility were arbitrarily determined as a percentage of beads associated with microtubules that were undergoing gliding movements. Similar inhibitory effects were observed with potassium aspartate.

crosslinking and sliding has not been determined. Sliding could be mediated by kinesin aggregates, bipolar filaments, or possibly by single kinesin molecules with independently acting heads. Recent observations (Navone *et al.*, 1992) have demonstrated that both the truncated kinesin head and tail regions, when independently expressed in transfected tissue culture cells, colocalize with MTs, suggesting that dual binding sites could act to dynamically crosslink MTs.

VI. Summary

Here we have described simple and reproducible methods to observe kinesin-mediated vesicle and microtubule movements under defined conditions using video microscopy. We are optimistic that this assay will provide a useful tool to study kinesin function, regulation, and dynamic physical interactions with membranous organelles and microtubules.

Acknowledgment

We thank Ms. Karen Anderson for her assistance in the printing and mounting of figures.

References

Albanesi, J. P., Hammer, J. A., III, and Korn, E. D. (1983). The interaction of F-actin with phosphorylated and unphosphorylated myosins IA and IB from *Acanthamoeba castellani*. *J. Biol. Chem.* **258**, 10176–10181.

Brenner, B., Schoenberg, M., Chalovich, J. M., Greene, L. E., and Eisenberg, E. (1982). Evidence for cross-bridge attachment in relaxed muscle at low ionic strength. *Proc. Natl. Acad. Sci. U.S.A.* **79**, 7288–7291.

Chalovich, J. M., and Eisenberg, E. (1982). Inhibition of actomyosin ATPase activity by troponin–tropomyosin without blocking the binding of myosin to actin. *J. Biol. Chem.* **257**, 2432–2437.

Cohn, S. A., Ingold, A. L., and Scholey, J. M. (1987). Correlation between the ATPase and microtubule translocating activities of sea urchin egg kinesin. *Nature* (London) **328**, 160–163.

Gibbons, I. R. (1966). Studies on the adenosine triphosphatase activity of 14 S and 30 S dynein from cilia of *Tetrahymena*. *J. Biol. Chem.* **241**, 5590–5596.

Hackney, H. D. (1988). Kinesin ATPase: Rate-limiting ADP release. *Proc. Natl. Acad. Sci. U.S.A.* **85**, 6314–6318.

Kachar, B., Albanesi, J. P., Fujisaki, H., and Korn, E. D. (1987). Extensive purification from *Acanthamoeba castellani* of a microtubule-dependent translocator with microtubule-activated Mg^{2+}-ATPase activity. *J. Biol. Chem.* **262**, 16180–16185.

Kuznetsov, S. A., and Gelfand, V. I. (1986). Bovine brain kinesin is a microtubule-activated ATPase. *Proc. Natl. Acad. Sci. U.S.A.* **83**, 8530–8534.

Mandelkow, E. M., Herrmann, M., Ruehl, U. *et al.* (1985). Tubulin domains probed by limited proteolysis and subunit-specific antibodies. *J. Mol. Biol.* **185**, 311–327.

Moos, C. (1972). Actin activation of heavy meromyosin and subfragment-1 ATPases: Steady state kinetics studies. *Cold Spring Harbor Symp. Quant. Biol.* **37**, 137–143.

Murofushi, H., Ikai, A., Okuhara, K., Kotani, S., Aizawa, H., Kumakura, K., and Sakai, H. (1988). Purification and characterization of kinesin from bovine adrenal medulla. *J. Biol. Chem.* **263**, 12744–12750.

Murphy, D. B., and Wallis, K. T. (1992). Mt-activated kinesin ATPase and organelle motility: Salt reduces the affinity of kinesin for Mts but does not alter the ATPase mechanism. Submitted for publication.

Navone, F., Niclas, J., Hom-Booher, N., Sparks, L., Bernstein, H. D., McCaffrey, G., and Vale, R. D. (1992). Cloning and expression of a human kinesin heavy chain gene: Interaction of the COOH-terminal domain with cytoplasmic microtubules in transfected CV-1 cells. *J. Cell Biol.* **117**, 1263–1275.

Schnapp, B. J., Reese, T. S., and Bechtold, R. B. (1991). Attachment of microtubule motors to vesicular organelles. *J. Cell Biol.* **115**, 40a.

Schroer, T. A., Schnapp, B. J., Reese, T. S., and Sheetz, M. P. (1988). The role of kinesin and other soluble factors in organelle movement along microtubules. *J. Cell Biol.* **107**, 1785–1792.

Shpetner, S., Pascal, B. M., and Vallee, R. B. (1988). Characterization of the microtubule-activated ATPase of brain cytoplasmic dynein (MAP IC). *J. Cell Biol.* **107**, 1001–1009.

Smith, A. D., and Winkler, H. (1967). A simple method for the isolation of adrenal chromaffin granules on a large scale. *Biochem. J.* **103**, 480–482.

Stein, L. A., Chock, P. B., and Eisenberg, E. (1981). Mechanism of the actomyosin ATPase: Effect of actin on the ATP hydrolysis step. *Proc. Natl. Acad. Sci. U.S.A.* **78**, 1346–1350.

Urrutia, R., and Kachar, B. (1992). An improved method for the purification of kinesin from bovine adrenal medulla. *J. Biochem. Biophys. Methods* **24**, 63–70.

Urrutia, R., McNiven, M. A., Albanesi, J. P., Murphy, D. B., and Kachar, B. (1991). Purified kinesin promotes vesicle motility and induces active sliding between microtubules *in vitro*. *Proc. Natl. Acad. Sci. U.S.A.* **88,** 6701–6705.

Vale, R. D. (1987). Intracellular transport using microtubule-based motors. *Annu. Rev. Cell Biol.* **3,** 347–378.

Vale, R. D., Reese, T., and Sheetz, M. (1985). Identification of a novel force-generating protein, kinesin, involved in microtubule-based motility. *Cell* (*Cambridge, Mass.*) **42,** 39–50.

von Hippel, P. H., and Schleich, T. (1969). Effects of neutral salts on the structure and conformational stability of macromolecules in solution. *In* "Structure and Stability of Biological Macromolecules" (S. N. Timasheff, and G. D. Fasman, eds.). Decker, New York.

Wagner, M. C., Pfister, K. K., Bloom, G. S., and Brady, S. T. (1989). Copurification of kinesin polypeptides with microtubule-stimulated Mg-ATPase activity and kinetic analysis of enzymatic properties. *Cell Motil. Cytoskel.* **12,** 195–215.

Williams, D. L., Jr., Greene, L. E., and Eisenberg, E. (1984). Comparison of effects of smooth and skeletal muscle tropomyosins on interactions of actin and myosin subfragment 1. *Biochemistry* **23,** 4150–4155.

CHAPTER 19

An Assay for the Activity of Microtubule-Based Motors on the Kinetochores of Isolated Chinese Hamster Ovary Chromosomes

A. A. Hyman and T. J. Mitchison

Department of Pharmacology
University of California, San Francisco
San Francisco, California 94143

I. Introduction
II. Methods for Assaying Motor Activity on the Kinetochore
    A. Chromosome Preparation
    B. Microscope Requirements
    C. Setup
    D. Recording Minus-End-Directed Movement
    E. Recording Plus-End-Directed Movement
    F. Variables in the Assay
III. Conclusions
    References

I. Introduction

During mitosis, chromosomes attach to mitotic spindle microtubules via specific structures called kinetochores. Following microtubule attachment, the kinetochore–microtubule attachment mediates complicated patterns of chromosome movement which lead eventually to correct chromosome movement. Many recent experiments have suggested that microtubule-based motors may be responsible in part for driving the movements of kinetochores on microtu-

METHODS IN CELL BIOLOGY, VOL. 39

bules (McIntosh and Pfarr, 1991); however, the complexity of mitosis *in vivo* has meant that it is difficult to examine the detailed role and function of microtubule-based motors in mitosis. This chapter describes a detailed method derived from a published procedure for examining the behavior of microtubule-based motors on the kinetochores of isolated chromosomes (Hyman and Mitchison, 1991b).

In brief, chromosomes are isolated from Chinese hamster ovary (CHO) cells arrested at prometaphase using microtubule inhibitors, by swelling followed by digitonin lysis, and are then collected on a sucrose gradient in a low-ionic-strength buffer. This buffer minimizes extraction of loosely bound kinetochore components. The isolated chromosomes are adsorbed to the glass slide of a perfusion chamber. To assay the microtubule-based motor activity of the kinetochores of these isolated chromosome, polarity-marked microtubules (see Chapter 7) are attached to the kinetochores. After addition of ATP, the movement of these microtubules over the kinetochores of the isolated chromosomes is followed by video microscopy.

II. Methods for Assaying Motor Activity on the Kinetochore

A. Chromosome Preparation

This procedure is based on Mitchison and Kirschner (1985).

1. Take 20 plates of CHO cells that are one division before confluence growing in minimum essential medium (MEM) + nucleosides + 10% supplemented calf serum (University of California, San Francisco, cell culture facility) and add 10 μg/ml vinblastin final concentration to each plate. Leave cells in drug for 8 hours.

2. Rinse the mitotically arrested cells from the plates using gentle squirts of medium with a Pasteur pipet. Overvigorous washing will remove cells in interphase.

3. Spin down the cells at 1000g for 5 minutes at room temperature. The cells will form a loose pellet.

4. Remove all the medium, and resuspend in swelling buffer (5 mM K-Pipes, 5 mM NaCl, 5 mM MgCl$_2$, 1 mM EGTA, pH 6.8), at 30°C for 5 minutes. We find that swelling at 30°C improves the release of chromosomes on lysis.

5. Spin down the cells at 1000g for 5 minutes at 30°C. Remove excess swelling buffer and resuspend in 7 ml lysis buffer [10 mM potassium, 1,4-piperazine-diethanesulfonic acid (K-Pipes), 1 mM MgCl$_2$, 1 mM ethylene glycol bis(β-

aminoethyl ether) N,N'-tetraacetic acid (EGTA), 0.1% digitonin from a 10% stock in dimethylsulfoxide (DMSO), pH 7.2] on ice. Immediately transfer to a 10-ml glass Dounce homogenizer with a tight pestle and administer 20 rapid strokes on ice. The assay of kinetochore activity relies on specific capture of microtubules by kinetochores, and trapping by chromosome arms obscures the assay. This trapping appears to be caused by adhesion of cytoskeletal fragments to the chromosome arms, not to chromatin. In general, the cleaner the chromosome preparation, the fewer the number of microtubules attached to the chromosome arms. We find that clean chromosomes are obtained only if the cells lyse within the first few strokes of the pestle.

6. Transfer the lysate to a sucrose step gradient made with 2-ml steps of 30, 40, 50, and 60% (w/v) sucrose in 10 mM Pipes, pH 7.2, 5 mM MgCl$_2$, 1 mM EGTA, prepared in a 15-ml glass Corex tube. Spin at 5K for 15 minutes at 4°C in a Sorval HB4 swinging bucket rotor. The chromosomes accumulate as flocculent white material at the 40–50 and 50–60% interfaces. Collect them in a minimum volume using a Pasteur pipet. Freeze in liquid nitrogen in 10-μl aliquots. Store at -80°C. Chromosomes remain functional for 3 months at this temperature.

B. Microscope Requirements

We use a Zeiss standard microscope equipped for epifluorescence, because simple microscopes such as these give the greatest light throughput. Simple Nikon or Olympus upright scopes should perform equivalently. The objective was a 60× DPlanApo 1.4 Olympus lens, but the equivalent Nikon lens also works. Increased magnification to the video camera is provided by a Zeiss 4× tube. The magnification is chosen so that one chromosome with its microtubules (ca. 40 μm wide) fills a field on the video camera, ensuring high resolution and allowing us to distinguish plus- from minus-end-directed movement. A silicon-intensified-target (SIT) camera is used to generate the images, four to eight frames are averaged with an image processor, and a Panasonic OMDR is used to record images. Kinetochores with rhodamine-labeled microtubules attached are very sensitive to light damage, and shuttering of the light is essential to record movement of the microtubules on the kinetochores. A fully functional antifade system is absolutely necessary to prevent oxygen radical-induced damage. We usually record images every 3 seconds. The whole system can be coordinated using an Image-1 AT image processor (Universal Imaging Corporation) or a different image processor with homewritten shutter control software. It is fairly easy even for beginning programmers to write shutter control and OMDR control into software that controls a simple averaging device, such as a Hamamatsu Argus 10. We use a Data Translation IO board (DT-2817) together with a Screw terminal plate (DT-758c) for controlling shutters and filter wheels.

C. Setup

We first describe a basic motility assay and then consider some of the variables associated with the assay. The assay is very sensitive to variables such as time of incubation and washing, so great care and attention must be paid these factors to obtain reproducible results. We try not to let the time vary more than 5 seconds at each stage. It is further recommended that all reagents be stored as small frozen aliquots which can be used fresh for each experiment. We are standardly able to prepare four individual chambers in one experiment using this protocol.

1. Make perfusion chambers using two strips of double-stick tape (3M No. 665) on a glass slide to support a 22-mm^2 coverslip. Cool the chambers to 0–4°C on a metal block placed on ice.

2. Thaw the chromosomes and dilute 1/300 in PME (10 mM K-Pipes, pH 7.4, 5 mM 4 mM MgCl$_2$, 1 mM EGTA, 100 μg/ml leupeptin, 100 μg/ml pepstatin, 100 μg/ml chymostatin, 5 mM dithiothreitol) containing 10% bovine serum albumin (BSA, Sigma No. A7638). Polyamines were used previously to stabilize chromosome arms, but in the real-time assay they cause the seeds to stick to the glass surface of the perfusion chamber and therefore must be avoided. The PME is warmed to 37°C for 10 minutes and returned to ice before addition of chromosomes, as this appears to help the BSA subsequently in blocking the binding of other proteins to the glass.

3. The diluted chromosomes are perfused into the chamber and left for 5 minutes (step 1, Fig. 1). The chromosomes tend to stick preferentially to the glass slide over the coverslip, and all subsequent experiments describe complexes attached to the glass slide. BSA prevents most other proteins from binding subsequently to the glass, although some free microtubule motors will bind to the glass. The slide, which forms the bottom surface of the chamber, gives inferior optical visualization compared with the coverslip undersurface, but it was used in all our assays because of its more favorable surface chemistry.

4. Stable, rhodamine-labeled microtubule seeds are made. Three steps (4–6) result in the attachment of segmented, polarity-marked microtubules (see Chapter 7) to the chromosomes attached to the glass slide. First, the brightly labeled seeds are attached to the kinetochores. To survive the washing steps, in which the unattached seeds are removed from the chamber, the seeds must be stabilized. Taxol cannot be used in these experiments because it will interfere with the subsequent growth of GTP–tubulin from the seeds. In our experiments, the seeds are made from GMPCPP–tubulin. Microtubules polymerized in the presence of GMPCPP are stable to dilution (Hyman *et al.*, 1992) and can therefore be attached to kinetochores and washed without having to worry about stabilization as with GTP microtubules. A protocol to synthesize GMPCPP has been described (Hyman *et al.*, 1992). To make the GMPCPP seeds, 40 μM rhodamine–tubulin and 0.15 mM GMPCPP are mixed together

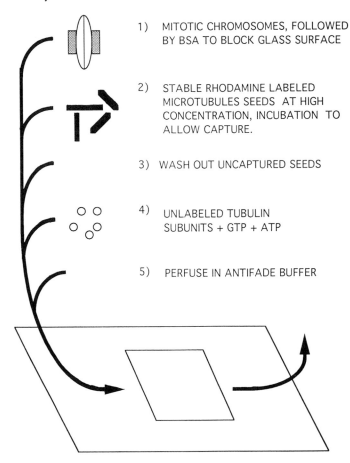

1) MITOTIC CHROMOSOMES, FOLLOWED BY BSA TO BLOCK GLASS SURFACE

2) STABLE RHODAMINE LABELED MICROTUBULES SEEDS AT HIGH CONCENTRATION, INCUBATION TO ALLOW CAPTURE.

3) WASH OUT UNCAPTURED SEEDS

4) UNLABELED TUBULIN SUBUNITS + GTP + ATP

5) PERFUSE IN ANTIFADE BUFFER

Fig. 1 Diagram to show the perfusion steps required in setting up the assay for motor activity on the kinetochores. For details see text.

(Hyman *et al.*, 1991). This is most conveniently made up as a large volume on ice and spun 30 psi in the airfuge at 4°C to remove aggregates of tubulin formed on thawing; the supernatant is frozen in 5-μl aliquots. Prior to use an aliquot is thawed and polymerized for 15 minutes at 37°C.

5. To attach the seeds to the kinetochores of the chromosomes that have been attached to the glass slide, the chamber is washed once (one wash is one chamber volume), in PMD (80 mM K-Pipes, pH 6.8, 1 mM EGTA, 4 mM MgCl$_2$, 5 mM dithiothreitol) + 1x protease inhibitors (10 μg/ml leupeptin, pepstatin, and chymostatin), and then the slide is transferred to an aluminum block at 37°C for 30 seconds. The seeds, diluted 1/30 into PMD + 1x protease inhibitors at 37°C, are then perfused into the chamber. The reaction is allowed to proceed for

5 minutes so that the seeds are captured by the kinetochores (step 2, Fig. 1). Uncaptured seeds are washed out by perfusing the chamber with PMD + 1 mM GTP four times at 37°C (step 3, Fig. 1).

6. To elongate more dimly labeled microtubule segments off the ends of the GMPCPP seeds, the chamber is washed twice with PMD + 1 mM GTP + 0.1x protease inhibitors + 3% polyethylene glycol (PEG) 3360, containing once-cycled tubulin (Hyman *et al.*, 1991) at 15 μM and rhodamine–tubulin (Hyman *et al.*, 1991) at 1.5 μM. The final stochiometry is approximately one rhodamine per 10 tubulin subunits. The 3% PEG 3360 is included in the polymerization buffer to ensure growth from the minus end and also to prevent adsorption of the rhodamine–tubulin to the glass surface of the chamber. Polymerization is allowed to proceed for 10 minutes at 37°C (step 4, Fig. 1).

7. The microtubules are stabilized by washing five times in an antifade buffer containing taxol (PMD + 0.2% 2-mercaptoethanol, 10 μM taxol, 5 μg/ml glucose oxidase, 10 μg/ml catalase, 10 mM glucose, 0.1x protease inhibitors) + 1 μg/ml Hoescht 33342 (step 5, Fig. 1). The glucose oxidase and catalase are most conveniently stored as a 100× stock in 50% glycerol at −20°C. The antifade buffer is essential to the assay. It works by scavenging oxygen, but a number of problems are associated with its use. First, the glucose tends to be used up if solutions are not kept in sealed tubes. Second, it essential that the temperature of solutions only increase, not decrease, subsequent to addition of the antifade component, because oxygen redisolving at lower temperatures uses up the glucose. Third, as the glucose oxidase works, its FAD cofactor becomes oxidized. This greatly reduces the strength of the fluorescent signal by an unknown mechanism, and may also increase the fluorescence background. It is therefore preferred that the antifade system be kept at 0°C until just prior to use. Fourth, glucose is converted to gluconic acid, potentially changing the pH of the buffer in weakly buffered solutions. Lowering of pH is particularly prevalent around bubbles if these are sealed under the coverslip.

8. If successful, the chromosomes should be attached to the glass slide, and their arms and primary constriction should be visible by Hoechst fluorescence. The degree of swelling of the chromatin arms depends on the concentration of MgCl$_2$ in the assay. Excessive swelling reduced our ability to discriminate the primary constriction from the rest of the chromosome by Hoechst fluorescence; however we also found that swelling tended to reduce background binding of microtubule to chromosomes arms and made kinetochore visualization in the rhodamine channel easier. Thus, our buffer represents a compromise. At the primary constriction, the two kinetochores should be visible as clusters of brightly labeled seeds, and in addition the kinetochores may bind some rhodamine–tubulin directly. If seeds are present all over the slide, then too many seeds were used in the capture reaction. The more dimly labeled portions of the microtubules should radiate away from the kinetochores. If rhodamine–tubulin is adsorbed to the glass surface, try increasing the concentration of PEG

slightly. The most common problem is sticking of multiple microtubules to chromosome arms. This is caused by cytoskeletal debris on the arms, and must be addressed by improving the chromosome isolation as discussed above. The chromosomes should be fairly sparse, one to four per field with a 60× objective, so that individual chromosomes do not obscure each other.

D. Recording Minus-End-Directed Movement

Minus-end-directed movement is the most difficult to record. The amount of minus-end growth rarely exceeds 10 μm, and because the minus-end motor moves at 40 μm per minute, only about 20 seconds elapses between addition of ATP and the microtubule running off the kinetochore. Although it is possible with practice to record minus-end-directed movement following perfusion of ATP into the chamber, for routine measurement it is easier to initiate movement by uncaging caged ATP.

1. Wash the chamber with 1 mM caged ATP in antifade buffer. It is important that caged ATP be protected from room light at all times, and we perform these manipulations under a sodium vapor safelight.

2. Using rhodamine fluorescence, select a nice chromosome, that is, a chromosome in which the two kinetochores can be distinguished as clusters of brightly labeled seeds and no microtubules are attached to the arms of the chromosomes.

3. With the video recording sequence running, locally activate the caged ATP. The easiest way to uncage the ATP on a simple microscope system is to employ a filter used in a Hoechst or DAPI filter set. With a typical 360-nm bandpass filter, a 100-W mercury arc illuminator, and a bright 60×, 1.4NA lens, complete uncaging in a small area takes about 0.3 second. Start recording the chromosomes at 3-second intervals with a 0.3-second shutter-open time during each recording. During the 3-second wait time between recordings, switch the filter holder to the Hoechst position, so that the next time the shutter opens, the field is irradiated with 360-nm light for 0.3 second, which uncages the ATP. Then switch back to rhodamine fluorescence to record any subsequent movement of the microtubules. Alternatively, more sophisticated systems using filter wheels can be employed.

E. Recording Plus-End-Directed Movement

The plus-end-directed motor will not move in an ATP concentration below 100 μM. It is difficult to observe this activity by activation of caged ATP because the activated ATP rapidly diffuses away from the observation field; however, because the movement is slower and the plus-end segments are longer, it is easy to record plus-end-directed movement after perfusing 1 mM ATP in antifade buffer into the chamber, refocusing, and beginning the recording.

F. Variables in the Assay

1. Switching between Plus- and Minus–End-Directed Movement at the Kinetochore Using ATPγS

The addition of ATPγS during the capture reaction (step A, Fig. 2) is sufficient to switch the kinetochores from plus- to minus-end-directed movement (Hyman and Mitchison, 1991b). Different chromosome preparations have different levels of endogenous plus- or minus-end-directed movement, although movement tends to be more minus end directed in the absence of ATPγS. Therefore, for each chromosome preparation, a dose–response curve must be constructed to determine the concentration of ATPγS required to fully turn on plus-end-directed movement. The concentration required for full switching varied between 10 and 500 μM. One reason that plus-end-directed movement was routinely observed in early work (Mitchison and Kirschner, 1985) may be that GTPγS was routinely included in the assay buffer because it was used in the microtubule seed stabilization step.

2. Washing the Kinetochores Prior to the Capture Reaction

A key variable in the assay is the wash between steps A and B (Fig. 2), as illustrated in Hyman and Mitchison (1991a). The relationship between washing and kinetochore motor activity varies from chromosome preparation to preparation. The general rule is that incubation of chromosomes at 37°C, without microtubules attached to their kinetochores, tends to inactivate the plus-end-directed movement system. A second reason that plus-end-directed movement was routinely observed in early work (Mitchison and Kirschner, 1985) may be that the chromosomes were not washed between microtubule capture and ATP addition. Presumably, the washing removes some weakly bound kinetochore components necessary for plus-end movement. We do not know whether this is due to inactivation of the motor itself or of the phosphorylation machinery.

3. Variation in Microtubule Number at the Kinetochore

The number of microtubules attached to the kinetochore determines the concentration of ATPγS required to activate the plus-end-directed motor (Hyman and Mitchison, 1991a). At fewer than three microtubules per kinetochore it is very difficult to initiate plus-end-directed movement, even in the presence of 1 mM ATPγS. At more than six microtubules per kinetochore, plus-end-directed movement may be observed even in the absence of ATPγS. The number of microtubules attached per kinetochore can be varied by varying the number of microtubules in the capture reaction (step B, Fig. 2) while keeping the time constant.

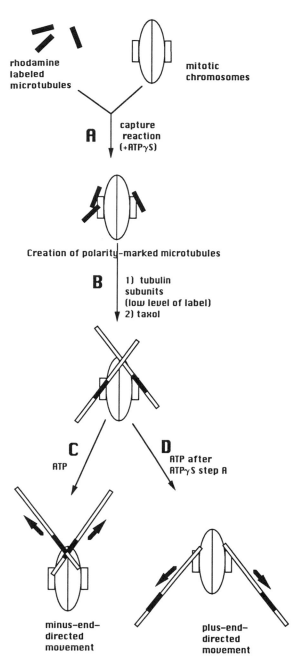

Fig. 2 Diagram of the results of ATP addition to the kinetochores of isolated chromosomes. Summary of the movements of microtubules on the kinetochores of isolated chromosomes. (A) The capture of the microtubule seeds by the kinetochores. (B) Tubulin grows from the ends of the captured seeds. After addition of ATP, microtubules move either by minus-end directed (C) or plus-end directed (D) movement.

III. Conclusions

We have described an assay in which the motor activity of the kinetochores of isolated mammalian chromosomes can be examined by video microscopy. The assay is technically challenging: Only small numbers of chromosomes can be assayed in each experiment, and the assay is very sensitive to the buffers used and the incubation times. Multiple perfusions must be performed in reproducible time intervals, and the sequence of temperature changes cannot be varied. Under a defined set of conditions, however, the assay can be made quite reproducible. It should prove very useful for testing specific inhibitory reagents to probe the identities of the motors and kinases involved in mammalian kinetochore function. The assay demonstrates the utility of modern fluorescence microscopy in the study of subcellular reactions and obtains quantitative biochemical information on their molecular mechanisms.

References

Hyman, A. A., and Mitchison, T. J. (1991a). Regulation of the direction of chromosome movement. *Cold Spring Harbor Symp. Quant. Biol.* **56,** 745–750.

Hyman, A. A., and Mitchison, T. J. (1991b). Two different microtubule-based motor activities with opposite polarites in kinetochores. *Nature (London)* **351,** 206–211.

Hyman, A. A., Drexel, D., Kellog, D., Salser, S., Sawin, K., Steffen, P., Wordeman, L., and Mitchison, T. J. (1991). Preparation of modified tubulins. *In* "Methods of Enzymology" (R. B. Vallee, ed.), Vol. 196, pp. 478–485. Academic Press, San Diego.

Hyman, A. A., Salser, S., Drechsel, D., Unwin, N., and Mitchison, T. J. (1992). The role of GTP hydrolysis in microtubule dynamics: Information from a slowly hydrolyseable analogue GMPCPP. *Mol. Biol. Cell* **3,** 1155–1167.

McIntosh, J. R., and Pfarr, C. M., (1991). Mitotic motors. *J. Cell Biol.* **115,** 577–587.

Mitchison, T. J., and Kirschner, M. W. (1985). Properties of the kinetochore in vitro. 2. Microtubule capture and ATP dependent translocation. *J. Cell Biol.* **101,** 767–777.

CHAPTER 20

The Diatom Central Spindle as a Model System for Studying Antiparallel Microtubule Interactions during Spindle Elongation *in Vitro*

Christopher J. Hogan, ★ **Patrick J. Neale,**†,[1] **Manlin Lee,**‡ **and W. Zacheus Cande**★

★Departments of Molecular and Cell Biology and †Plant Biology
University of California, Berkeley
Berkeley, California 94720

‡Institute of Life Science
National Tsing Hua University
Hsingchu, Taiwan 30043, Republic of China

I. Introduction
II. Diatom Cultures
   A. Growth and Maintenance of *Cylindrotheca fusiformis*
   B. Synchronization of Bulk Cultures
   C. Accumulation of Metaphase Spindles
III. *In Vitro* Models
   A. General Considerations
   B. Permeabilized Cell Model
   C. Isolated Spindles
   D. Spindle Elongation Assay
IV. Concluding Remarks
   References

[1] Present address: Smithsonian Environmental Research Center, Edgewater, Maryland 21037.

I. Introduction

Mitotic spindle elongation is one of two primary motility events responsible for chromosome separation during anaphase in eukaryotes. This aspect of karyokinesis is accomplished by a combination of pulling forces and by interactions between antiparallel microtubules (MTs) that cause these MTs to slide apart and drive the respective poles to which they are anchored in opposite directions [for reviews of various mechanisms that may contribute to spindle elongation, see Cande and Hogan (1989) and Hogan and Cande (1990)]. Although this is an unusual motility event, where MT bundles of mixed polarity are used to separate cargoes (the poles), the mechanism may be widespread throughout nature and operative in spindles as diverse as those of diatoms, yeast, and mammals. Few model systems have been developed to study this motility *in vitro* that feature the necessary spindle components to function outside the cell yet are simple enough to analyze. The diatom central spindle has thus far proved unique in both respects as an *in vitro* model and has made analysis of the mechanochemistry of spindle elongation possible. The ultimate goal in using such models is to characterize the molecular components and the assemblages they constitute that are integral to both the establishment of a bipolar spindle and to the separation of spindle poles (and the chromosomes tethered to them) during karyokinesis. This chapter describes the system our laboratory currently uses to study spindle elongation *in vitro*.

The diatom spindle is an excellent choice for analysis of spindle elongation *in vitro* because of its unique and highly ordered spindle morphology (Pickett-Heaps and Tippet, 1978; McDonald and Cande, 1989). The MTs that participate in anaphase B (spindle elongation), that is, the MTs of the central spindle, are spatially separated from the kinetochore MTs involved in anaphase A (chromosome-to-pole movement). Microtubules of each half-spindle are nucleated from platelike pole complexes and lie parallel to each other in a large bundle. They are of approximately uniform length and MTs from opposing half-spindles interdigitate to form a well-defined midzone where antiparallel MTs overlap. Moreover, the MT plus ends distal to the poles are clustered at the periphery of the overlap zone, thus making the overlap zone visible even with a light microscope (Pickett-Heaps and Tippet, 1978; Cande and McDonald, 1985). These exceptional structural features have permitted the mechanism of central spindle elongation *in vitro* to be determined in isolated central spindles from the centic diatom *Stephanopyxis turris* (Cande and McDonald, 1985, 1986; Masuda *et al.*, 1988). Unfortunately, a large number of spindles per unit culture volume cannot be obtained with this organism and spindle function is relatively labile under extended experimental conditions. The system described herein makes use of the pennate, marine diatom *Cylindrotheca fusiformis*. *C. fusiformis* has an array of attributes that make its spindles suitable both for an *in vitro* motility model and for biochemical analysis of spindle components; these include ease of growing bulk quantities with respect to medium, cost, and space; high log phase cell densities; simple culture synchronization and metaphase spindle collection scheme; and robustness of central spindles so that a high proportion remain functional not only during detergent permeabilization of cells or spindle isolation, but also during extended experimental manipulations. Although several other systems have been useful for studying spindle elongation *in vitro,* including lysed mammalian cells (Cande, 1982), isolated sea urchin spindles (Rebhun and Palazzo, 1988; Palazzo *et al.,* 1991), and permeabilized fission yeast cells (Masuda *et al.,* 1990), *C. fusiformis* spindles represent the most reliable and readily interpretable model for assaying spindle elongation *in vitro.*

II. Diatom Cultures

A. Growth and Maintenance of *Cylindrotheca fusiformis*

Cylindrotheca fusiformis is a diatom typical of estuarine and marine benthic habitats. It will grow in seawater to very high densities in continuously stirred bulk cultures when provided with nutrient and carbon enrichment. *C. fusiformis* clones are available from the University of Texas Culture Collection of Algae, Clone Nos. 2081–2087 (see Starr and Zeikus, 1987), and from the Center for the Culture of Marine Phytoplankton, Bigelow Laboratory for Ocean Sciences, McKown Point, West Boothbay, Maine (Clone MNC).

Clones obtained from these culture collections will aggregate rapidly to form a plate of cells at the bottom of a flask if left unstirred. These cells will not readily resuspend if the flask is stirred. The cells secrete a sticky "slime" (probably a mucopolysaccharide) that bind cells together in irregular clumps, a behavior typical of benthic diatoms. It is crucial both for bulk culturing of diatoms and for the spindle isolation procedure presented below that so-called "slimeless" clones be selected for. Slimeless clones can be isolated by plating diatoms on agar and selecting for nonclumping colonies (as judged under an inverted microscope). Nonclumping colonies will typically appear dull, small, and round on agar. Colonies containing cells that secrete slime tend to appear shiny, spread out, and jagged-edged owing to greater motility of the cells. Settled cells of slimeless clones in seawater should easily disperse when stirred. In stirred seawater they grow as small chains of cells that appear to have midthecal connections. Carboy production has focused on using these "slimeless" cells, which are almost planktonic in habit.

Desirable clones can be stored on agar slants (made with sterile seawater) in culture tubes at 16–22°C in the dark for 6-month stretches before requiring transfer. As a ready source for bulk cultures, sterile seeder flasks are maintained on a weekly basis. These are small flasks containing ~400 ml medium prepared from autoclaved seawater and f/2 nutrient medium (Guillard, 1975) sterilized by 0.22-μm filtration. Transfers to and from seeder flasks (except in the case of a terminal flask used to seed a carboy) should be done in a sterile transfer hood.

Seeder flasks are kept on an orbital shaker rotating at 100 rpm at 20–22°C on a 12/12 day/night light cycle. It takes about 1 week for a seeder flask to get to a useful density if it is inoculated with ~30 ml of plateau density culture (~3 × 10^6 cells ml^{-1}) from a previous seeder.

B. Synchronization of Bulk Cultures

Cylindrotheca fusiformis will grow well both in continuous light and in defined light/dark regimes. A pulsed synchronization method is used to provide a division "burst" at a specific time. The pulsed synchronization scheme employs three incubation periods: an initial growth period in the light, a growth-arresting dark period, and a second light period to bring about the division burst. A typical cell density curve over the incubation period is illustrated in Fig. 1. During the first incubation period the cells grow in continuous light with an average growth rate of 0.08 h^{-1}. Under these conditions cell division is occurring continuously and the doubling time of cell density is 8.6 hours. During the second, dark incubation period, cells continue to divide for about 6 hours. This represents the proportion of cells that have satisfied the light-dependent part of the cell cycle and can proceed with division (Vaulot and Chisholm, 1987). After an extended period in darkness all cells presumably are arrested at the same point in the cell

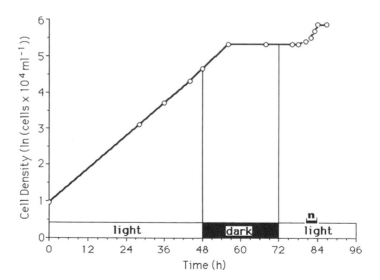

Fig. 1 Growth curve and synchronization scheme for *Cylindrotheca fusiformis* cultures. Log phase cultures were grown for 2 days in continuous light. Cell growth was arrested by placing the cells in the dark for 24 hours. After 6 hours of a second light period the cultures divided synchronously. Metaphase spindles are collected by treating cultures with nocodozole for 5 hours (n) during this division period.

cycle. The original synchronization scheme (Darley and Volcani, 1969, 1971; Lewin *et al.*, 1966) specified a 24-hour dark period. We have found that in practice the dark regime can be as short as 12 hours with satisfactory results. This flexibility in dark period is convenient as it allows for an adjustment of the first light period to compensate for any lags that may occur in growth of the cells. In the third incubation period, a second light pulse allows for restart of cell growth and a subsequent burst of cell division. In the original synchronization scheme the light intensity used in the second light period was much greater than that used in the first light period (Darley and Volcani, 1971). This also has been found to be unnecessary so that our standard scheme employs the same light intensity for the pulse as is used during growth. This intensity is 200 μmol (photons) m^{-2} s^{-1}.

To prepare a bulk culture twelve, 8-liter polycarbonate carboys are filled with 7–8 liters of filtered seawater. A large stir bar is added and carboys are autoclaved. Before inoculation 100 ml sterile f/2 medium is added. In normal circumstances exponential growth of *C. fusiformis* in 8-liter carboys will continue until a density of between 2 and 3 $\times$ 10^6 cells ml^{-1} is achieved. Cell density usually doubles during the dark period and again during the second light pulse. Thus cells should be seeded with a target density of 5 $\times$ 10^5 cells ml^{-1} by the end of the growth light period. Usually, carboys are seeded at a density of 2.5 $\times$ 10^4 cells ml^{-1} and grown in the continuous-light growth phase for 36–48 hours before turning the lights off for a 12- to 24-hour dark period. An initial lag period before exponential growth sometimes occurs and adjustments in duration of the initial growth phase can be made after determining cell density at 36 hours. During all three incubation phases carboys are vigorously stirred in a growth chamber maintained at 22°C and fitted with wide-spectrum fluorescent lights (GE F40PL/AQ, General Electric). The carbon source for cell growth is supplied by continuous bubbling of sterilely filtered 3% carbon dioxide into carboys from an external tank at a (low) rate such that cultures are maintained at ~pH 8.

C. Accumulation of Metaphase Spindles

After the initiation of the second light period cells reenter the cell cycle, and by the sixth hour of illumination approximately 80–85% of the cells divide over the following 6 hours (Darley and Volcani, 1971). To collect cells with metaphase spindles, cultures are made 10^{-7} *M* nocodozole by adding 10 μl/liter 10^{-2} *M* nocodozole in dimethylsulfoxide (DMSO) directly to the carboys. Spindles are allowed to accumulate for 5 hours (hours 7–11 of the second light period) before harvesting (Fig. 1). This concentration of nocodozole disrupts interphase arrays of MTs (Figs. 2b, c); however, central spindles continue to form and existing spindles will not depolymerize. During the drug treatment, spindles reorient from perpendicular to parallel with respect to the long axis of the cell (Figs. 2a–c) and appear slightly longer and thinner than in untreated cells. The spindles after drug treatment are variable in length but on average are

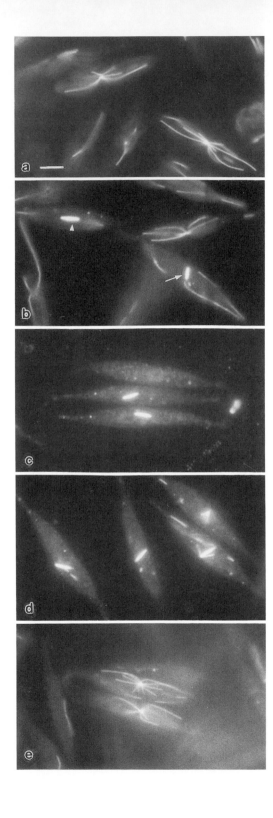

4–5 μm long. Preliminary electron microscopy suggests that half-spindles contain at least 10 MTs and that the zone of MT overlap may be about 0.5 μm. This estimate of the size of the zone of MT overlap is supported by our observations of biotinylated tubulin incorporation into the spindle. Two bands of newly incorporated tubulin in the spindle midzone mark the positions of plus ends of overlapping MTs. The two bands of incorporated tubulin are about 0.5 μm apart (Hogan *et al.,* 1992). Nocodozole treatment also retards anaphase and inhibits cytokinesis (Figs. 3a, b) so that mostly metaphase spindles accumulate during drug treatment.

After 5 hours in nocodozole a 1-ml cell sample is collected, pelleted in the microfuge 30 seconds, and resuspended in 30 μl PMEG (see below for formulation) containing 0.5% Triton X-100, 3% DMSO, and 1 μg ml^{-1} DAPI. To determine the proportion of spindle-containing cells prior to harvest, this sample is viewed under the fluorescence microscope and scored for percentage of elongate or dumbell-shaped nuclei. These nuclear configurations indicate the presence of a spindle and typically 40–60% of the cells contain spindles at this point (Fig. 3). At harvest, cell density is 1.9–2.5 ×10^6 cells ml^{-1} (Fig. 1). If cultures are allowed to continue to grow in nocodozole, spindles remain stable for several hours thereafter (Fig. 3a); however, if the drug is washed out, spindles reorient normal to the cell axis and undergo spindle elongation, karyokinesis is completed, chromosomes decondense, and interphase arrays of microtubules return (Figs. 2d and e, 3b). Cytokinesis does not occur, however, as cell populations in which the drug was washed out show a dramatic decrease in mitotic index, while the cell density remains unchanged (Fig. 3b). This suggests that nocodozole treatment can be used to accumulate metaphase spindles while eliminating interphase MT arrays and that it is unnecessary to remove the drug prior to harvest.

III. *In Vitro* Models

A. General Considerations

Two types of central spindle models can be made from synchronized *C. fusiformis* cells: detergent-permeabilized cells and isolated spindle prepara-

---

Fig. 2 Effect of nocodozole on different populations of microtubules and on spindle orientation. (a) At the time of nocodozole treatment most of the population exhibits an interphase array of MTs that radiate from the centrosome. Spindles are oriented perpendicular to the cell's long axis at this time (not shown). (b) One hour after nocodozole addition interphase arrays begin to disappear. Spindles begin to accumulate and reorient from perpendicular (arrow) to parallel (arrowhead) with respect to the long axis of the cell. (c) Most spindles show a parallel orientation, and no interphase arrays remain after 3 hours in nocodozole. (d, e) If nocodozole is washed out of the cultures, spindles return to the parallel orientation (d) and interphase arrays return (e) within 1 hour. Bar = 5 μm.

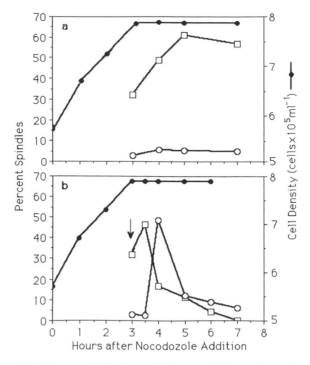

Fig. 3 Effect of nocodozole on cell division, spindle accumulation, and spindle orientation. At the initiation of nocodozole treatment cytokinesis slows, as can be seen from the cell density curves (●), and finally ceases 3 hours after drug treatment while spindles (□ and ○) accumulate to 30–40% (a, b). If nocodozole is left in cultures (a) spindles with a parallel orientation (□) continue to accumulate for 2 hours and then remain at a constant level. If nocodozole is removed after 3 hours (b, arrow) spindles change their orientation from parallel (□) to perpendicular (○) and most cells lose spindles by 2 hours after drug reversal (b). Although karyokinesis occurs (not shown), cytokinesis never takes place.

tions. Spindles in permeabilized cells as well as isolated spindles are able to undergo ATP-dependent elongation. A larger percentage of spindles are functionally competent, and in general, assessment of reactivation of spindle elongation is easier to determine in the permeabilized cell model than in isolated spindle fractions. Thus for most types of studies in which spindle function is being assessed with respect to pharmacological sensitivities, various function-blocking treatments, function-restoration treatments, and so on, the permeabilized cell model is preferable to isolated spindles. Isolated spindles have the advantage of being enriched in tubulin, and by inference spindle-associated protein, by ~25× over extracts from permeabilized cells (Hogan *et al.,* 1992). For correlating loss or gain of function with specific polypeptides or, for instance, for identifying spindle-specific kinase substrates, isolated spindle fractions are the most desirable preparation to use.

In either case synchronized bulk cultures first must be concentrated to a manageable volume and then seawater replaced with buffer. Cells can be harvested and concentrated relatively quickly using tangential flow filtration. We use a Pellicon filtration system (Millipore Corp., Bedford, MA) with nine sets of polyethylene spacers, each containing a 0.45-μm Durapore filter (total surface area = 4.5 ft^2). At a filtrate flow rate of 1.25 liter min^{-1} and back-pressure of 5 psi, 80–100-liters of culture can be concentrated to 1 liter in 45–60 minutes with minimal cell loss. At this point cells can be pelleted by spinning 5 minutes at 2500g in GSA tubes (Sorvall, Dupont Co., Wilmington, DE), and the remaining seawater discarded. Cells should then be washed twice in 1 liter of an isotonic buffer (Cande and McDonald, 1986) that contains 10 mM EGTA. Although so-called "slimeless" clones are used, we have found no clone in which secretions of a sticky matrix have been completely eliminated. Not only do these washes serve to eliminate remaining traces of seawater, but also the chelation of Ca^{2+} at this stage seems to greatly reduce the secreted matrix.

To obtain functional spindles it is essential that cells be permeabilized or homogenized in the presence of phosphatase inhibitors. Functional spindle models prepared from the diatom *Stephanopyxis turris* also exhibit this requirement (Wordeman and Cande, 1987). A high phosphorylation state of spindle proteins may be essential for spindle elongation. When *C. fusiformis* cells are thiophosphorylated during permeabilization, a specific and consistent pattern of thiophosphate incorporation can subsequently be seen with an antibody that binds to thiophosphorylated epitopes (Wordeman and Cande, 1987). Thiophosphate is selectively incorporated into components located in the spindle midzone and at the poles (Hogan *et al.*, 1992). Sites of thiophosphate incorporation in *C. fusiformis* spindles are the same as those seen in diatom spindles isolated from *S. turris* (Wordeman and Cande, 1987; Wordeman *et al.*, 1989). Table I illustrates the dramatic difference found when phosphatase inhibitors are used during spindle preparation. One notable difference between *C. fusiformis* and *S. turris* is that β-glycerophosphate is much more effective than ATPγS for preserving spindle function in *C. fusiformis*.

B. Permeabilized Cell Model

After isotonic washing, the cell pellet derived from 80–100 liters of culture (typically 20–25 g wet wt of cells) is resuspended in a GSA tube with 500 ml permeabilization buffer that contains PMEG [50 mM 1,4-piperazinediethanesulfonic acid (Pipes), pH 7.0, 5 mM MgSO$_4$, 5 mM ethylene glycol bis (β-aminoethyl ether) N, N'-tetraacetic acid (EGTA), 40 mM β-glycerophosphate, 100 μM *rac*-6-hydroxy-2,5,7,8-tetramethylchromane-2-carboxylic acid (TROLOX), a proteinase inhibitor cocktail, Masuda *et al.*, 1988; plus 1 mM phenylmethylsulfonyl fluoride (PMSF) and 1 mM dithiothreitol] containing 3% DMSO and 1% Triton X-100. The tube is capped and cells are permeabilized on ice while

Table I

Effect of Phosphate Inhibitors on Spindle Elongation *in Vitro*

Cell permeabilization buffer[a]	Spindle reactivation buffer[b]	% Spindles elongated[c]
PME	—	4
PME	PME, ATP	4
PME, 50 μM ATPγS	PME, ATP	25
PME, 50 μM ATPγS	PMEG, ATP	45
PMEG	PMEG, ATP	85

[a] P, 50 mM Pipes, pH 7.0; M, 5 mM MgSO$_4$; E, 5 mM EGTA; G, 40 mM β-glycerophosphate.

[b] All reactivation buffers included 1 mM ATP. Spindles were reactivated for 5 minutes.

[c] Spindle elongation was determined by counting the number of spindles that contained gaps (see text). At least 500 spindles were scored for each set of conditions.

shaking for 15 minutes. Cells are then pelleted by centrifugation at 2500g, 5 minutes, 4°C, and the supernatant is discarded. The supernatant should be dark green because of the chlorophyll liberated from permeabilized chloroplasts. The procedure is repeated two more times. Following the third permeabilization wash, the supernatant should be pale green. We use chlorophyll depletion as an indication that most detergent-extractable protein has been washed from the cells. The permeabilized cells should then be washed twice in 500 ml permeabilization buffer without Triton X-100 and the final pellet resuspended in 50 ml of the same buffer. At this point cells can be aliquoted and then frozen and stored in liquid nitrogen. Cells can be stored in this manner for at least 2 years without any appreciable loss in the ability of spindles to elongate *in vitro*.

C. Isolated Spindles

To prepare an isolated spindle fraction, diatoms are collected and washed in isotonic buffer as described above. The pellet of cells from the final isotonic wash is resuspended in ~250 ml of a buffer conducive to both MT stabilization and preservation of spindle function that consists of PMEG containing 3% DMSO. The resuspended cells are then added to a prechilled 350-ml Bead Beater homogenization chamber (Bead Beater, Biospec Prod., Bartlesville, OK) that has been filled to approximately one-third capacity with 0.1-mm glass beads. Buffer is added until the chamber is full to avoid foaming during homogenization. During homogenization the chamber should be kept in a cooling chamber filled with ice water. The two chambers are placed on the Bead Beater base and cells are disrupted by four, 30 second bursts. Chamber contents should be allowed to cool for 30 seconds between each burst. To separate the glass beads from the homogenate, the chamber contents are poured over 30-μm Nitex (Tekto Inc., Elmsford, NY) and the filtrate is collected. To remove ruptured diatom frustules and other large debris, the filtrate is spun at 200g, 10 minutes, 4°C. At this point the vast majority of liberated spindles remain in the supernatant along with interphase nuclei, chloroplasts, and other organelles (Figs. 4a and b). Chromatin is no longer attached to the central spindles after homogenization. The greatest loss of the spindles' ability to elongate *in vitro* occurs during homogenization. Sixty percent of the spindles retain the ability to elongate compared with spindles in permeabilized cells (Table II). Much of the loss of function is probably due to breakage of spindles into component half-spindles because a substantial number of the MT bundles observed by immunolocalization are approximately half the size (2–3 μm) of metaphase spindles seen in permeabilized cells. Centrifugation of the 200g supernatant at 2600g, 25 minutes, 4°C, pellets nuclei and large organelles while leaving spindles in the supernatant with little loss of spindle function (Figs. 4c and d, Table II). Spindles can be stored in liquid nitrogen for up to a year in this state and maintain functional

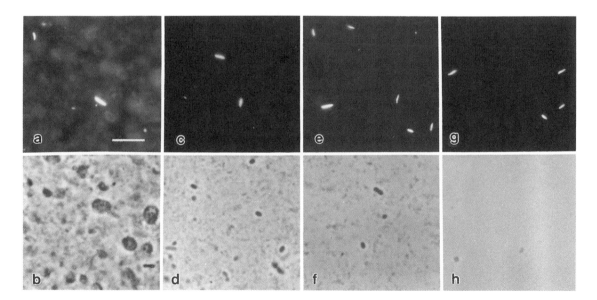

Fig. 4 Fractions during spindle isolation. (a, c, e, g) Antitubulin immunofluorescence of a random microscope field from four different steps during spindle isolation. (b, d, f, h) Same fields as a, c, e, g, respectively, viewed with phase optics. (a, b) Filtered homogenate. (c, d) 2600g supernatant. (e, f) Minitan tangential flow filtration retentate. (g, h) Spindle fraction from the buffer/sucrose interface. Bar = 10 μm.

Table II
Functional Spindles Retained during Spindle Isolation

Fraction	% Spindles Retained	ATP-Dependent Elongation[a] (% control[b])
Filtered homogenate	100	63
2600*g* supernatant	60	39
Filtration retentate	30	39
Spindle fraction	30	39

[a] Fractions were incubated in 1 mM ATP for 5 minutes and percentage gaps were determined for 500 spindles in each fraction.

[b] An aliquot of cells from cultures used to isolate spindles was permeabilized and spindles were reactivated in 1 mM ATP for 5 minutes. Percent reactivation was determined and this value was used as control. From Hogan *et al.* (1992).

integrity. To solubilize remaining membrane vesicles the 2600*g* supernatant is made 1% Triton X-100 and gently stirred for 20 minutes at 4°C. If spindles are centrifuged at this stage aggregates of extraneous proteins will copellet with and contaminate the spindle pellet. Removal of many soluble proteins from the spindle-containing fraction can now be accomplished by tangential flow filtration. Spindles are loaded into a Minitan tangential flow filtration unit (Millipore Corp.), and soluble proteins are eliminated from the spindle fraction by filtration through 0.2-μm Durapore filters (Milipore Corp.). This process is continued by washing spindles with 500–700 ml of PMEG containing 3% DMSO and 1% Triton X-100 followed by 200 ml buffer without Triton X-100 (Figs. 4e,f). The spindle-containing retentate (20–50 ml) is then underlayered with 20% sucrose (in PMEG) and spun at 8000*g*, 10 minutes, 4°C, in an HB-4 swinging bucket rotor (Sorval, DuPont). This concentrates spindles at the buffer/cushion interface and allows remaining large particles in the spindle fraction to pellet free of the spindles (Figs. 4g, h). Approximately 30% of the spindles collected at the interface are capable of undergoing elongation *in vitro* (Table II).

D. Spindle Elongation Assay

For reactivation of spindle elongation *in vitro*, 0.5 ml of permeabilized cells (either fresh or thawed from liquid nitrogen) are pelleted through 10 ml of PMEG at 2500*g*, 3 minutes, 4°C, to wash DMSO from cells (and to supply fresh dithiothreitol and proteolytic inhibitors for cells that have been stored frozen). The pellet is resuspended in 2 ml of PMEG and 10- to 30-μl aliquots are spun at 2500*g*, 3 minutes, 4°C, through 3 ml of PMEG onto poly-L-lysine-treated (0.5 mg ml^{-1}) coverslips. Coverslips are then drained of excess buffer by blotting onto a Kimwipe. Spindles are stable, both structurally and functionally, so that pre-treatments (e.g., antibodies, inhibitors, extractions) of up to 30 minutes at room temperature can be tolerated. To assay spindles for the ability to elongate, the pretreatment solutions are rinsed from the coverslip by dipping in PMEG, excess buffer is drained, and coverslips are incubated in PMEG containing 1 mM ATP for 5 minutes. Coverslips are then fixed either for 10 minutes in aldehyde (0.5% paraformaldehyde, 0.1% gluteraldehyde in PMEG) or for 5 minutes in −20°C CH$_3$OH. Spindles are visualized by immunolocalization using an antibody to tubulin followed by an appropriate fluorochrome-conjugated secondary antibody. Forty-five-minute antibody incubations are required for adequate antibody penetration.

In the case of isolated spindles, 20–50 μl of the buffer/cushion interface is spun through 3 ml of PMEG at 15,000g, 10 minutes, 4°C onto acid-cleaned coverslips. Pretreatment, ATP reactivation, fixation, and spindle immunolocalization are performed as described above except only 10-minute antibody incubation steps are required.

Spindles can be observed under the fluorescence microscope and scored for morphological changes that result from ATP treatment. Whereas spindles that are not incubated in ATP or that did not elongate appear as a fluorescent bar (Fig. 5a), reactivated spindles exhibit a gap or a bend and a gap between half-spindles (Figs. 5b, c). Typically 75–90% of spindles in permeabilized cells and 30% of isolated spindles will show gaps after ATP treatment. Gap formation between half-spindles result from antiparallel microtubules that have slid apart during spindle elongation (Cande and McDonald, 1985, 1986). After ATP addi-

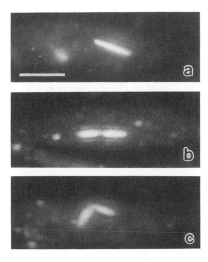

Fig. 5 Spindle elongation *in vitro*. Permeabilized cells stained with antitubulin both before (a) and after (b, c) ATP addition. After incubation in 1 mM ATP for 5 minutes, spindles typically exhibit a gap (b) or a bend and a gap (c). Bar = 5 μm.

tion most of the MTs slide past one another out of the zone of MT overlap; however, some antiparallel interactions remain between the longest MTs in the overlap zone that allow the gap, which is visible in the light microscope, to develop between half-spindles (Cande and McDonald, 1986). That gaps are not formed by poleward depolymerization of half-spindle MTs has been demonstrated by marking the free MT (plus) ends with exogenously added biotinylated tubulin and then following the marked ends throughout subsequent ATP reactivation (Masuda *et al.*, 1988; Hogan *et al.*, 1992). Before ATP addition the marked tubulin is seen as two zones flanking the zone of MT overlap. After ATP incubation the two marked zones have merged into one in the spindle midzone, indicating that MTs from each half-spindle slide apart through the MT overlap zone during ATP treatment. Moreover, there is a strong correlation between the percentage of spindles that show ATP-dependent sliding in the above experiment and ATP-dependent gap formation in spindles that are not (plus) end labeled (Hogan *et al.*, 1992). The "gap assay" is thus a relatively simple and unambiguous indicator of spindle elongation *in vitro* in diatom central spindles.

The percentage of spindles that form gaps during ATP incubation depends on ATP concentration and on incubation time (Hogan *et al.*, 1992). Thus if under identical conditions of time and nucleotide concentration a different percentage of gaps are formed, variation in the rate of spindle elongation is indicated. Under such conditions it is possible to study the effects that various inhibitors, enhancers, nucleotides, and other compounds have on the spindle elongation machinery by comparing relative rates of elongation as measured by percentage of gaps formed after different incubation periods.

IV. Concluding Remarks

The structural complexity and functional diversity of events that constitute mitosis preclude a simple analysis. It is clear that many of the motility events that occur during mitosis may be mediated by MT-stimulated mechanochemical enzymes. In developing an *in vitro* assay for spindle elongation we have attempted to dissect one aspect of mitosis from the overall process with the goal of identifying and characterizing components and assemblages involved in the antiparallel sliding of MTs. Diatom central spindles were used because they comprise the main component of the spindle apparatus in diatoms, they are segregated from other mitotic components, and they represent the structural unit (a bundle of antiparallel MTs) that is functional during spindle elongation. Moreover, the central spindle of *C. fusiformis* has proved to be an extremely stable structure, making it amenable to a wide range of *in vitro* experiments.

The gap assay described here is simple, reliable, and easily interpretable. Using it we have been able to characterize enzymological characteristics of spindle elongation that include sensitivity to MT motor inhibitors as well as nucleotide triphosphates and ATP analogs that support spindle elongation at variable rates (Hogan *et al.*, 1992). Preincubation of spindles with function-blocking antibodies to specific motors or other spindle components should also yield valuable mechanistic information. In the future it may also be possible to inactivate the native spindle motor(s) and add back purified components that reconstitute function.

References

Cande, W. Z. (1982). Nucleotide requirements for anaphase chromosome movements in permeabilized mitotic cells: Anaphase B but not anaphase A requires ATP. *Cell (Cambridge, Mass.)* **28,** 15–22.

Cande, W. Z., and Hogan, C. J. (1989). The mechanism of anaphase spindle elongation. *BioEssays* **11,** 5–9.

Cande, W. Z., and McDonald, K. L. (1985). *In vitro* reactivation of anaphase spindle elongation using isolated diatom spindles. *Nature (London)* **316,** 168–170.

Cande, W. Z., and McDonald, K. L. (1986). Physiology and ultrastructural analysis of elongating mitotic spindles reactivated *in vitro. J. Cell Biol.* **103,** 593–604.

Darley, W. M., and Volcani, B. E. (1969). Role of silicon in diatom metabolism: A silicon requirement for deoxyribonucleic acid synthesis in the diatom Cylindrotheca fusiformis Reimann et Lewin. *Exp. Cell Res.* **58,** 334–342.

Darley, W. M., and Volcani, B. E. (1971). Synchronized cultures: Diatoms. *In* "Methods in Enzymology" (A. S. Pietro, ed.), Vol. 23, Part A, pp. 85–96. Academic Press, New York.

Guillard, R. R. L. (1975). Culture of phytoplankton for feeding marine invertebrates. *In* "Culture of Marine Invertebrate Animals" (W. L. Smith and M. H. Chanley, eds.), pp. 29–60. Plenum, New York.

Hogan, C. J., and Cande, W. Z. (1990). Antiparallel microtubule interactions: Spindle formation and anaphase B. *Cell Motil. Cytoskel.* **16,** 99–103.

Hogan, C. J., Stephens, L., Shimizu, T., and Cande, W. Z. (1992). Physiological evidence for the involvement of a kinesin-related protein during anaphase spindle elongation in diatom central spindles. *J. Cell Biol.* **119,** 1277–1286.

Lewin, J. C., Reimann, B. E., Busby, W. F., and Volcani, B. E. (1966). Silica shell formation in synchronously dividing diatoms. *In* "Cell Synchrony, Studies in Biosynthetic Regulation" (I. L. Cameron, and G. M. Padilla, eds.), pp. 169–188. Academic Press, New York.

Masuda, H., McDonald, K. L., and Cande, W. Z. (1988). The mechanism of anaphase spindle elongation: Uncoupling of tubulin incorporation and microtubule sliding during *in vitro* spindle reactivation. *J. Cell Biol.* **107,** 623–633.

Masuda, H., Hirano, T., Yanagida, M., and Cande, W. Z. (1990). *In vitro* reactivation of spindle elongation in fission yeast *nuc2* mutant cells. *J. Cell Biol.* **110,** 417–425.

McDonald, K. L., and Cande, W. Z. (1989). Diatoms and the mechanism of anaphase spindle elongation, *In* "Algae as Experimental Systems" (A. W. Coleman, L. J. Goff, and J. R. Stein-Taylor, eds.), pp. 3–18. Liss, New York.

Palazzo, R. E., Lutz, D. A., and Rubhun, L. I. (1991). Reactivation of isolated mitotic apparatuses: Metaphase vs anaphase spindles. *Cell Motil. Cytoskel.* **18,** 304–318.

Pickett-Heaps, J. D., and Tippit, D. H. (1978). The diatom spindle in perspective. *Cell (Cambridge, Mass.)* **14,** 455–467.

Rebhun, L. I., and Palazzo, R. E. (1988). *In vitro* reactivation of anaphase B in isolated spindles of the sea urchin egg. *Cell Motil. Cytoskel.* **10,** 197–209.

Starr, R. C., and Zeikus, J. A. (1987). UTEX—The culture collection of algae at the University of Texas at Austin. *J. Phycol.* **23,** Suppl., 1–47.

Vaulot, D., and Chisholm, S. W. (1987). A simple model of the growth of phytoplankton populations in light/dark cycles. *J. Plankton Res.* **9,** 345–366.

Wordeman, L., and Cande, W. Z. (1987). Reactivation of spindle elongation *in vitro* is correlated with the phosphorylation of a 205 kd spindle-associated protein. *Cell (Cambridge, Mass.)* **50,** 535–543.

Wordeman, L., Davis, F. M., Rao, P. N., and Cande, W. Z. (1989). The distribution of phosphorylated spindle-associated proteins in the diatom *Stephanopyxis turris. Cell Motil. Cytoskel.* **12,** 33–41.

INDEX

A

Actin
   filaments
      load introduction, 39–42
      sliding, 42–48
   polystyrene bead-labeled, 6
   purification, 6
   rhodamine–phalloidin-labeled, 28–29
Adsorption, proteins onto flow cell surfaces,
      141–142
Algae, extruded cytoplasm preparations, 180–
      181
Antibodies, myosin-isoform-specific, 39
Artifacts, motility assays in *Xenopus* extracts,
      222
ATP
   concentrations in motility assays, 157
   microtubule sliding in axonemes
      Chlamydomonas, 91–93
      sea urchin sperm, 94–97
ATP analogs
   adenine-modified, 170
   chemistry, 169–173
   precautions, 174
   for probes of conformational change, 174–
      175
   purification, 168–169
   ribose-modified, 170–172
   for substrate specificity determination, 175–
      176
   triphosphate-modified, 173
ATPase, motor proteins expressed in
      Escherichia coli, 124–125
Axonemes, *Chlamydomonas reinhardtii*
   polarity assay, 70–73
   preparation, 157
Axoplasm, squid
   extrusion, 192–194
   vesicle motility
      buffers for, 194–195
      limitations, 197–201
      quantification of moving particles, 197–
         201
      video microscopy, 195–197

B

Bacterial expression vectors
   pET system, 118–119
   pGEX system, 119
   plasmid construction, 116–118
Buffers, physiological, for motility assays,
      194–195
Bundling, microtuble-associated
   in cell-free cytoplasmic extracts, 247–249
   materials, 246
   slide preparation, 246–247

C

Cell cycle, status, assay in *Xenopus* extracts,
      211, 216
Chara corallina, extruded cytoplasm
      preparations, 180–181
Chinese hamster ovary cells
   chromosome kinetochores, motor activity
         assay
      chromosome preparation, 268–269
      microscope requirements, 269
      procedure, 273
      recording techniques, 273
      variables, 274–276
   mitotic chromosomes, isolation, 154–156
Chlamydomonas reinhardtii
   axonemes, preparation, 157
   flagellar axonemes, polarity assay with,
      70–73
CHO, *see* Chinese hamster ovary cells
Chromaffin granules, isolation, 255–256
Chromatography, thin-layer and HPLC, ATP
      analog purity checking, 169
Chromosomes
   CHO kinetochores, motor activity assay
      chromosome preparation, 268–269
      microscope requirements, 269
      procedure, 270–273
      recording techniques, 273
      variables, 274–276
   mitotic CHO cells
      isolation, 154–156

motility assay, 158–160
Clam, egg cytoplasm extracts
    microtubule-associated motility and
        crosslinking, 245–249
    preparation, 243–245
Conformation, molecular motors, ATP analog
    probes for, 174–175
Coverslips, preparation, 28
Crosslinking, microtubule-associated
    in cell-free cytoplasmic extracts, 247–249
    materials, 246
    slide preparation, 246–247
Cylindrotheca fusiformis, cultures
    bulk, synchronization, 280–281
    growth and maintenance, 279–280
    metaphase spindles, accumulation, 281–283
Cytoplasm, algal
    extruded preparations, 180–181
    myosin-mediated organelle movements
        DIC microscopy, 181–183
        in vitro observations, 183–186

D

Depolymerization, microtubule, induction of
    chromosome and vesicle motions, 158–160
Diatoms
    cultures
        bulk, synchronization, 280–281
        growth and maintenance, 279–280
        metaphase spindle accumulation, 281–283
    spindle elongation assay, 288–290
DIC microscopy, *see* Differential interference
    contrast microscopy
Differential interference contrast microscopes,
    25–26
Differential interference contrast microscopy,
    video-enhanced
    motility and crosslinking, microtubule-
        associated, in clams and sea urchins,
        245–249
    motility in interphase/metaphase *Xenopus*
        extracts, 213–214, 220
    myosin-mediated organelle movements,
        181–183
    nanometer movements of single motor
        molecules, 134–135
    real-time motility *in vitro,* 158–162
    tubulovesicular networks, 233
Directionality assay, motility in interphase/
    metaphase *Xenopus* extracts, 217–219

Dynein
    cytoplasmic
        depletion from supernatant, 232
        force production
            microtubule gliding assay, 67–70
            polarity, 70–73
        microtubule mobility, real-time video
            microscopy, 86–87
        purification, 132
        single molecules, nanometer movements,
            133–135
    flagellar, microtubule mobility, real-time
        video microscopy, 87
    14 S and 22 S, substrate specificity, 174–175
    microtubule translocation, dark-field
        microscopy, 97–102

E

Egg extracts
    clam, cytoplasmic
        equipment and reagents, 243
        microtubule-associated motility and
            crosslinking, 245–249
        protocol, 243–245
    sea urchin, cytoplasmic
        equipment for, 238–240
        protocol, 241–243
        reagents for, 240
    Xenopus laevis
        interphase
            membrane motility assay, 213–219
            preparation, 207–210
        metaphase
            membrane motility assay, 213–219
            preparation, 210
Endoplasmic reticulum, tubulovesicular
    network formation
    cell lines for, 239
    difficulties in, 233–234
    harvesting of cell and membrane vesicles,
        239–240
    microtubule motors
        depletion, 232
        purification, 232–233
    microtubule preparation, 230
    motor dependence of, 235
    peripheral membrane protein extraction, 232
    procedure, 231–232
    video-enhanced DIC microscopy, 233
Enzymes, motor, *see specific enzymes*

Escherichia coli, expression of microtubule
    motor proteins in, 119–120
    ATPase activity, 124–125
    binding assays, 124
    expression vectors, 118–119
    motility assays, 123
    plasmids for, 116–118
    procedure, 119–120
    purification of proteins, 120–122
N-Ethylmaleimide, myosin modification, 5
Expert vision system, for actin filament
    sliding, 42–45
Expression vectors, bacterial, *see* Bacterial
    expression vectors

F

Fixation, pellicle-initiated microtubules, 163
Flow cells
    for kinesin-driven microtubule movement
        adsorption of proteins, 141
        construction, 140
        solution exchange in, 140–141
        spacers and coverglasses, 142–143
    for myosin-specific adaptation of motility
        assay, 28
Fluorescence microscopy, motility in
    interphase/metaphase *Xenopus* extracts,
    220–221
Force production
    microtubule gliding assay, 70–73
    myosin–actin interaction, 12–13

G

Glass, nanometer movement on, 130–135
Golgi membranes
    gradient purification, 234
    tubulovesicular network formation
        cell lines for, 239
        difficulties in, 233–234
        harvesting of cell and membrane vesicles,
          239–240
        microtubule motors
          depletion, 232
          purification, 232–233
        microtubule preparation, 230
        motor dependence of, 235
        peripheral membrane protein extraction,
          232

        procedure, 231–232
        video-enhanced DIC microscopy, 233
Gradient purification, Golgi membranes, 234

H

Histone H1 kinase assay, for cell cycle status,
    211

I

Image processing
    transfer of video signals to film, 222–223
    vesicle motility in squid axoplasm, 195–197
Image processors, for myosin motility, 27
Intensifier silicon intensifier target camera, 26

K

Kinesin
    depletion from supernatant, 232
    force production
        microtubule gliding assay, 67–70
        polarity, 70–73
    microtubule mobility, real-time video
        microscopy, 83–86
    nanometer movements of single molecules,
        133–135
    preparation, 139
    purification, 131–132, 254–255
    single molecule, microtubule motility driven
        by
        flow cells for, 139–143
        microscopy, 139–144
        motion analysis, 143–144
        sufficiency, 145–146
        in vitro assay, 144–145
    substrate specificity, determination with
        ATP analogs, 174–175
Kinetochores, motor activity assay
    chromosome preparation, 268–269
    microscope requirements, 269
    procedure, 270–273
    recording techniques, 273
    variables, 274–276

L

Laser trap technique, *see* Optical trap
    technique
Light microscopy, with DIC microscopy, for
    motility observation, 158–162

Lipid membranes, myosin I motility
    evaluation, 59–61
    formation, 55–57
    recording, 57–59
    vesicle preparation, 54–55
Load, introduction onto actin filaments, 39–42
Luciferin–firefly luciferase assay, for ATP
    contamination, 169

M

Membranes, *see also* Golgi membranes; Lipid
    membranes
    tubular networks, formation in *Xenopus*
        extracts, 221–222
Meromyosin, heavy
    preparation, 5
    quick freeze, 5
    substrate specificity, 174–175
Methylcellulose, in actin filament motility
    assay, 31
Microscopy, *see specific types*
Microtubules
    aster formation
        kinesin-mediated sliding during, 263–264
        in vitro, 258
    ATP-induced sliding in flagellar axonemes
        Chlamydomonas, 91–93
        sea urchin sperm, 94–97
    gliding assay
        polarity of force production, 70–73
        protocols, 67–70
    for interphase/metaphase *Xenopus* samples,
        214–215
    motility assay with computer-assisted video
        microscopy
        cytoplasmic dynein, 86–87
        flagellar dynein, 87
        kinesin activity, 83–86
        ncd protein, 86
        procedures and modifications, 81–83
    motor protein expression in *Escherichia coli*
        ATPase determination, 124–125
        binding assays, 124
        limitations, 125
        motility assays, 123
        procedure, 119–120
        purification for motility assays, 120–122
    movement, single kinesin molecule-related
        flow cell for, 139–142
        microscopy, 139–144
        motility assay *in vitro*, 144–145

    motion analysis, 143–144
    sufficiency of single molecules, 145–146
pellicle-initiated, fixation, 163
phosphocellulose-purified, preparation, 230
polarity-marked
    bright seeds preparation, 106–108
    labeling, 106
    procedure, 109–110
    reagents for, 106
    visualization, 110–112
MKLP-1, microtubule sliding activity assay,
    160–162
Motility assays
    ATP-induced microtubule sliding in
        axonemes
        Chlamydomonas, 91–93
        sea urchin sperm, 94–97
    dynein-driven microtubule translocation,
        97–102
    interphase/metaphase *Xenopus* extracts
        cell cycle status, 211, 216
        image transfer to film, 222–223
        measurements, 221–222
        membrane movements, 215–216
        microscope perfusion chamber for, 213–
            214
        microscopy, 220–221
        microtubules for, 214–215
        plus-end- and minus-end-directed motor
            activity, 217–219
        soluble motor activity, 216
        troubleshooting, 219
    kinesin-mediated
        dynamics, 258–260
        intact cell phenomena and, 260–263
        microtubule aster assay, 258
        microtubule sliding during aster
            formation, 263–264
        salt concentration effects, 263
        vesicular transport assay, 257–258
    microtubules
        in clam and sea urchin egg cytoplasm
            extracts, 245–249
        MKLP-1 sliding activity, 160–162
        nanometer movements by single
            molecules
            procedures, 130–133
            recording and DIC microscopy, 133–
                135
        real-time video microscopy
            cytoplasmic dynein activity, 86–87
            flagellar dynein activity, 87

hardware/software setup, 76–80
kinesin activity, 83–86
ncd protein activity, 86
single kinesin molecule-driven, 144–145
single motor molecule-driven
   procedures, 130–133
   recording and DIC microscopy, 133–135
tethered to *Tetrahymena* pelicles
   chromosome and vesicle motions, 158–160
   sliding activity of motor enzymes, 160–162
motor proteins expressed in *Escherichia coli*
   assays, 123
   partial purifications, 120–122
myosin
   antibody-assisted assays, 39
   equipment for, 25–27
   force production, 12–13
   load introduction methods, 39–42
   with native thick filaments, 33–36
   with nonmuscle myosin II, 37
   procedure, 27–31
   with smooth muscle myosin, 37
   with soluble myosin, 32–33
   with subfragments, 32–33
   with synthetic thick filaments, 32
   velocity
      digital image analysis, 8
      extraction of motor events, 8–12
      features, 7–8
   with very fast myosin, 38–39
   with very slow myosin, 37–38
myosin–actin interaction
   components, preparation, 4–7
   force production, 12–13
   velocity
      digital image analysis, 8
      extraction of motor events, 8–12
      features, 7–8
myosin I
   evaluation, 59–61
   on planar lipid membranes, 54–55
   planar membrane formation, 55–57
   recording techniques, 57–59
for thin-filament regulatory systems, 39
vesicles in squid axoplasm
   buffers for, 194–195
   limitations, 197–201
   quantification of moving particles, 197–201
   video microscopy, 195–197

Motor enzymes, *see specific enzymes*
Myosin
   motility assays
      antibody-assisted, 39
      equipment for, 25–27
      force production, 12–13
      load introduction methods, 39–42
      with native thick filaments, 33–36
      with nonmuscle myosin II, 37
      procedure, 27–31
      with smooth muscle myosin, 37
      with soluble myosin, 32–33
      with subfragments, 32–33
      with synthetic thick filaments, 32
      velocity
         digital image analysis, 8
         extraction of motor events, 8–12
         features, 7–8
      with very fast myosin, 38–39
      with very slow myosin, 37–38
   organelle movements in characean
      cytoplasm
      direct visualization methods, 181–183
      in vitro observations, 183–186
   purification, 4–5
   quick freeze, 5
   soluble pool, identification in *Nitella*
      cytoplasm, 180–181
   step size, 3–4
Myosin I, motility on planar lipid membranes
   evaluation, 59–61
   lipid vesicles, 54–55
   planar membrane formation, 55–57
   recording techniques, 57–59

N

Nanometer movements
   recording, 133–134
   video-enhanced DIC microscopy, 134
ncd protein, microtubule mobility, real-time
   video microscopy, 86
Nitella flexilis, cytoplasm
   extruded, preparation, 180–181
   soluble myosin pool, identification, 180–181
Nuclei, demembranated *Xenopus* sperm, 211

O

Optical trap technique
   condensers, 17
   detection system, 17–18

feedback control, 18
illumination and fluorescence, 16–17
lasers
    beam modulation and deflection, 15–16
    types, 15
microscopes
    choice of, 14–15
    stage, 18
objectives, 16
Organelles, cytoplasmic
    direct visualization methods, 181–183
    myosin-mediated movements *in vitro*, 183–186

P

Perfusion chamber, for motility observation, 213–214
Photobleaching, fluorescently labeled microtubules, 110
Plasmids, for bacterial expression systems, 116–118
Polarity assays, microtubule marking
    bright seeds preparation, 106–108
    labeling, 106
    procedure, 109–110
    reagent for, 106
    visualization, 110–112

R

Recording techniques
    for CHO kinetochore motor activity, 273
    for myosin I motility, 57–59
    for myosin motility, 276
Rhodamine, tubulin labeling with, 106–108

S

Salt, effects on kinesin motility, 263
Sea urchin
    egg cytoplasm extracts
        microtubule-associated motility and crosslinking, 245–249
        preparation, 238–243
Silicon intensifier target camera, 26
Sperm, demembranated *Xenopus* nuclei for cell cycle status assay, 211
Spindle elongation
    assay for, 288–290
    in vitro models
        isolated spindle, 286–288
        permeabilized cell, 285–286

Squid, axoplasm, vesicle motility, 195–197
Step size, myosin, estimation, 3–4
Substrate specificity, determination with ATP analogs, 175–176

T

Tetrahymena thermophila, pellicles, preparation, 151–153
Transport, vesicular, *see* Vesicular transport
Tubular membrane networks, *see* Membranes, tubular networks
Tubulin
    bovine brain, purification, 153–154
    preparation, 139
    purification for vesicle motility assay, 254–255
    rhodamine-labeled, 106–108
Tubulovesicular networks, formation
    cell lines for, 239
    difficulties in, 233–234
    harvesting of cell and membrane vesicles, 239–240
    microtubule motors
        depletion, 232
        purification, 232–233
    microtubule preparation, 230
    motor dependence of, 235
    peripheral membranes, protein extraction, 232
    procedure, 231–232
    video-enhanced DIC microscopy, 233

V

Vesicular transport
    kinesin-mediated, in biochemically defined assay
        dynamics *in vitro*, 258–260
        intact cell phenomena and, 260–263
        protocol, 257–258
        reagents and components, 254–257
        salt concentration effects, 263
    myosin-mediated, in characean cytoplasm
        direct visualization methods, 180–181
        in vitro observations, 183–186
Video-enhanced differential interference contrast microscopy, *see* Differential interference contrast microscopy, video-enhanced
Video imaging systems, for myosin motility, 26

Video microscopy
  computer-assisted
    computer for, 78
    microtubule motility assay, 81–83
    optics, 77–78
    superimposition of video signals, 78–79
    velocity program for, 79–80
  dark-field
    ATP-induced microtubule sliding in
      axonemes
        Chlamydomonas, 91–93
        sea urchin sperm, 94–97
    cleaning of slides for, 143
    dynein-driven microtubule translocation,
      97–102
  vesicle motility in squid axoplasm, 195–197
  video-enhanced DIC, *see* Differential
    interference contrast microscopy,
    video-enhanced

X

Xenopus laevis
  cell cycle status assay, 211, 266
  demembranated sperm nuclei, 211
  egg extracts
    interphase
      membrane motility assay, 213–219
      preparation, 207–210
    metaphase
      membrane motility assay, 213–219
      preparation, 210
  membrane fractions and supernatants
    motility assay, 215–219
    preparation, 211–213

VOLUMES IN SERIES

Founding Series Editor
DAVID M. PRESCOTT

Volume 1 (1964)
Methods in Cell Physiology
Edited by David M. Prescott

Volume 2 (1966)
Methods in Cell Physiology
Edited by David M. Prescott

Volume 3 (1968)
Methods in Cell Physiology
Edited by David M. Prescott

Volume 4 (1970)
Methods in Cell Physiology
Edited by David M. Prescott

Volume 5 (1972)
Methods in Cell Physiology
Edited by David M. Prescott

Volume 6 (1973)
Methods in Cell Physiology
Edited by David M. Prescott

Volume 7 (1973)
Methods in Cell Biology
Edited by David M. Prescott

Volume 8 (1974)
Methods in Cell Biology
Edited by David M. Prescott

Volume 9 (1975)
Methods in Cell Biology
Edited by David M. Prescott

Volume 10 (1975)
Methods in Cell Biology
Edited by David M. Prescott

Volume 11 (1975)
Yeast Cells
Edited by David M. Prescott

Volume 12 (1975)
Yeast Cells
Edited by David M. Prescott

Volume 13 (1976)
Methods in Cell Biology
Edited by David M. Prescott

Volume 14 (1976)
Methods in Cell Biology
Edited by David M. Prescott

Volume 15 (1977)
Methods in Cell Biology
Edited by David M. Prescott

Volume 16 (1977)
Chromatin and Chromosomal Protein Research I
Edited by Gary Stein, Janet Stein, and Lewis J. Kleinsmith

Volume 17 (1978)
Chromatin and Chromosomal Protein Research II
Edited by Gary Stein, Janet Stein, and Lewis J. Kleinsmith

Volume 18 (1978)
Chromatin and Chromosomal Protein Research III
Edited by Gary Stein, Janet Stein, and Lewis J. Kleinsmith

Volume 19 (1978)
Chromatin and Chromosomal Protein Research IV
Edited by Gary Stein, Janet Stein, and Lewis J. Kleinsmith

Volume 20 (1978)
Methods in Cell Biology
Edited by David M. Prescott

Advisory Board Chairman
KEITH R. PORTER

Volume 21A (1980)
Normal Human Tissue and Cell Culture, Part A: Respiratory, Cardiovascular, and Integumentary Systems
Edited by Curtis C. Harris, Benjamin F. Trump, and Gary D. Stoner

Volume 21B (1980)
Normal Human Tissue and Cell Culture, Part B: Endocrine, Urogenital, and Gastrointestinal Systems
Edited by Curtis C. Harris, Benjamin F. Trump, and Gary D. Stoner

Volume 22 (1981)
Three-Dimensional Ultrastructure in Biology
Edited by James N. Turner

Volume 23 (1981)
Basic Mechanisms of Cellular Secretion
Edited by Arthur R. Hand and Constance Oliver

Volume 24 (1982)
The Cytoskeleton, Part A: Cytoskeletal Proteins, Isolation and Characterization
Edited by Leslie Wilson

Volume 25 (1982)
The Cytoskeleton, Part B: Biological Systems and *in Vitro* Models
Edited by Leslie Wilson

Volume 26 (1982)
Prenatal Diagnosis: Cell Biological Approaches
Edited by Samuel A. Latt and Gretchen J. Darlington

Series Editor
LESLIE WILSON

Volume 27 (1986)
Echinoderm Gametes and Embryos
Edited by Thomas E. Schroeder

Volume 28 (1987)
***Dictyostelium discoideum:* Molecular Approaches to Cell Biology**
Edited by James A. Spudich

Volume 29 (1989)
Fluorescence Microscopy of Living Cells in Culture, Part A: Fluorescent Analogs, Labeling Cells, and Basic Microscopy
Edited by Yu-Li Wang and D. Lansing Taylor

Volume 30 (1989)
Fluorescence Microscopy of Living Cells in Culture, Part B: Quantitative Fluorescence Microscopy—Imaging and Spectroscopy
Edited by D. Lansing Taylor and Yu-Li Wang

Volume 31 (1989)
Vesicular Transport, Part A
Edited by Alan M. Tartakoff

Volume 32 (1989)
Vesicular Transport, Part B
Edited by Alan M. Tartakoff

Volume 33 (1990)
Flow Cytometry
Edited by Zbigniew Darzynkiewicz and Harry A. Crissman

Volume 34 (1991)
Vectorial Transport of Proteins into and across Membranes
Edited by Alan M. Tartakoff

Selected from Volumes 31, 32, and 34 (1991)
Laboratory Methods for Vesicular and Vectorial Transport
Edited by Alan M. Tartakoff

Volume 35 (1991)
Functional Organization of the Nucleus: A Laboratory Guide
Edited by Barbara A. Hamkalo and Sarah C. R. Elgin

Volume 36 (1991)
***Xenopus laevis:* Practical Uses in Cell and Molecular Biology**
Edited by Brian K. Kay and H. Benjamin Peng

Volume 37 (1993)
Antibodies in Cell Biology
Edited by David J. Asai

Volume 38 (1993)
Cell Biological Applications of Confocal Microscopy
Edited by Brian Matsumoto

Volume 39 (1993)
Motility Assays for Motor Proteins
Edited by Jonathan M. Scholey

Volume 40 (1994)
A Practical Guide to the Study of Ca^{2+} in Living Cells
Edited by Richard Nuccitelli